Methods of Environmental Impact Assessment

Methods of
Environmental
Impact
Assessment

Edited by
Peter Morris
Riki Therivel
Oxford Brookes University

UBCPress / Vancouver

First published in 1995 by UCL Press.
The name of University College London (UCL) is a registered trade mark
used by UCL Press with the consent of the owner.

Published in Canada by UBC Press.

University of British Columbia
6344 Memorial Road
Vancouver, BC V6T 1Z2

ISBN 0-7748-0526-9

Canadian Cataloguing in Publication Data
Main entry under title:
Methods of environmental impact assessment

ISBN 0-7748-0526-9
1. Environmental impact analysis.   2. Environmental impact
statements.   I. Morris, Peter, 1934-   II. Therivel, Riki, 1960-
TD194.6.M47 1995   333.7'14   C95-910159-4

Typeset in Times New Roman And Omega.
Printed and bound in Canada by Friesens.

For Angie and (again) Tim

Contents

Preface and acknowledgements

The idea of a book on methods of environmental impact assessment first arose during the writing of *Introduction to Environmental Impact Assessment* (Glasson et al. 1994). We realized that very few books exist on how EIA should be carried out for specific environmental components such as air, flora, or socio-economics, and that, of these, none was written for the UK/EU context since the implementation of EC Directive 85/337 on EIA. Rather than crowd this information into that book – with its focus on EIA principles and procedures, process, practice and prospects – we decided to put together this sister volume which considers the nuts and bolts of EIA as carried out for particular environmental components by the individual environmental specialists who contribute to the process. Together the books provide a comprehensive coverage of the theory and practice of EIA in the UK and EU in the 1990s.

This book is aimed at people who organize, review, and make decisions about EIA; at environmental planners and managers; at students taking first degrees in planning, ecology, geography and related subjects with an EIA content; and at postgraduate students taking courses in EIA or environmental management. It explains what the major concerns of the individual environmental specialists are, how data on each environmental component are collected, what standards and regulations apply, how impacts are predicted, what mitigation measures can be used to minimize or eliminate impacts, what some of the limitations in these methods are, and where further information can be obtained. It does not aim to make specialists out of its readers; to do so would require at least one book per environmental component. Instead it aims to foster better communication between experts, a better understanding of how EIAs are carried out, and hopefully better EIA-related decisions. Like its sister volume, this book emphasizes best practice – what ideally should happen – as well as minimal regulatory requirements. EIA is a constantly evolving and improving process, and best practice today is likely to be the minimal regulatory requirement of tomorrow.

The basis of this book is a unit on Oxford Brookes University's MSc course in Environmental Assessment and Management entitled Methods of Environmental Assessment. The unit is taught by a range of staff from across the university who have practical expertise in EIA, and most of the chapters of this book are written by these specialists. As joint course leaders for the MSc, and joint organizers for the unit, we are grateful to our colleagues for their excellent contributions prepared under tight time constraints. Further expert input was kindly provided by members of Hampshire County Council and Ove Arup, an engineering and environmental consultancy.

We are also grateful for the help of Colin English (Arup Acoustics) and Owen McIntyre (Oxford Brookes University Law Unit) with Chapter 4; Gordon Baker (SIAS Ltd., Edinburgh) and Martin Baxter (Institute of Environmental Assessment, East Kirkby) with Chapter 5; Roger Barrowcliffe (Environmental Resources Management, London) who also provided Figures 8.1 and 8.2, David J. Carruthers (Cambridge Environmental Research Consultants, Cambridge), D. Owen Harrop (Aspinwall and Company, Baschurch, Shrewsbury), Ian S. McCrae (Transport Research Laboratory, Crowthorne), Frank Pavy (Rotork Analysis, Faringdon, Oxfordshire), and Roger J. Timmis (Warren Spring Laboratory, Stevenage) with Chapter 8; and Richard Frost (Oxford Brookes University), who commented on Chapter 15, and assisted in organising the bibliography. Thanks are due also to Derek Whiteley and Rob Woodward (both of Oxford Brookes University) for the line drawings.

Finally, we thank our editor Roger Jones for his enthusiasm, encouragement, and practical support.

RIKI THERIVEL
PETER MORRIS
January 1995

List of contributors

Jeremy Biggs, Antony Corfield, David Walker, Mericia Whitfield and *Penny Williams* are members of Pond Action, an independent non-profit group initiated in 1986 with the support of the World Wide Fund for Nature and Oxford Brookes University, which promotes the conservation of freshwater ecosystems.

Nicola Bourdillon is a Lecturer in Environmental Assessment and a Research Associate in the School of Planning, Oxford Brookes University.

Rosemary Braithwaite works in the Archaeology Team of Hampshire County Council's Planning Department.

Andrew Chadwick is a Research Associate with the Impacts Assessment Unit in the School of Planning, Oxford Brookes University.

Derek Elsom is Professor of Climatology in the Geography Unit, Oxford Brookes University.

Roger France is a Senior Lecturer in Historic Conservation in the School of Planning, Oxford Brookes University.

John Glasson is Professor of Planning and Head of the School of Planning at Oxford Brookes University, and the Research Director of the university's Impacts Assessment Unit.

Brian Goodey is Professor of Urban Landscape Design in the School of Planning, Oxford Brookes University.

Martin Hodson is a Senior Lecturer in Environmental Biology in the School of Biological and Molecular Sciences, Oxford Brookes University.

David Hopkins works in the Archaeology Team of Hampshire County Council's Planning Department.

Aidan Hughes is an Associate with Ove Arup in their Birmingham office.

Peter Morris is a Principal Lecturer in Ecology in the School of Biological and Molecular Sciences, Oxford Brookes University.

Agustin Rodriguez-Bachiller is a Senior Lecturer in Quantitative Methods in the School of Planning, Oxford Brookes University.

Tim Shreeve is a lecturer in Ecology in the School of Biological and Molecular Sciences, Oxford Brookes University.

Riki Therivel is a Senior Lecturer in Environmental Assessment and a Research Associate in the School of Planning, Oxford Brookes University.

Stewart Thompson is a Lecturer in Environmental Biology in the School of Biological and Molecular Sciences, Oxford Brookes University.

David Thurling is a Principal Lecturer in Environmental Biology in the School of Biological and Molecular Sciences, Oxford Brookes University.

Abbreviations and units used in the text

Abbreviations

AAI	area of archaeological importance
ACAO	Association of Country Archaeological Officers
ADAS	Agricultural Development and Advisory Service
ADMS	atmospheric dispersion modelling system
AE	actual evapotranspiration
AONB	Area of Outstanding Natural Beauty
AOSP	Area of Special Protection for birds
ASPT	average score per taxon
ASSI	Area of Special Scientific Interest
BATNEEC	best available technique not entailing excessive cost
BES	British Ecological Society
BHS	British Herpetological Society
BMWP	Biological Monitoring Working Party
BOD	biological oxygen demand
BPEO	best practicable environmental option
BRC	Biological Records Centre
BRE	Building Research Establishment
BS	British Standard
BSI	British Standards Institute
BTO	British Trust for Ornithology
BUTT	Butterflies Under Threat Team
CEAA	Canadian Environmental Assessment Agency
CC	Countryside Commission
CCW	Countryside Council for Wales (member of JNCC)
CEC	Commission of the European Communities
CFC	chlorofluorocarbon
CH_4	methane
CIMAH	Control of Industrial Major Accident Hazards
CIPFA	Chartered Institute of Public Finance and Accountancy
CIRIA	Construction Industry Research and Information Association
CO	carbon monoxide
CO_2	carbon dioxide
CPRE	Council for the Protection of Rural England
DoE	Department of the Environment
DoT	Department of Transport
DTI	Department of Trade and Industry

EC	European Community (now EU)
EH	English Heritage
EIA	environmental impact assessment
EIS	environmental impact statement
EN	English Nature (member of JNCC)
EPA	Environmental Protection Agency (US)
ES:CW	Environment Service: Countryside and Wildlife (Northern Ireland)
ESA	Environmentally Sensitive Area
ET	evapotranspiration
EU	European Union (formerly EC)
EUN	enhanced urban network
FBA	Freshwater Biological Association (now IFE)
GIS	geographical information system
GRO	General Register Office
HCFC	hydrochlorofluorocarbon
HCI	Herpetofauna Conservation International
HGV	heavy goods vehicle
HMIP	Her Majesty's Inspectorate of Pollution
HMSO	Her Majesty's Stationery Office
IAU	Impacts Assessment Unit, Oxford Brookes University
IEA	Institute of Environmental Assessment
IFE	Institute of Freshwater Ecology (formerly FBA)
IH	Institute of Hydrology
IHT	Institution of Highways and Transportation
ISC	industrial source complex (–ST short term, –LT long term)
ITE	Institute of Terrestrial Ecology
IRPTC	International Register of Potentially Toxic Compounds
IUCN	International Union for the Conservation of Nature
JNCC	Joint Nature Conservation Committee
LA	local authority
LAD	local authority district
LBRC	Local Biological Records Centre
LEA	local education authority
LIS	land information system (for GIS)
LNR	Local Nature Reserve
MAFF	Ministry of Agriculture, Fisheries and Food
MNR	Marine Nature Reserve
MO	Meteorological Office
MoA	measure of agreement
N	nitrogen
NCA	nature conservation authority/body (EN, CCW, ES:CW, and SNH)
NCC	Nature Conservancy Council (former entity; now split into the NCAs)
NERC	Natural Environment Research Council
NETCEN	National Environmental Technology Centre
NFBR	National Federation for Biological Recording
NGO	Non-Government Organisation
NHA	National Heritage Area
NI	Northern Ireland

NNR	National Nature Reserve
N_2O	nitrous oxide
NO_2	nitrogen dioxide
NO_x	nitrogen oxides
NOMIS	National Online Manpower Information Service
NPC	National Parks Commission
NPP	net primary production
NRA	National Rivers Authority
NRC	National Research Council (US)
NSCA	National Society for Clean Air and Environmental Protection
NVC	National Vegetation Classification
NWC	National Water Council
O_3	ozone
OPCS	Office of Population Censuses and Surveys
OS	Ordnance Survey
P	phosphorus
PAH	polycyclic aromatic hydrocarbons
PCA	principal components analysis
PCB	polychlorinated biphenyls
PCG	Pond Conservation Group
PE	potential evapotranspiration
Pn	precipitation
PPG	Planning Policy Guidance Note
QUARG	Quality of Urban Air Review Group
RCS	river corridor survey
RICS	Royal Institute of Chartered Surveyors
RIVPACS	river invertebrate prediction and classification system
RPB	River Purification Board
RSNC	Royal Society for Nature Conservation (Wildlife Trusts Partnership)
RSPB	Royal Society for the Protection of Birds
RWA	regulatory water authority
SAC	Special Area of Conservation
SDD	Scottish Development Department
SERPLAN	South East Regional Planning Conference
SI	Statutory Instrument
SIA	socio-economic impact assessment
SIC	Standard Industrial Classification
SMD	soil moisture deficit
SMR	Sites and Monuments Record
SNH	Scottish Natural Heritage (member of JNCC)
SO	Scottish Office
SO_2	sulphur dioxide
SPA	Special Protection Area
SRL	Sound Research Laboratories Ltd
SSA	standard spending assessment
SSSI	Site of Special Scientific Interest
SWQO	statutory water quality objective
TOMPS	toxic organic micro-pollutants

TPP	Transport Policies and Programmes
TRL/TRRL	Transport Research Laboratory/Transport and Road Research Laboratory
TTWA	travel to work area
UAFHA	Urban Advisors to the Federal Highway Commission (US)
UK	United Kingdom
UNECE	United Nations Economic Commission for Europe
UNEP	United Nations Environment Programme
UNESCO	United Nations Educational, Scientific and Cultural Organisation
USA	United States of America
USACE	US Army Corps of Engineering
USDA-SCS	US Department of Agriculture, Soil Conservation Service
USDC	US Department of Commerce
USEPA	US Environmental Protection Agency
USHUD	US Department of Housing and Urban Development
VAT	value added tax
VOC	volatile organic compound
WCED	World Commission on Environment and Development
WHO	World Health Organisation
WII	Wetland of International Importance (Ramsar site)
WO	Welsh Office
WQO	water quality objective
WTL	water table level
WWF	World Wide Fund for Nature

Units

cm	centimetre
cumec	cubic metres per second
dB	decibel
g	gram
ha	hectare
hr	hour
Hz	hertz
kg	kilogram
l^{-1} or /l	per litre
m	metre
mg	milligram
min	minute
mm	millimetre
MW	megawatt
ppb	parts per billion
ppm	parts per million
s	second
yr	year
µg	microgram

CHAPTER 1
Introduction

Riki Therivel & Peter Morris

1.1 The current status of EIA

Since the first system of environmental impact assessment (EIA) was established in the USA in 1970, EIA systems have been set up worldwide and become a powerful environmental safeguard in the project planning process. EC Directive 85/337 on EIA (CEC 1985) was adopted in 1985, and since then the individual EC (now EU) member states have implemented the Directive through their own regulations. In the UK, the production of the resulting environmental impact statements (EISs) increased more than tenfold between the early 1980s and the early 1990s as a result of the Directive; more than 350 EISs are now prepared annually in the UK, and many more in other member states.

However, despite the rapid growth in EIA, most of the parties involved in the EIA process in the UK are still relatively inexperienced: most environmental consultancies have prepared only one or two EISs, most local authorities have received only one or two EISs, and most environmental groups are uncertain about what they can expect of EIA. EIAs are often poorly carried out and presented, the resulting predictions are often incorrect, and it is unclear to what extent they are used in decision-making. This book aims to improve EIA practice by helping to disseminate information about how EIAs are, and should be, carried out. Although it focuses on the UK context in its discussion of policies and standards, the techniques it discusses apply universally. This introductory chapter summarizes the legislative background to EIA in the UK and EU, and explains the book's structure.

1.2 EIA legislation

Several important internationally accepted principles underlie the recent rapid growth in EIA. The World Commission on Environment and Development

1

espoused the principle of sustainable development in its report of 1987 (WCED 1987). The United Nations Conference on Environment and Development of 1992 established an objective to adapt human activities to nature's carrying capacity. European policy is guided by four principles laid down in the EU Action Programmes on the Environment, namely (a) preventive action is better than remedial measures, (b) environmental damage should be rectified at the source, (c) the polluter should pay the costs of measures taken to protect the environment, and (d) environmental policies should form a component of other EU policies. EIA is a clear example of this emphasis on preventive, holistic, strategic approaches to environmental protection which acknowledge environmental limits.

EC Directive 85/337 requires that, for a specified list of project types (Annex I of the Directive), EIA *must* be carried out; and that EIA *may* be carried out for projects in another list (Annex II) if they are likely to have significant environmental impacts. The required contents of the EIS, providing the regulatory basis for the methods used in EIA, are given in Annex III of the Directive. These are as follows:

1. Description of the project, including in particular:
   - a description of the physical characteristics of the whole project and the land-use requirements during the construction and operational phases
   - a description of the main characteristics of the production processes
   - an estimate, by type and quantity, of expected residues and emissions resulting from the operation of the proposed project.
2. Where appropriate, an outline of the main alternatives studied by the developer and an indication of the main reasons for his choice, taking into account the environmental effects.
3. Description of the aspects of the environment likely to be significantly affected by the proposed project, including, in particular, population, fauna, flora, soil, water, air, climatic factors, material assets, including the architectural and archaeological heritage, landscape and the inter-relationship between the above factors.
4. Description[1] of the likely significant effects of the proposed project on the environment resulting from:
   - the existence of the project
   - the use of natural resources
   - the emission of pollutants, the creation of nuisances and the elimination of waste; and the description by the developer of the forecasting methods used to assess the effects on the environment.
5. Description of the measures envisaged to prevent, reduce and where possible offset any significant adverse effects on the environment.
6. Non-technical summary of information provided under the above headings.

---

1. This description should cover the direct effects and any indirect, cumulative, short, medium and long-term permanent and temporary, positive and negative effects of the project.

7. An indication of any difficulties encountered by the developer in compiling the required information (CEC 1985).

In the UK, the Directive is implemented through 20 regulations, the core of which are the *Town and Country Planning (Assessment of Environmental Effects) Regulations 1988*. The requirements of each regulation differ slightly, but all are essentially variants of these core regulations. The regulations divide the contents required by the EC Directive into two lists, one mandatory and the other discretionary. In essence, the consideration of the use of natural resources, the emission of pollutants, alternatives to the project and difficulties encountered are discretionary whereas most of the other requirements listed above remain mandatory. Appendix A shows these requirements.

Further guidance on the content of EIAs is given in the DoE booklet *Environmental assessment: a guide to the procedures* (DOE 1989). This gives a much longer checklist of points which could be discussed in an EIA; this is also shown at Appendix A. Glasson et al. (1994) discuss how these requirements affect the overall organization, preparation, presentation, review and monitoring of EIAs.

1.3 Book structure

This book discusses how EIAs are carried out for the environmental components listed in the EC Directive and the member states' EIA regulations. Table 1.1 shows how the book chapters correspond to these legislative requirements. The book also includes some components not specifically listed in the regulations but often discussed in practice, namely noise and traffic. Chapters 2 and 3 deal with socio-economic impacts. Chapters 4–7 deal with impacts that are partly socio-economic and partly physical: noise, landscape, traffic, and archaeology. Chapters 8–10 cover the physical environment in terms of air and climate, soils and geology, and water. Chapters 11–14 consider ecological impacts, i.e. biological components (fauna and flora) in the context of **ecosystems**: terrestrial, freshwater, and coastal. Finally, interactions between the impacts associated with various components are discussed in Chapter 15.

Impacts on agriculture are primarily covered in Chapter 9 on soils, but are further addressed in the chapters on socioeconomics and impact interactions. The book does not cover impacts to health and safety or risk analysis, because these are generally not considered in an EIA, are covered by separate legislation, and require quite different methods from other types of impacts. The book also does not discuss the organization of an EIA team, project management, or broader aspects of environmental management, as these subjects deserve a book of their own. The term "methods" is used here to mean the practical steps involved in analyzing impacts; other authors sometimes refer to this as "techniques" or "methodologies".

Table 1.1   This book's coverage of the requirements of EC Directive 85/337 and subsequent legislation.

Environmental component (CEC 1985, DoE 1988)	Chapter number and title
Human beings, population	2 Economic impacts 3 Social impacts 4 Noise 5 Traffic
Fauna and flora	11 Ecology – overview 12 Terrestrial ecology 13 Freshwater ecology 14 Coastal ecology
Soil	9 Soils and geology
Water	10 Water 13 Freshwater ecology 14 Coastal ecology
Air	8 Air and climate
Climatic factors	8 Air and climate
Material assets	2 Economic impacts 7 Archaeological and other material and cultural assets
Architectural, archaeological and cultural heritage	3 Social impacts 7 Archaeological and other material and cultural assets
Landscape	6 Landscape
Interactions between the above	Various chapters 15 Interaction between impacts (including those in agriculture)

1.4 Chapter structure

Each chapter broadly follows the main steps involved in analyzing the impact of a development on an environmental component. As illustrated in Figure 1.1, these are as follows:

1. A *preliminary review* is carried out of the environmental component and the proposed project in order to decide whether to carry out an EIA (screening), and what key impacts, issues and alternatives to consider (scoping). Scoping therefore involves a preliminary assessment of the area likely to be affected (the *impact area*), likely impacts of the development, and potential mitigation measures. It is essential that screening and scoping are carried out early in the EIA process.

2. Relevant *parameters are selected* for describing the environmental component and gauging the impacts of the project (scoping continued). If associated parameters are not being investigated by other team members, it may
be necessary to consider these, e.g. information on geology and soils to
assess ecological or hydrological impacts.

3. *Baseline data* are obtained on each of the selected parameters by (a) collecting existing information from appropriate sources, and (b) undertaking
field surveys and sampling where additional information is required. The
aim is to: assess the value of the baseline environment (e.g. does it contain

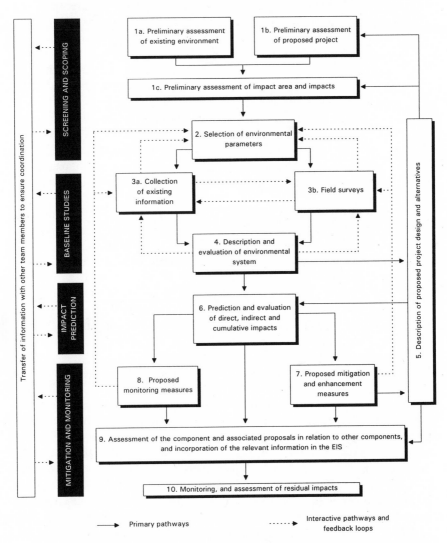

Figure 1.1 Procedures in the assessment of an environmental component for an EIA.

features of international, national or regional importance?); provide data needed to predict likely changes resulting from the proposed development; and provide a baseline for monitoring the component during and after construction of the development.

4. The results of the baseline survey are analyzed, and *description and evaluation of the baseline environmental systems* is carried out. Future conditions without the project are predicted if possible.

5. The *proposed project and alternatives* are described in relation to the environmental component.

6. Potential *impacts of the project on the component* are predicted and evaluated. These may include: (a) direct (primary) impacts; (b) indirect (secondary) impacts, i.e. knock-on effects occurring within or between components; and (c) cumulative impacts that arise from different impact factors and/or other developments. All such impacts may be positive or negative, short, medium, or long-term, reversible or irreversible, and permanent or temporary.

7. *Mitigation measures* are proposed to avoid or minimize the adverse impacts of the project (and, where relevant, to enhance the value of existing features), and are incorporated into the project description.

8. If considered desirable, *monitoring* during construction, operation and decommissioning phases of the project is proposed and included in the project description.

9. An *assessment of the component* (including the mitigation and monitoring proposals) is carried out in relation to other components, to assess aspects such as its relative importance, possible conflicts of interest, legislative limits, carrying capacities, and public concerns; and the relevant EIS section is prepared.

10. *Monitoring* is carried out. This is not strictly part of the EIA process, and is not statutory in the UK. However, it is included in Figure 1.1. because it is the only means of validating the effectiveness of mitigation measures, and assessing *residual impacts*; and lack of monitoring is a serious deficiency in current EIA practice.

A limitation of a model like Figure 1.1 is that it tends to overemphasize the stepwise nature of EIA. In fact (as the feedback loops are intended to indicate) the process is iterative. For example, whereas impact predictions and mitigation proposals will be finalized when all the pertinent information has been gathered, preliminary appraisals should have been made at the scoping stage and reassessed as the study progressed. Similarly, analysis of the data (step 4 et seq.) may reveal the need for further baseline studies, or even for the selection of additional parameters. However, it may then be too late the achieve this, and a policy of continuous reappraisal and adjustment of the study programme should be adopted.

Each chapter of the book has similar main sections, namely:

- introduction (importance of the component, links to other components)

6

- definitions and concepts
- legislative background and interest groups
- baseline studies, and predictions without the development
- impact prediction
- mitigation measures
- monitoring.

These subjects cannot all be discussed in depth in a book of this size. Some aspects are more pertinent to some components or impacts than others, and have thus been discussed in greater depth in those chapters. The chapters' authors have emphasized the topics that they feel are most important in assessing their particular component.

The chapters are meant to be both an overview of the subject and a spring-board for further reading. Because of the broad coverage in relation to the size of the book there is some emphasis on general literature such as texts containing details of specific techniques and/or extensive bibliographies. Other useful contacts and sources of further information are given in Appendices B and F. A glossary is provided of terms used in the book, each of which is highlighted in bold when it first appears in the text.

1.5 The broader context and the future of EIA methods

As is discussed in Glasson et al. (1994), the procedures outlined in Figure 1.1 take place within a broader context of project planning and decision-making. They would be preceded by feasibility studies carried out by the developer, and would in turn provide a basis for a decision about whether the project should proceed. In the future, a project EIA may be set in the context of a strategic EIA of the sectoral or regional policies, plans and programmes which provide its basis.

EIA methods in the UK and EU are evolving rapidly, spurred on by the requirements of EC Directive 85/337. The Directive's five-year review (CEC 1993) is likely to trigger further changes, including mandatory scoping, more thorough analysis of alternatives to the proposed project, and a greater emphasis on monitoring (see Glasson et al. 1994). Revisions to the Directive are expected shortly.

Other worldwide trends in EIA include an increasing focus on indirect and cumulative impacts (see Ch. 15), transboundary impacts (e.g. where pollution from one country has an impact in another country), and the links between sustainability, carrying capacity, and strategic EIA (Therivel et al. 1992). These in turn are triggering research into how to deal more effectively with uncertainty and the use of much larger databases than have been used to date. EIA methods are likely to rely increasingly on geographical information systems (see Appendix D) and computer models, and to be more closely linked to other sources of environmental information such as environmental audits and data gathered under the Environmental Protection Act 1990. Concern about wider distributional impacts, for instance about whether some countries are "importing" sustainability at the cost of making environmental conditions in other countries unsustainable, is likely to lead to even more complex forms of public participation and political negotiations, but ultimately to a more equitable approach to development and the environment.

CHAPTER 2

Socio-economic impacts 1: overview and economic impacts

John Glasson

2.1 Introduction

Major projects have a wide range of impacts on a locality – including biophysical and socio-economic – and the trade-off between such impacts is often crucial in decision-making. Major projects may offer a tempting solution to an area's, especially a rural area's, economic problems, which however may have to be offset against more negative impacts such as pressure on local services and social upheaval, in addition to possible damage to the physical environment. Socio-economic impacts can be very significant for particular projects and the analyst ignores them at his/her peril. Nevertheless, they have often had a low profile in EIA.

This chapter begins with an initial overview of socio-economic impacts of projects/developments, which explains the nature of such impacts. Economic impacts, including the direct employment impacts and the wider, indirect impacts, on a local and regional economy are then discussed in more detail. The chapter dovetails with Chapter 3, which focuses on related impacts such as changes in population levels and associated effects on the social infrastructure, including accommodation and services. Several of the methods discussed straddle the two chapters and will be cross-referenced to minimize duplication. Chapters 2 and 3 draw in particular on the work of the Impacts Assessment Unit (IAU) in the School of Planning at Oxford Brookes University, which has undertaken many research and consultancy studies on the socio-economic impacts of major projects.

2.2 Definitions and concepts: socio-economic impacts

2.2.1 Origins and definitions

Socio-economic impact assessment (SIA) developed in the 1970s and 1980s mainly in relation to the assessment of the impacts of major resource development projects, such as nuclear power station in the USA, hydroelectric schemes in Canada and the UK's North Sea oil- and gas-related developments. The growing interest in socio-economic impacts, partly stimulated by the introduction of the US National Environmental Policy Act of 1969 and subsequent amendments of 1977, generated some important studies and publications, including the works of Carley & Bustelo (1984), Finsterbusch (1980, 1985), Lang & Armour (1981), and Wolf (1974). It also led to considerable debate on the nature and rôle of SIA. Some authors refer to socio-impact assessment; others refer to socio-economic impact assessment. Some see SIA as an integral part of EIA, providing the essential "human elements" complement to the often narrow biophysical focus of many EISs ". . . from the perspective of the social impact agenda, this meant: valuing people 'as much as fish' . . ." (Bronfman 1991). Others see SIA as a separate field of study, a separate process, and some authors raise the legitimate concern that SIA as an integral part of EIA runs the risk of marginalization and superficial treatment. Chapters 2 and 3 of this text focus on the wider definition of socio-economic impacts, *within* the EIA process.

Wolf (1974), one of the pioneers of SIA, adopted the wide-ranging definition of SIA as "the estimating and appraising of the conditions of a society organized and changed by the large scale application of high technology". Bowles (1981) has a similarly broad definition: "the systematic advanced appraisal of the impacts on the day to day quality of life of people and communities when the environment is affected by development or policy change". A more lighthearted, but often relevant, approach to definition can be typified as the "grab bag" (Carley & Bustelo 1984) or "Heineken" approach – with SIA including all those vitally important but often intangible impacts which other methods cannot reach.

2.2.2 Socio-economic impacts in practice: the poor relation?

The early recognition, by some analysts, of the importance of socio-economic impacts in the EIA process and in the resultant EISs has been partly reflected in legislation. The EC Directive (CEC 1985), outlined in Chapter 1, requires a description of possible impacts on human beings. Furthermore the UK government has produced guidance which suggests that "certain aspects of a project including numbers employed and where they will come from should be considered within an environmental statement" (DoE 1989). Yet despite some legislative impetus, the consideration of social and economic impacts has continued to be the poor relation in EIA and in EISs (Glasson & Heaney 1993). There may be several reasons for this which can be summed up by the general perceptions that socio-economic impacts seldom occur, they are invariably negative, and cannot easily be measured. Such perceptions are, of course, a gross distortion. Socio-economic impacts invariably follow from a development, they are often positive, they can be measured and they are important. Indeed the key trade-offs in the decisions on projects often revolve around the balancing of socio-economic benefits (usually employment) against biophysical costs. Socio-economic impacts are important because the economic fortunes and lifestyles and values of people are important.

In a review of the coverage of socio-economic impacts in EISs produced in the UK between 1988–92, Glasson & Heaney showed that, from a sample of 110 EISs, only 43% had considered socio-economic impacts at all. Coverage was better than (a low) average for power station, mixed development and mineral extraction projects. Within those EISs that included socio-economic impacts, there was more emphasis on economic impacts (particularly direct employment impacts) than social impacts. Both operational and construction stages of projects were considered, although with more emphasis on the former. The geographical level of analysis was primarily local, with only very limited coverage of the wider regional scale and no consideration of impacts at the national level. There was very limited use of techniques; where they were included they were primarily economic or employment **multipliers**. Quality was also generally unsatisfactory; only 36% of EISs that considered socio-economic impacts were considered to deal with the economic impacts adequately or better. For social impacts, the figure was only 15%. Socio-economic impacts merit a higher profile. A recent United Nations study of EIA practice in a range of countries advocated changes in the EIA process and in the EIS documentation (UNECE 1991). These included giving greater emphasis to socio-economic impacts in EIA. A starting point is to clarify the various dimensions of socio-economic impacts.

2.2.3 The scope of socio-economic impacts

A consideration of socio-economic impacts needs to clarify the type, duration, spatial extent and distribution of impacts; that is, the analyst needs to ask the questions what to include, over what period of time, over what area and who will be affected?

An overview of *what to include* is outlined in Table 2.1. There is usually a functional relationship between impacts. Direct economic impacts have wider indirect economic impacts. Thus, direct employment on a project will generate expenditure on local services (e.g. for petrol, food and drink). The ratio of local to non-local labour on a project is often a key determinant of many subsequent impacts. A project with a high proportion of inmigrant labour will have greater implications for the demography of the locality. There will be an increase in population, which may also include an influx of dependants of the additional

Table 2.1   What to include? – types of socio-economic impacts.

1. Direct economic
   - local – non-local employment
   - characteristics of employment (e.g. skill group)
   - labour supply and training
   - wage levels

2. Indirect/wider economic/expenditure
   - employees' retail expenditure
   - linked suppliers to main development
   - labour market pressures
   - wider multiplier effects

3. Demographic
   - changes in population size; temporary and permanent
   - changes in other population characteristics (e.g. family size, income levels, socio-economic groups)
   - settlement patterns

4. Housing
   - various housing tenure types
   - public and private
   - house prices
   - homelessness and other housing problems

5. Other local services: public and private sector
   - educational services
   - health services; social support
   - others (e.g. police, fire, recreation, transport)
   - local finances

6. Sociocultural
   - life styles/quality of life
   - social problems (e.g. crime, illness, divorce)
   - community stress and conflict; integration and alienation

employees. The demographic changes will work through into the housing market and will impact on other local services and infrastructure (for example, on health and education services), with implications for both the public and private sector (see Fig. 2.1).

In some cases, population changes themselves may be initiators of the causal chain of impacts; new small settlements (often primarily for commuters) would fit into this category. Development actions may also have sociocultural impacts. A new settlement of 15,000 people may have implications for the lifestyles in a rural, small village based environment. The introduction of a major project, with a construction stage involving the employment of several thousand people over several years, may be viewed as a serious threat to the quality of life of a locality. Social problems may be associated with such development, which may generate considerable community stress and conflict. In practice, such sociocultural impacts are usually poorly covered in EISs, being regarded as more intangible and difficult to assess.

The question of *what period of time* to consider in SIA raises in particular the often substantial differences between impacts in the construction and operational

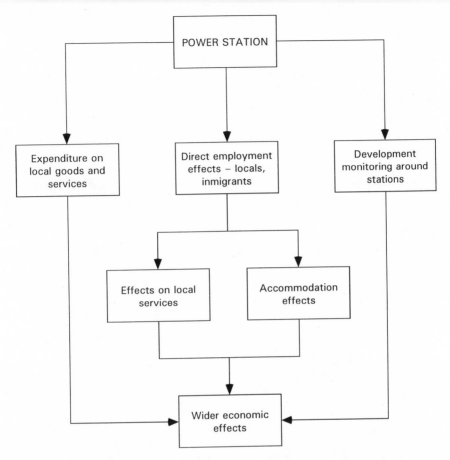

Figure 2.1 Example of linkages between socio-economic impacts for a power-station project.

stages of a project. Major utilities (such as power stations and reservoirs) and other infrastructure projects, such as roads, may have high levels of construction employment but much lower levels of operational employment. In contrast, manufacturing and service industry projects often have shorter construction periods with lower levels of employment, but with considerable employment levels over project lives that may extend for several decades. The closure of a project may also have significant socio-economic impacts; unfortunately these are rarely covered in the initial assessment. Socio-economic impacts should be considered for all stages of the life of a development. Even within stages, it may be necessary to identify substages, for example peak construction employment, to highlight the extremes of impacts which may flow from a project. Only through monitoring can predictions be updated over the life of the project under consideration.

What area to cover in SIA raises the often contentious issue of where to draw the boundaries around impacts. Boundaries may be determined by several factors. They may be influenced by estimates of the impact zone. Thus, for the construction stage of a major project, a subregional or regional boundary may be taken, reflecting the fact that construction workers are willing to travel long distances daily for short-term, well paid employment. On the other hand, permanent employees of an operational development are likely to locate much nearer to their work. Other determinants of the geographical area of study may include the availability of data (e.g. for counties and districts in the UK), and policy issues (e.g. providing spatial impact data related to the areas of responsibility of the key decision-makers involved in a project). Different socio-economic impacts will often necessitate the use of different geographical areas, reflecting some of the determinants already discussed. As noted earlier, EISs in practice have focused on local areas. This may provide a very partial picture; economic impacts often have wider regional, and occasionally national and international implications.

The question of *who will be affected* is of crucial importance in EIA, but is very rarely addressed in EISs. The differential effects of development impacts do not fall evenly on communities; there are usually winners and losers. For example, a new tourism development in a historic city in the UK may benefit visitors to the city and tourism entrepreneurs, but may generate considerable pressures on a variety of services used by the local population. Distributional impacts can be analyzed by reference to geographical areas and/or to groups involved (for example local and non-local, age groups, socio-economic groups, employment groups).

There are of course many other dimensions to impacts besides the areas discussed here, including adverse and beneficial, reversible and irreversible, quantitative and qualitative, and actual and perceived impacts (see Glasson et al. 1994). All are relevant in SIA. The distinction between actual and perceived impacts raises the distinction between more "objective" and more "subjective" assessments of impacts. The impacts of a development perceived by residents of a locality may be significant in determining local responses to a project. They can constitute an important source of information to be considered alongside more "objective" predictions of impacts.

2.3 Baseline studies: direct and indirect economic impacts

2.3.1 Understanding the project/development action

Socio-economic impacts are the outcome of the interaction between the characteristics of the project/development action and the characteristics of the "host" environment. As a starting point, the analyst must assemble baseline information on both sets of characteristics.

The assembling of relevant information on the characteristics of the project would appear to be one of the more straightforward steps in the process. However, projects have many characteristics and for some, relevant data may be limited. The drafting of a *direct employment labour curve* is the key initial source of information (see Fig. 2.2). This shows the anticipated employment requirements of the project. To be of maximum use it should include several dimen-

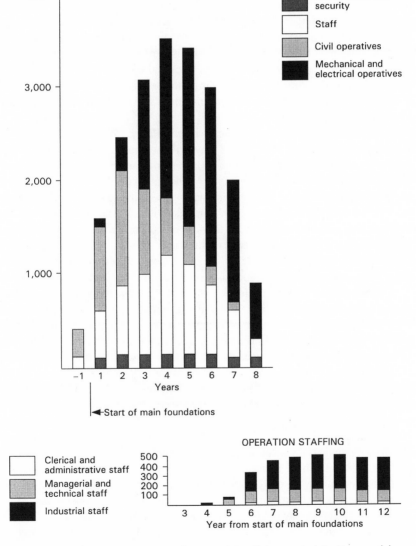

Figure 2.2 Labour requirements for a project disaggregated in time and by employment category.

sions, including in particular the duration and categories of employment. The labour curve should indicate the anticipated labour requirements for each stage in the project life-cycle. For the purposes of prediction and further analysis, there may be a focus on certain key points in the life-cycle. For example, an SIA of peak construction employment could reveal the maximum impact on a community; an analysis of impacts at full operational employment would provide a guide to many continuing and long-term impacts. The labour curve should also indicate requirements by employment or skill category. These may be subdivided in various ways according to the nature of employment in the project concerned, but often involve a distinction between managerial and technical staff, clerical and administrative staff and project operatives. For a construction project, there may be a further significant distinction in the operatives category between civil works and mechanical and electrical operatives. A finer disaggregation still would focus on the particular trades or skills involved, including levels of skills (e.g.skilled/semi-skilled/unskilled) and types of skills (e.g. steel erector, carpenter, electrician).

Projects also have *associated employment policies* which may influence the labour requirements in a variety of ways. For example, the use/type of shift working and the approach to training of labour may be very significant in determining the scope for local employment. An indication of likely wage levels could be helpful in determining wider economic impacts into the local retail economy. An indication of the main developer's attitude/policy to subcontracting can also be helpful in determining the wider economic impacts for the local and regional manufacturing and producer services industries.

Hopefully, the initial brief from the developer will provide a good starting point on labour requirements and associated policies. But this is not always the case, particularly where the project is a "one-off" and the developer cannot draw on comparative experience from within the firm involved. In such cases the analyst may be able to draw on EISs of comparative studies. However, many major projects are at the forefront of technology and there may be few national, or even international, comparators available. There may be genuine uncertainty on the relative merits of different designs for a project, and this may necessitate the assessment of the socio-economic impacts of various possibilities. For example, an assessment by the IAU at Oxford in 1987 for the Hinkley Point C power station proposal, considered the socio-economic impacts for both pressurized water reactor and advanced gas-cooled reactor designs (Glasson et al. 1987). Projects also tend to change their characteristics through the planning and development process; these may have significant socio-economic implications. For example, the discovery during the early stages of project construction of major foundation problems may necessitate a much greater input of civil works operatives. Major projects also tend to have many contractors, and it may be difficult to forecast accurately without knowledge of such subcontractors, and indeed of the main contractor. Such uncertainties reinforce the necessity of regular monitoring of project characteristics throughout project planning and development.

2.3.2 Establishing the economic environment baseline

Defining the "host" economic environment area depends to some extent on the nature of the project. Some projects may have significant national or even international employment implications. The construction of the Channel Tunnel had wide-ranging interregional economic impacts in the UK, bringing considerable benefits to areas well beyond Kent and the South East region of England, for example to the West Midlands (Vickerman 1987). Many projects have regional or subregional economic impacts, and almost all have local economic impacts. As noted in §2.2, it can be useful to make a distinction between the anticipated construction and operational daily commuting zones for a project. The former is invariably much larger in geographical area than the latter, possibly extending up to 90 minutes one-way daily commuting time from the project. For these areas, and for the wider region and nation as appropriate, it is necessary to assemble data on current and anticipated labour market characteristics, including size of labour force, employment structure, unemployment and vacancies, skills and training provision.

The size of the labour force provides a first guide to the ability of a locality to service a development. Information is needed on the economically active workforce (i.e. those males and females in the 16 to retirement age bands). This then needs disaggregation into industrial and/or occupational groups to provide a guide to the economic activities and employment types in the study area(s). An industrial disaggregation would identify, for example, those in agriculture, types of manufacturing and services. In the UK, the Standard Industrial Classification (SIC) provides a template of categories (Table 2.2). An occupational disaggregation indicates particular skill groups (Table 2.3). Data on unemployment and vacancies provide indicators of the pressure in the labour market and the availability of various labour groups. It should be disaggregated by length of unemployment, as well as by skill category and location. Data should also be collected on the provision of training facilities in an area. Such facilities may be employed to enhance the quality of labour supply.

Table 2.2 UK broad Standard Industrial Classification (SIC).

Division of industry

---

0 Agriculture, forestry, fishing
1 Energy and water supply
2 Extraction of minerals and ores other than fuels, manufacture of metals and chemicals
3 Metal goods, engineering and vehicles industries
4 Other manufacturing industries
5 Construction
6 Distribution, hotels and catering, repairs
7 Transport and communications
8 Banking, finance, insurance, business services and leasing
9 Public administration and other services

Source: Census of Employment/Department of Employment.

Table 2.3 UK broad occupational categories.

- Managers and administrators
- Professional
- Associate professional and technical
- Clerical and secretarial
- Personal and protective services
- Sales
- Craft and skilled manual
- Plant and machine operatives
- Other occupations (largely low-skill manual)

Source: Department of Employment, Labour Force Survey.

In the UK, the provision of labour market data comes from various, and changing, sources. The national Department of Employment is a primary source, and a guide to available data is provided in Table 2.4. The National Online Manpower Information Service (NOMIS) computerized database is a particularly useful source of employment and unemployment data at various geographical levels. Department of Employment regions may also provide useful annual and more regular reviews of the employment situation in their region. A basic geographical area for the Department of Employment data is the Travel to Work Area (TTWA). Another important UK source of data is the Census of Population. The results of the 1991 census are now largely available and include information on the economic activity, workplace and transport to work of the population. The statutory local and structure plans for the area under consideration also provide valuable employment data; this may be complemented by data in advisory regional strategies and reviews/studies which are well developed in some regions (e.g. South East via SERPLAN).

Table 2.4 Major UK Department of Employment data sources.

Employment Gazette – published monthly. This is the major source on employment. At the regional level there is monthly information on employment, redundancies, vacancies, unemployment and Regional Development Grants, and annual information on number of employees (age/sex/SIC), activity rates, seasonal unemployment and new employment data. Breakdowns by travel-to-work areas, Assisted Areas and Parliamentary constituencies are also available. There are also occasional labour force projections (male/female/total) by region.
New Earnings Survey – produced annually since 1971. It relates to earnings of employees by industry, occupation, region, etc., at April each year. Part E of the six parts includes detailed analysis of earnings (weekly, hourly) by occupation and industry for regions.
Skills and Enterprise Network – introduced in 1991 and produced quarterly. It provides an important data source on skills, employment and training.
Labour Market Quarterly Report – provides a commentary, including tables and charts, on current labour market trends and the implications for training, employment, unemployment, and includes special features on particular labour market topics. It includes some regional data.
National Online Manpower Information System (NOMIS) (with University of Durham) – a pay-as-you-use nationally networked information system offering rapid access and integrated analysis for data on employment, unemployment, job vacancies, population and migration, at various geographical levels.

Source: Glasson (1992)

In some areas, the sources noted may be enhanced by various one-off studies, including for example skills audits which seek to establish the current and latent skills provision of an area. In the UK, a network of Training and Enterprise Councils or Local Enterprise Companies exist in each region, and provide a useful contact, particularly on training information. Predictably, the various data sources do not use the same geographical bases; in particular the discrepancy between TTWAs and local authority areas can cause problems for the analyst. The latter should also be aware of the influence of "softer" data – for example information on possible developments in other major projects in a locality which may have labour market implications for the project under consideration. Data on other "host" area economic characteristics – such as wage levels, characteristics of the retail economy, detailed characteristics of local manufacturers – may be more limited, although some local authorities do produce very useful industrial directories.

Local economic impacts may also be influenced by the policy stance(s) of the host area. For many localities the possibility of employment and local trade gains from a project may be the only perceived benefits. There may be a desire to maximize such gains and to limit the **leakage** of multiplier benefits (see §2.5). This may result in an authority taking a policy stance on the percentage of "local" labour to be employed on a project. For example, in an extreme case, Gwynedd County Council negotiated, through the use of an Act of Parliament, a very high percentage of local labour for the construction of the Wylfa nuclear power station on Anglesey. A local position may also be taken on the provision of training facilities. There may be concern about the possible local employment "boombust" scenario associated with some major projects, which may of course bring caution into the setting of high local employment ratios.

2.3.3 Clarifying the issues

Consideration of project and host environment characteristics can help to clarify key issues. Denzin (1970) and Grady et al. (1987) remind us that issue specification should be rooted in several sources, and they advocate the use of the philosophy of "triangulation" for data (the use of a variety of data sources), for investigators (the use of different sets of researchers), for theory (the use of multiple perspectives to interpret a single set of data) and for methods (the use of multiple methods). Thus, the use of quantitative published and semi-published data, as outlined, should be complemented by the use of key informant interviews, working groups (e.g. of developer, local planning officers, councillors, and representatives of interest groups) and, possibly, public meetings.

Although many direct and indirect employment impacts will be specific to the case in hand, the following key questions tend to be raised in most cases:
(a) What proportions of project construction and operation jobs are likely
   to be filled by local workers, as compared to inmigrants, and what are

the likely origins of the inmigrant workers?

(b) What is likely to be the magnitude of the secondary (indirect and induced) employment resulting from project development? What proportions of these jobs will be filled by local workers?

(c) How will local businesses be affected by rapid growth resulting from a major project? For example, will development provide opportunities for expansion or will local firms experience difficulty competing with new chain stores and in attracting and retaining quality workers? (Murdock et al. 1986).

2.4 Impact prediction: direct employment impacts

2.4.1 The nature of prediction

Prediction of socio-economic impacts is an inexact exercise. Ideally the prediction of the direct employment impacts on an area would be based on information relating to the recruitment policies of the companies involved in the development, and on individuals' decisions in response to the new employment opportunities. In the absence of firm data on these and related factors, predictions need to be based on a series of *assumptions* related to the characteristics of the development and of the locality. These could for example include the following: "the labour requirement curves for construction and operation will be as provided by the client; local recruitment will be encouraged by the developer with a target of 50%; employment on the new project will be attractive to the local workforce by virtue of the comparatively high wages offered."

Predictive approaches may use *extrapolative* methods, drawing on trends in past and present data. In this respect, use can be made of comparative situations and the study of the direct employment impacts of similar projects. Unfortunately the limited monitoring of impacts of project outcomes reduces the value of this source, and primary surveys may be needed to obtain such information. Predictive approaches may also use *normative* methods. Such methods work backwards from desired outcomes to assess whether the project, in its environmental context, is adequate to achieve them. For example, the desired direct employment outcome from the construction stage of a major project may be "X"% local employment.

Underpinning all prediction methods should be some clarification of the *cause–effect relationships* between the variables involved. Figure 2.3 provides a simplified flow diagram for the local socio-economic impacts of a power station development. Prediction of the local (and regional as appropriate) labour recruitment ratios is the key step in the process. Non-local workers are, by definition, not based in the study area. Their inmigration for the duration of a project will have a wider range of secondary demographic, accommodation, services and

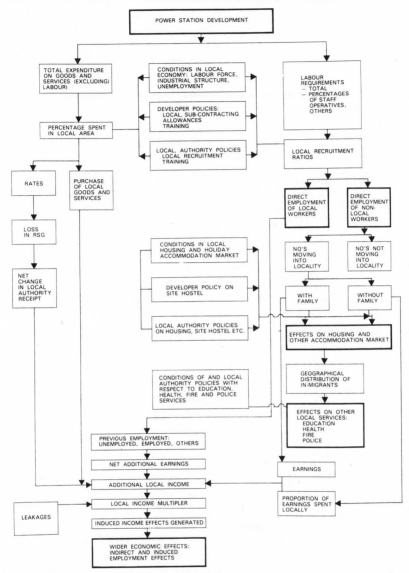

Figure 2.3 A cause–effect diagram for the local socio-economic impacts of a power station proposal (*source:* Glasson et al. 1987).

sociocultural impacts (as discussed in Ch. 3). The wider economic impacts on, for example, local retail activity, will be discussed further in this chapter. The key determinants of the local recruitment ratios are the labour requirements of the project, the conditions in the local economy, and relevant local authority and developer policies on topics such as training, local recruitment and travel allowances. It is possible to quantify some of the cause–effect relationships, and

21

various *economic impact models*, derived from the multiplier concept, can be used for predictive purposes. These will be discussed further in §2.5.

Whatever prediction method is used, there will be a degree of *uncertainty* attached to the predicted impacts. Such uncertainty can be partly handled by the application of probability factors to predictions, by sensitivity analysis, and by the inclusion of ranges in the predictions (see Glasson et al. 1994).

2.4.2 Predicting local (and regional) direct employment impacts

Disaggregation into project stages, geographical areas and employment categories is the key to improving the accuracy of predictions. For example, the construction stage of major projects will usually involve an amalgam of professional/managerial staff, administrative/secretarial staff, local services staff (e.g. catering, security) and a wide range of operatives in a variety of skill categories. Most projects will involve civil-works operatives (e.g. plant operators, drivers), and most will also include some mechanical and electrical activity (e.g. electricians, engineers). For each employment category there is a labour market, with relevant supply and demand characteristics. Guidance on the mix of local/non-local employment for each category can be obtained from comparative studies and from the best estimates of the participants in the process (e.g. from the developer, from the local employment office). Hopefully, but in practice not often, guidance will be informed by the monitoring of direct employment impacts in practice.

As a normal rule, the more specialist the staff, the longer the training needed to achieve the expertise, the more likely that the employee will not come from the immediate locality of the project. Specialist professional staff and managerial staff are likely to be brought in from outside the study area; they may be transferred from other sites, seconded from headquarters or recruited on the national or international market. Only a small percentage may be recruited from the local market, which may simply just not have the expertise available in the numbers necessary. On the other hand, local services staff (e.g. security, cleaning, catering), and to a slightly lesser extent secretarial and administrative staff, may be much more plentiful in most local labour markets, and the local percentage employed on the project may be quite high, and in some cases very high. Other skill categories will vary in terms of local potential according to the degree of skill and training needed. There may be an abundance of general labourers, but a considerable shortage of coded welders.

Comparative analysis of the disaggregated employment categories is likely to produce broad bands for the level of local recruitment. These can then be refined with reference to the conditions applicable to the particular project and locality under consideration. For example, high levels of unemployment in particular skill categories in the locality may boost local recruitment in those categories. Normative methods may also come into play. The developer may introduce training programmes to boost the supply of local skills. Table 2.5 provides an example of the sort of estimates which may be derived. Although the predictions may still use ranges, a prediction from the disaggregated analysis is much more robust than taking employment as an homogenous category.

Table 2.5 Example of predicted employment of local and non-local labour for the construction stage of a major project.

	Total labour requirements	Local		Non-local	
		%	range	%	range
Site services, security and clerical staff	300	90	250–290	10	10–50
Professional, supervisory and managerial staff	430	15	50–80	85	350–380
Civil operatives	500	55	250–300	45	200–250
Mechanical and electrical operatives	1520	40	550–670	60	850–970
Total	2750	44	1100–1340	56	1410–1650

Local labour: employees already in residence in the construction daily commuting zone before being recruited on site. Non-local labour: all other employees.

A further level of micro-analysis would be to predict the employment impacts for particular localities within the study area, and for particular groups, such as the unemployed. A further level of macro-analysis, used in some EISs, would include an estimate of the total person days of employment per year generated by the project (e.g. 10,000 employment days in 1997).

2.5 Impact prediction: wider economic impacts

2.5.1 The range of wider economic impacts

In addition to the direct local (and/or regional) employment effects, major projects have a range of secondary or indirect impacts. The workforce, which may be very substantial (and well paid) in some stages of a project, can generate considerable retail expenditure in a locality, on a whole range of goods and services. This may be a considerable boost for the local retail economy; for example, IAU studies of the impact of power station developments suggest that retail turnover in adjacent medium and small towns may be boosted by at least 10% (Glasson et al. 1982). The projects themselves require supplies ranging from components from local engineering firms, to provisions for the canteen. These can also boost the local economy. Such demands create employment, or sustain employment, additional to that directly created by the project. As will be discussed in Chapter 3, the additional workforce may demand other services locally (e.g. health, education), and housing, which may generate additional construction. These

23

demands will create additional employment. Training programmes associated with a project may bring other economic benefits in terms of a general upgrading of the skills. Overall, the net effect may be considerably larger than the original direct injection of jobs and income into a locality, and such wider economic impacts are invariably regarded as beneficial.

However, there can be wider economic costs. Existing firms may fear the competition for labour which may result from a new project. They may lose skilled labour to high wage projects. There may be inflationary pressures on the housing market and on other local services. Major projects may be a catalyst for other development in an area. A road or bridge can improve accessibility and increase the economic potential of areas. But major projects may also cast a shadow over an area in terms of alternative developments. For example, large military projects, nuclear power stations, mineral extraction projects and others, may have a deterrent impact on other activities, such as tourism – although the construction stage and the operation of many projects can be tourism attractions in themselves, especially when aided by good interpretation and visitor centre facilities.

2.5.2 Measuring wider economic impacts: the multiplier approach

The analysis of the wider economic effects of introducing a major new source of income and employment into a local economy can be carried out using several different techniques (Brownrigg 1971, Glasson 1992, Lewis 1988, McNicholl 1981). The three methods most frequently used are (a) the economic base multiplier model, (b) the input–output model, and (c) the Keynesian multiplier, although it should be added that the percentage of EISs including such studies is still small.

The *economic base multiplier* is founded on a division of local (and/or regional) economies into basic and non-basic activities. Basic activities (local/ regional supportive activities) are seen as the "motors" of the economy; they are primarily orientated to markets external to the area. Non-basic activities (regional dependent activities) support the population associated with the basic activities, and are primarily locally orientated services (e.g. retail services). The ratio of basic to non-basic activities, usually measured in employment terms, is used for prediction purposes. Thus, an "X" increase in basic employment may generate a "Y" increase in non-basic employment. The model has the advantages, and disadvantages, of simplicity (Glasson 1992).

Input–output models provide a much more sophisticated approach. An input–output table is a balancing matrix of financial transactions between industries or sectors. Adapted from national input–output tables, regional or local tables can provide a detailed and disaggregated guide to the wider economic impacts resulting from changes in one industry or sector. However, unless an up-to-date table exists for the area under study, the start-up costs are normally too great for most

EIA exercises. See Batey et al. (1993) for an interesting example of the use of input–output analysis to assess the socio-economic impacts of an airport development.

For several reasons – primarily related to the availability of appropriate data at a local level – the *Keynesian multiplier approach* has been used in several studies and is discussed in further detail here. The basic theory underlying the Keynesian multiplier is simple: "a money injection into an economic system, whether national or regional, will cause an increase in the level of income in that system by some multiple of the original injection" (Brownrigg 1974). Mathematically this can be represented at its most simple as:

$$Y_r = K_r J \qquad (1)$$

where Y_r is the change in the level of income in region r, J is the initial income injection (or multiplicand), and K_r is the regional income multiplier.

If the initial injection of money is passed on intact at each round, the multiplier effect would be infinite. The £X million initial injection would provide £X million extra income to workers, which in turn would generate an extra income of £X million for local suppliers, who would then spend it, and so on, ad infinitum. But the multiplier is not infinite because there are obvious leakages at each stage of the multiplier process. Five important leakages are:

s the proportion of additional income saved (and therefore not spent locally);

t_d the proportion of additional income paid in direct taxation and National Insurance contributions;

m the proportion of additional income spent on imported goods and services;

u the marginal transfer benefit:income ratio (representing the relative change in transfer payments, such as unemployment benefits, which result from the rise in local income and employment);

t_i the proportion of additional consumption expenditure on local goods which goes on indirect taxation (e.g. VAT)

The multiplier can therefore be formulated as follows:

$$K_r = \frac{1}{1-(1-s)\,(1-t_d-u\,)(1-m\,)(1-t_i)} \qquad (2)$$

Substituting (2) into (1) then gives:

$$Y_r = \frac{1}{1-(1-s\,)(1-t_d-u\,)\,(1-m)\,(1-t_i)}\,J \qquad (3)$$

Thus, when applied to the multiplicand J, the multiplier K_r gives the accumulated wider economic impacts for the area under consideration, as in equation (3). The Keynesian multiplier can be calculated in income or employment terms. The various leakages normally reduce the value of local and regional multipliers in practice to between 1.1 and 1.8; in other words, for each £1 brought in directly by the project, an extra £0.10–0.80 is produced indirectly. The size of the import

25

leakage is a major determinant, since the larger the leakage, the smaller the multiplier. Leakages increase as the size of the study area declines, and decrease as the study area becomes more isolated. Thus, of the UK regions, Scotland has the highest regional multiplier (Steele 1969). Local (county and district level multipliers) normally vary between approximately 1.1 and 1.4.

In practice, EIA studies will probably limit such analyses to gross estimates of the wider economic impacts at perhaps the peak construction and full operation stages. But it is possible to disaggregate also with reference to the various employee groups. A study of the predicted local socio-economic impacts of the construction and operation of the proposed Hinkley Point C nuclear power station illustrates the variations, with higher multipliers associated with inmigrants with families (1.3–1.5) than with unaccompanied inmigrants (1.05–1.11) (Glasson et al. 1988). The Keynesian multiplier model, with modifications as appropriate, is well suited to the assessment of the wider economic impacts of projects. But it can only be as good as the information sources on which it is based to construct both the multiplicand and the multiplier. Predictive studies of proposed developments are more problematic in this respect than studies of existing developments, although knowledge of the latter can inform prediction.

2.5.3 Assessing significance

Socio-economic impacts, including the direct employment and wider economic impacts, do not have recognized standards. There are no easily applicable "state of local society" standards against which the predicted impacts of a development can be assessed. Although a reduction in local unemployment may be regarded as positive, and an increase in local crime as negative, there are no absolute standards. Views on the significance of economic impacts, such as the proportion and types of local employment on a project, are often political and arbitrary. Nevertheless, it is sometimes possible to identify what might be termed threshold or step changes in the socio-economic profile of an area. For example, it may be possible to identify predicted impacts which threaten to swamp the local labour market, and which may produce a "boom–bust" scenario. It may also be possible to identify likely high levels of leakage of anticipated benefits out of a locality, which may be equally unacceptable.

In the assessment of significance, the analyst should be aware of the philosophy of "triangulation" noted earlier. Multiple perspectives on significance can be gleaned from many sources, including the local press, which can be very powerful as an opinion former, other key local opinion formers (including local councillors and officials), surveys of the population in the host locality, and public meetings. All can help to assess the significance, perceived and actual, of various socio-economic impacts. A very simple analysis might measure the column-centimetres of local newspaper coverage of certain issues in the planning stage of the project; a survey of local people might seek to calculate simple measures of agreement (MOA) with certain statements relating to economic impacts. MOA is defined as the number of respondents who agree with the statement, minus the number who disagree, divided by the total numbers of respondents. Thus, a MOA of 1 denotes full agreement; –1 denotes complete disagreement.

2.6 Mitigation and enhancement

Most predicted economic impacts are normally encouraged by the local decision-makers. However, there may be concern about some of the issues already noted, such as the poaching of labour from local firms, the swamping of the local labour market, or the shadow effect on other potential development. In such cases, there may be attempts to build in formal and/or informal controls, such as "no poaching agreements". The fear of the "boom–bust" scenario may lead to requirements for a compensatory "assisted area" package for other employment with the demise of employment associated with the project in hand. But in general the focus for economic impacts is more on measures to enhance benefits. When positive impacts are identified there should be a concern to ensure that they do happen and do not become diluted. The potential local employment benefits of a project can be encouraged through appropriate skills training programmes for local people. Targets for the proportion of local recruitment may be set. Various measures, such as project open days for potential local suppliers and a register of local suppliers, may help to encourage local links and to reduce the leakage of wider economic impacts outside the locality. The use of good management practices, including a local liaison committee which brings together the operator and community representatives, and a responsive complaints procedure, can help mitigation and enhancement. A commitment to monitoring, and the publication of monitoring data, can also make a major contribution to effective mitigation.

2.7 Monitoring

Previous stages in the EIA process should be designed with monitoring in mind. Key indicators for monitoring direct employment impacts include: levels and types of employment, by local and non-local sources and by previous employment status; trends in local and regional unemployment rates; and the output of training programmes. All these indicators should be disaggregated to allow analysis by employment/skill category. Relevant data sources include developer/contractor returns, monthly unemployment statistics, and training programme data; these can be supplemented by direct survey information. Key indicators of the wider economic impacts include: trends in retail turnover, the fortunes of local companies, and development trends in the locality. Some guidance on such indicators may be gleaned from published data. The project developer may also provide information on the distribution of subcontracts, but surveys of, for example, workforce expenditure, and the linkages of local firms with a project, may be necessary to gain the necessary information for useful monitoring.

Monitoring is currently (1994) not mandatory for EIA in the UK. There are few comprehensive studies to draw on. The work of the IAU at Oxford on monitoring the local socio-economic impacts of the construction of Sizewell B (Glasson et al. 1988–93) provides one of the few examples of a longitudinal study of socio-economic impacts in practice. It shows the significance of direct employment and wider economic impacts for the local economy. At peak over 2000 local jobs were provided, but with a clear emphasis on the less skilled jobs. Local skills have been upgraded through a major training programme, and although some local companies have experienced recruitment difficulties as a result of Sizewell B, the impact did not appear to be too significant. A group of about 30 to 40 mainly small local companies have benefited substantially from contracts with the project. Although the actual level of project employment was higher than predicted, many of the predictions made at the time of the public inquiry have stood the test of time, and the key socio-economic condition of encouraging the use of local labour has been fulfilled.

2.8 Conclusions

Socio-economic impacts are important in the EIA process. They have traditionally been limited to no more than one EIS chapter, and often a small late chapter, if they have been included at all. Our placing of such impacts early in this text, and in two chapters, emphasizes our concern to indicate their importance in a comprehensive EIA. The discussion has outlined the broad characteristics of such impacts and discussed economic impacts in more detail, with a particular focus on approaches to establishing the information baseline and to prediction. Some predictive methods can become complex. This may be appropriate for major studies; for smaller studies, some of the simpler methods may be more appropriate. The non-local:local employment ratio associated with a project has been identified as a key determinant of many subsequent socio-economic effects.

CHAPTER 3
Socio-economic impacts 2: social impacts

Andrew Chadwick

3.1 Introduction

Chapter 2 discussed how the workforce involved in the construction and operation of any major project is likely to be drawn partly from local sources (within daily commuting distance of the project site) and partly from farther afield. Those employees recruited from beyond daily commuting distance can be expected to move into the locality, either temporarily during construction or permanently during operation. Some of these employees will bring families into the area. Inmigrant employees and their families will have several effects on the locality:

- They will result in an increase in the *population* of the area and possibly in changes to the age and sex structure of the local population.
- They will require *accommodation* within reasonable commuting distance of the project site.
- They will place additional demands on a range of *local services*, including schools, health and recreational facilities, police and emergency services.
- They will have *financial implications* for the local authorities in the area, with additional costs of service provision set against an increase in revenues.
- They will have *other social impacts*, such as changes in the local crime rate or in the social mix of the area's population.

3.2 Definitions

The geographical extent of social impacts – what might be termed the "impact zone" – will depend on the residential location of inmigrant workers and their families. Inmigrant employees can be expected to move into accommodation within reasonable commuting distance of the project site, although the definition of what constitutes a reasonable distance will depend on the project stage (construction or operation), as well as local settlement patterns and the local transport network. Ideally, monitoring data from similar projects elsewhere should be used to indicate the likely extent of daily commuting and thus the likely boundaries of the impact zone. These boundaries can be defined in various ways, for example in terms of a fixed distance or radius from the project site or, more usually, in terms of administrative or political areas such as local authority districts (LADs), health authority areas or school catchment areas.

3.3 Baseline studies

3.3.1 Demography

Establishing the existing baseline
The demographic impact of any development will depend on the project-related changes in population in relation to the existing population size and structure in the impact zone. It is therefore necessary to establish the existing population baseline in the impact zone (i.e. size and age/sex structure). The most useful source of population data in Great Britain, particularly for small geographical areas, is the *Census of population*. This is carried out once every ten years, most recently in April 1991. Since all households in Great Britain are included in the census, reliable information is available at all geographical levels, from census enumeration districts (covering 150–200 households) upwards. Census data are available in a wide range of formats, both published and unpublished:

- The series of *County monitors* and more detailed *County reports*, published by the Office of Population Censuses and Surveys (OPCS), provide data for counties and LADs in England and Wales. Similar reports are produced for health authority areas, and data for LADs and regions in Scotland are published by the General Register Office (GRO).
- For areas smaller than a LAD, or for user-defined areas, the census *Small area statistics* or more detailed *Local base statistics* should be used. These are available direct from the OPCS/GRO, from commercial census agencies and via on-line computerized databases such as NOMIS (see Appendix B).

Further details on the various types of census data available can be found in Leventhal et al. (1993), OPCS (1992), OPCS (forthcoming) and Rhind (1983).

The great strengths of the census are its comprehensiveness and the availability of data for small or user-defined areas. Its main weakness is that it is undertaken only once every ten years. Given the delay in the processing and publication of results, the latest data are sometimes more than a decade out of date. Between censuses, it is therefore necessary to consult other sources to obtain an up-to-date picture of population size and structure in the impact zone. The most often used of these sources are the official *mid-year population estimates*, published annually by the OPCS/GRO. These estimates, disaggregated by age and sex, are available for individual LADs and health authority districts (health board areas in Scotland). The OPCS publishes estimates for the previous June at the end of each year in *Key population and vital statistics, local and health authority areas*. In addition, most local authorities produce their own population estimates, both for the authority as a whole and its constituent parts (i.e. wards or parishes). These estimates may be based on either primary data collection or the use of various proxy measures of population change since the latest census, such as changes in the electoral roll or doctors' registrations (see England et al. 1985, Healey 1991). Some commercial market analysis companies also produce census-based population estimates for small or user-defined areas. Details of these companies are contained in Leventhal et al. (1993).

Projecting the baseline forwards
The data sources outlined above allow the existing population baseline in the impact zone to be established. But it will also be necessary to project this baseline forwards to the expected times of peak construction and full operational activity of the proposed development. Several data sources are available to guide this process. National population projections are published every two years by the OPCS (for England), the Welsh Office and the GRO (Scotland). The projections are available for local and health authority areas, but not for individual LADs in England and Wales (other than London boroughs and metropolitan districts). This means that, if the impact zone has been defined in terms of LADs, it may be necessary to use the projected population change in the relevant county as a rough indicator of likely changes at the district level. This assumes that population change will be distributed evenly throughout each county. It therefore ignores the potential rôle of local planning policies and other factors in producing different rates of migration into and out of each district. An alternative approach would be to use the population projections and forecasts produced by local authorities themselves. These are used by authorities as inputs to their land-use planning work (e.g. structure plan preparation) and to estimates of future service requirements (e.g. school places). Projections are usually available for LADs and in some cases for wards or parishes (see Congdon & Batey 1989, England et al. 1985, Woods & Rees 1986). Some commercial market analysis companies also produce population projections for small areas.

These various sources have limitations as means of projecting forwards the population baseline for relatively small geographical areas. Projections for smaller areas (e.g. LADs) tend to be less reliable than those for larger areas (e.g. counties or regions). This is because net migration is usually a more important determinant of population change for smaller areas; and migration flows are much more difficult to predict than the number of births and deaths. The sources also differ in the extent to which they simply project forwards past trends in an unmodified way. For example, the OPCS stress that their population projections are not "forecasts", in that they take no account of the potential effects of changes in local planning policies. These are often designed to counteract past trends, for example to slow down the rate of population and housing growth in an area. Local authority forecasts are much more likely to incorporate such anticipated policy effects and may therefore be preferable, although of course the intended policy effects may not materialize in practice.

3.3.2 Accommodation

Establishing the existing baseline

The 1991 census, as well as providing population data, is also the most useful source of data on the housing stock in small geographical areas. The census provides two measures of the housing stock in an area. The first is the number of "household spaces" in the area. This indicates the total amount of accommodation in which households can live, regardless of whether it is currently occupied. Vacant accommodation, second homes and non-permanent accommodation (such as caravans and mobile homes) are therefore included. Secondly, the census shows the number of "dwellings" in the area. The definition of a dwelling is slightly more restrictive, in that non-permanent accommodation is excluded, and household spaces that share the same entrance into a building and are not self-contained are counted as a single (shared) dwelling. The number of dwellings in an area is therefore always slightly less than the number of household spaces. The census provides a breakdown of both household spaces and dwellings, according to their tenure (i.e. whether they are owner-occupied, privately rented, rented with a job or business, or rented from a housing association, local authority, New Town or Scottish Homes). The amount of vacant accommodation is identified, as is accommodation not used as a main residence – this includes second homes and some holiday accommodation (e.g. self catering cottages).

All of this information, although providing a very detailed picture of the available housing stock, relates to the position in April 1991 and will therefore need to be updated. This can be achieved by using several sources. The DoE and the Welsh and Scottish Offices publish data on the number of new houses completed (by the private and public sector), and existing homes renovated and demolished, for each LAD. These data are published each quarter in *Local housing statistics* (for England and Wales) and *Scottish housing statistics*. This information, perhaps supplemented by more detailed development control data from local authorities themselves, should allow changes in the overall size of the housing stock since the latest census to be estimated. Changes in the tenure of this accommodation, vacancy levels, the number of second homes and the subdivision of properties, are much more difficult to determine. It may therefore be necessary to assume that these characteristics of the housing stock have remained largely unchanged since the latest census.

At any given time, only a small proportion of the total housing stock will be available on the market for sale or rental. A figure often quoted is that 3–4% of the owner-occupied stock will be on the market at any one time. However, this proportion will vary from one locality to another and over time, according to the state of the housing market (e.g. it will tend to be higher at a time of depressed housing market conditions). Local estate agents should be able to provide information on the number of residential properties for sale or to rent, although some double counting may occur in the case of properties listed by more than one agent. Estate agents are also a useful source of data on current house price and rent levels. House price data are also published by some of the larger building societies, down to the level of individual LADs.

During the construction stage of any development, some inmigrant employees are likely to move into bed and breakfast establishments, hotels, caravans or other types of *tourist accommodation*. It is therefore necessary to establish how much and what type of such accommodation is available in the impact zone, and the extent to which it is currently occupied. Any unoccupied accommodation (e.g. outside the peak tourist season) could be used by inmigrant employees without affecting the availability of accommodation for other – existing – users. Regional tourist boards, county and district councils and tourist information centres all maintain lists of accommodation establishments within their areas of jurisdiction. Details of each individual establishment are often available, including the location, number of rooms and charges/tariffs. A detailed picture of the existing stock of accommodation can therefore be obtained. When combining lists prepared by different organizations for the same geographical area, care should be taken to avoid the double counting of establishments.

Information on existing occupancy levels in tourist accommodation is more limited. The English Hotel Occupancy Survey, produced by the British Tourist Authority, shows the monthly bed and room occupancy rates for a sample of hotels (but not for other types of accommodation). This information is published in *Tourism fact sheets* and is available for tourist board regions and subregions (i.e. individual counties or groupings of adjacent counties). Data for smaller geographical areas may be available from the many area-specific tourism studies carried out by the regional tourist boards, local authorities, academics and consultants.

Projecting the baseline forwards
Forecasting the net change in the available housing stock which would be expected in the absence of the proposed development requires separate estimates of:

(a) *The number of new houses likely to be completed.* The simplest approach would be to assume that future completions would continue at the same annual rate as experienced in recent years in the impact zone (i.e. *simple trend projection*). The annual rate of housebuilding can be calculated from the data published in *Local housing statistics* and *Scottish housing statis-*

tics. This method assumes that there will be no changes in the state of the national or local housing markets, local rates of population and household growth, or local planning policies. An alternative approach would be to use estimates of future growth in the population and number of households in the area to predict the likely demand for new houses. *Local authority population and household forecasts* are likely to be particularly relevant. High and low estimates of household growth are usually made by local authorities, using different assumptions about net migration, employment and household formation (see England et al. 1985, Field & MacGregor 1987, King 1987). Of course, the anticipated increase in the number of households in an area may not be met by an equivalent increase in the housing stock. This is because local planning policies may be intended to meet only part of the projected increase in households. The extent, phasing and location of new housebuilding envisaged by local planning authorities is indicated by the *housing allocations in approved structure plans and adopted local plans*.

(b) *The number of existing houses likely to be demolished*.

(c) *The number of conversions from within the existing stock* (e.g. a single dwelling converted into several self-contained flats, or vice versa).

(d) *Changes in the amount of vacant accommodation, second homes, substandard and other accommodation not available for use*.

All of these changes are difficult to estimate accurately. It may be necessary to assume that recent rates of demolition will continue, and that no net changes will occur in categories (c) and (d) above (see Field & MacGregor 1987, for some possible approaches).

Likely changes in the stock of tourist accommodation are also difficult to predict, although regional tourist boards and local authorities may be able to indicate the scale of any additional provision, either already under construction, having gained planning approval or awaiting approval.

3.3.3 Local services

Inmigrant employees and their families will place demands on a wide range of services provided by local authorities, health authorities and other public bodies. In the space available, it is not possible to discuss each of these service areas in detail. The bulk of this section examines one service area – local education services – as an example of how the existing service baseline might be established and projected forwards. Other service areas are briefly discussed at the end of the section.

The number and type of schools and further education colleges within the impact zone can be obtained directly from local education authorities (i.e. county councils, metropolitan districts and London boroughs in England and Wales) or – for grant maintained and independent schools – from the Department for Education (in England). LEAs will be able to indicate the existing number of pupils on school rolls and the total available capacity (in permanent and temporary accommodation), both for the LEA area as a whole and for each individual school. An age breakdown of existing pupils should also be available. This information can be used to determine the extent to which the available capacity in LEA schools is currently being utilized, across the authority as a whole and for individual schools, high school catchment areas or age groups. Current pupil: teacher ratios can be obtained direct from the LEA or from the Department for Education's annual publication *Statistics of education: schools.*

Information on planned changes in school capacity as a result of the closure, amalgamation or enlargement of existing schools and the opening of new schools should be obtained from the LEA concerned. All LEAs also produce forecasts of future pupil numbers, both for the authority as a whole and for individual schools. These are derived in some cases from the authority's own population and household projections, and should incorporate the effects of anticipated non-project inmigration (see Jenkins & Walker 1985). These data sources allow any significant anticipated changes in pupil numbers and the utilization of capacity within the impact zone to be identified.

Information on other local services, such as recreation, police, fire and social services, should be obtained directly from the relevant local authority department. For health services, Family Health Service Authorities and District Health Authorities will be able to provide a wide range of data on existing medical, dental and pharmacy services, as well as hospital facilities in the impact zone.

3.4 Impact prediction

3.4.1 Demography

Estimating the direct population increase
The direct increase in population caused by a major project will consist of inmigrant employees and any other family members brought into the locality. Several separate estimates are therefore required to determine the population changes directly due to the project: (a) the total number of employees moving into the impact zone; (b) the proportion of these inmigrant employees bringing their family; and (c) the characteristics of these families (i.e. their size and age structure).

The total number of employees moving into the impact zone　　Chapter 2 has outlined the methods available for predicting the mix of local and inmigrant employees associated with the construction and operation of major projects. *During the construction stage*, the build-up in the number of inmigrant workers will reflect the build-up of the construction workforce and changes in the local labour percentage. At the end of the construction stage, most inmigrant workers will move out of the impact zone and return to their original address or another construction project elsewhere. However, a small proportion may establish local ties, especially during a lengthy construction project, and may decide to remain in the area. A construction project spanning several years may therefore result in a small permanent increase in the local population. *During operation*, the main flow of inmigrant employees will usually occur at a relatively early stage, with subsequent inmigration limited to that caused by the normal turnover of employees.

The proportion of inmigrant employees bringing their family During the *construction stage*, only a minority of inmigrant employees – mainly those on long-term contracts – are likely to bring their family into the area. The precise proportion will depend on various factors:

- the length of the construction programme (for projects lasting only a few months, it is likely to be negligible; for projects spanning several years, the proportion may reach at least 10–20%)
- the location of the project site, which will determine the relative merits of weekly commuting and family relocation
- conditions in the national and local housing markets (a depressed national housing market or sharp interregional house price differentials may discourage house and family relocation)
- the availability of suitable family accommodation in the locality.

All of these factors are either specific to particular projects or locations or subject to change. Precise prediction is therefore not possible. Ideally, monitoring data from previous construction projects of various durations and in different parts of the country should be used, as a guide to the likely proportion of employees bringing families. The resulting predictions should incorporate a range of high and low estimates, based on the available monitoring data.

During the operational stage, the vast majority of inmigrant employees will relocate permanently to the area, although there may be some initial delay while suitable accommodation is found and existing properties are sold. Those employees with partners or children can be expected to bring them into the area (with the exception of a few weekly commuters). The precise proportion of employees with families will depend on the age and sex profile of the inmigrant workforce. For example, a younger workforce will contain a higher proportion of single, unattached employees who will not bring families into the area. One crude way of estimating the number of inmigrant families would therefore be to determine the likely mix of married and unmarried employees. This can be approximated by using the latest census data. The census shows the percentage of people in specific age and sex categories who are married or single, both at the national level and for smaller geographical areas. Of course, in practice, some unmarried employees may have partners and/or children, and some allowance should be made for this in any estimates.

The characteristics of inmigrant families   Once the likely number of inmi-grant families has been determined, it is necessary to estimate the average size and broad age structure of these families. The usual approach to estimating the size of inmigrant families is to use detailed census data on household headship. The census shows the average size of households of different types, classified according to the age, sex and marital status of the head of household. Therefore, if it was considered likely that most inmigrant families would contain a married, male head of household, aged 20–59 years, the average size of this type of house-hold – either nationally or in the impact zone – could be calculated. For projects with a younger anticipated workforce, the average size of households with mar-ried male heads aged, say 20–44 years, could be calculated instead. This method assumes that the household characteristics at the time of the 1991 census will remain largely unchanged; it also requires some knowledge (or guesswork) about the age and sex profile of the inmigrant workforce.

Let us assume that the method outlined above suggests that each inmigrant family will contain an average of 3.2 persons. It could then be assumed that each of these families would consist of two adults of working age (the inmigrant employee and partner) and an average of 1.2 other family members – mainly dependent children up to 18 years old, but also including a small proportion of "adult" children (over 18 years old) still living with their parents and perhaps some elderly relatives. The precise proportion of adult children and elderly rel-atives should ideally be derived from monitoring data, but – in the absence of such information – a rough guestimate may be required. Information on the age structure of the 0–18 year old population is available from several sources, and this can be used as the basis of predictions of the ages of dependent children brought into the area. The current age breakdown of 0–18 year olds is provided by the 1991 census, the mid-year population estimates and local authority popu-lation estimates. The projected future age breakdown of this group can be obtained from the various population projections and forecasts outlined in §3.3.1. The census also provides an age breakdown of children (and others) moving into particular LADs or counties during the 12 months to April 1991.

The precise age distribution of dependent children will of course depend on the age profile of their parents. For example, a younger workforce will tend to have a higher proportion of pre-school children than might be suggested by the data sources above, whereas an older workforce may have higher proportions of secondary school children. Some fine-tuning of the age distribution revealed by the data sources above may therefore be required, to take account of the expected age profile of the project workforce. The age breakdown of the workforce should ideally be estimated by obtaining information on the age of employees on similar projects elsewhere. Such information should be readily available to the project developer (for operation) or its contractors (for construction).

Estimating the indirect population increase

As well as the direct population increase as a result of the arrival of inmigrant project employees and families, the development may give rise to indirect population impacts. These impacts can arise in two main ways. First, some locally recruited project employees will leave local employers to take up jobs on the project. This will result in local job vacancies, some of which may be filled by inmigrants. Indirect employment may also be created in local industries supplying or servicing the project, or in the provision of project-related infrastructure. Again, some of these jobs may be taken by inmigrant employees. The scale of the resulting additional inmigration is very difficult to estimate, but its possible existence should at least be acknowledged (see Clark et al. 1981, for some possible estimation methods).

A second source of indirect impacts arises from the fact that some locally recruited project employees might have migrated out of the impact zone if the project had not gone ahead, especially if alternative job opportunities locally were limited. The project may therefore lead to a reduction in outmigration from the area. Again, the extent of any reduction is difficult to predict. It is likely to be significant only in areas experiencing static or declining population, net outmigration and limited or declining employment opportunities.

Assessing significance

The significance of project-related population changes will depend on three main factors: (a) the existing population size and structure in the impact zone (i.e. the baseline); (b) the geographical distribution of the inmigrant population; and (c) the timing of the population changes. Put simply, if inmigrants are few relative to the existing population and have a similar age and sex structure, are distributed over a wide area and do not all arrive at once, then the impacts are unlikely to be significant. The first step in assessing significance is therefore to *express the estimated project-related population increase as a percentage of the baseline population* in the impact zone. The predicted age structure of inmigrants should be compared with the baseline age structure, and any significant differences outlined.

The next step is to estimate *the likely geographical distribution of inmigrants*. Population changes may be quite localized, rather than being evenly distributed throughout the impact zone. However, in the absence of information from monitoring studies, the precise distribution of inmigrants is difficult to predict. The simplest approach would be to assume that the number of employees moving into a particular settlement would be a positive function of that settlement's size and a negative function of its distance from the project site. In practice, the predictions derived from this type of model would need to be modified to allow for the characteristics of the particular locality. These could include the expected location of future housebuilding in the impact zone; differences in the availability and price of various types of housing; and the attractiveness of each settlement in terms of school and other facilities and general environment.

The timing of the arrival of inmigrant employees and the associated population changes will largely follow the expected build-up in the project workforce. However, during the construction stage, most inmigrant families are likely to arrive in the early stages, given that families will tend to be brought by those employees on long-term contracts for the duration of the project.

The nature and significance of population impacts will change as the project progresses through the various stages of its life cycle. Inmigrant employees and their families will become older. In addition, during the operational stage – which may span several decades – there may be some natural increase from the original inmigrant population. These changes can be estimated by using a simple "cohort survival" method, applying age-specific birth and death rates to the original population (see Field & MacGregor 1987). Some allowance may also need to be made for the turnover of employees on the project. As older employees retire, they will tend to be replaced by younger employees, with younger families. This process will counteract, but not completely reverse, the tendency for the inmigrant population to become older.

3.4.2 Accommodation

Estimating the amount of accommodation required
The total amount of accommodation required will be determined by the size of the inmigrant workforce and the extent to which accommodation is shared. Methods to estimate the total number of inmigrant employees were outlined in Chapter 2. Sharing of accommodation is likely to be minimal amongst the permanent *operational workforce*, since most inmigrant employees will be accompanied by their families. However, there may be a limited amount of sharing amongst younger, single employees. To use census terminology, the number of household spaces required will be slightly less than the number of inmigrant employees seeking accommodation. *During the construction stage*, sharing may be much more significant, especially amongst those employees using rented, caravan and perhaps B&B accommodation. Estimates of the likely extent of sharing should be incorporated into any predictions of the demand for accommodation by the construction workforce. Otherwise, the number of household spaces required is likely to be overestimated, perhaps significantly. Monitoring studies of other construction projects are the only reliable guide to the likely extent of sharing (see, for example, Glasson et al. 1992).

Estimating the type and location of accommodation required
The vast majority of inmigrant *operational employees* are likely to relocate permanently to the impact zone. Most will wish to purchase a property in the area, although a small proportion may prefer private rented accommodation. The latter group will include younger, single employees and a few weekly commuters not relocating their family. There may also be some demand for council rented

39

accommodation. The likely mix between owner-occupied, private and council rented accommodation requirements can be roughly estimated by using census data – the census provides information on the tenure of all households moving address during the 12 months to April 1991. Separate tenure patterns can be identified for different types of move (e.g. moves within the same LAD, inter-county or interregional moves). This information is also available for different age groups, according to the age of the head of household. These data can be combined with the expected age profile of the operational workforce, to produce estimates of the likely tenure patterns of inmigrant households.

Predicting the likely mix of accommodation used by inmigrant *construction workers* is a more complicated exercise. A wider range of accommodation is likely to be suitable, including B&B, caravan and other types of tourist accommodation. A further complication is that the developer may decide to provide accommodation specifically for the workforce. The extent of such provision will have important implications for the take up of other types of accommodation. Because the local supply of different types of accommodation and the extent of developer provision will vary from one locality and project to another, the precise mix of accommodation used can vary considerably from project to project. Monitoring data, even if they are available, may therefore provide only a rough indication of the likely take up of each type of accommodation.

In the absence of developer provision, the vast majority of inmigrant construction workers are likely to use private rented, B&B/lodgings or caravan accommodation. The use of each type of accommodation can be roughly estimated by drawing on the available monitoring data from other construction projects, adjusted to allow for the particular supply characteristics in the impact zone – i.e. the amount of each type of accommodation available, its location, cost and existing occupancy levels (see §3.3.2). For example, if the local supply of tourist accommodation is very limited, concentrated in highly priced hotels at some distance from the project site and is usually fully occupied, the proportion of employees using such accommodation is likely to be relatively low. Some construction workers may wish to purchase properties in the locality. The number is likely to be minimal during construction projects lasting only a few months, but may be more significant (at least 10%) in cases where construction activity spans several years. The proportion of inmigrant employees buying properties will be closely linked to the proportion bringing families into the impact zone. However, since some families will prefer to use rented accommodation, the number of owner-occupied properties required is likely to be lower than the total number of inmigrant families.

In certain cases, the project developer may decide to make specific accommodation provision for the construction workforce. This may involve negotiations with the local planning authority over the provision of additional caravan sites or the expansion of existing sites. In other cases, the developer may wish to provide purpose-built hostel accommodation, located on or adjacent to the construction site. This typically consists of single bedrooms and associated catering, recreational and other facilities. To the extent that such provision is made, the proportion of inmigrant employees using other types of accommodation will be lower than would otherwise have been the case. It is recommended that estimates of the demand for different types of accommodation should be made using various alternative scenarios – for example, without any hostel or additional caravan provision, with a small hostel or with a larger hostel. Such estimates should help to clarify the need for such developer provision. The *geographical distribution* of the accommodation taken up by inmigrant employees is difficult to predict. §3.4.1 outlined a possible approach.

Assessing significance

The project-related demand for local accommodation is likely to result in a net change in the amount of accommodation available in the impact zone. On the one hand, the availability of accommodation will be reduced by the take up of local accommodation by project employees and their families. This accommodation would otherwise have been available to local residents and non-project inmigrants. On the other hand, to the extent that project-related demands are met by the release of unoccupied or underoccupied accommodation and/or the bringing forward of speculative housebuilding development, the amount of accommodation available locally will be higher than would otherwise have been the case. The balance between these two types of change will represent the net change due to the project. This should then be expressed as a percentage of the existing (or projected) stock of accommodation in the impact zone. Similar calculations can be made for each separate type of accommodation and for particular settlements or areas within the impact zone.

In extreme cases, the net decline in the availability of accommodation caused by the project may be such that the project-related and non-project demands for accommodation may outstrip the available local supply. Assessment of such pressures requires projections of the following:

- the likely project-related demand for accommodation (as outlined earlier in the section)
- the likely non-project demand for accommodation by local residents and non-project inmigrants (derived from the projected growth in population and households, as outlined in §3.3.2), and
- likely changes in the local supply of accommodation, including project-induced changes, such as the release of unoccupied and underoccupied accommodation and the bringing forward of speculative development (see §3.3.2).

Cases in which the project results in a shortfall in the local supply of accommodation are likely to require the consideration of mitigation measures. However, in practice, pressure on one locality is likely to be relieved by the diversion of demand (both project and non-project) into adjacent localities. Unless this is seen as undesirable, this may eliminate the need for mitigation measures.

41

3.4.3 Local services

Estimating the demand for local services
Inmigrant employees and their families will place demands on a wide range of services provided by local authorities and other public bodies. The demand for these services will largely reflect the age and sex distribution of the inmigrant population (see §3.4.1). For example, in the case of health and personal social services, the number of young children and elderly people will be a critical determinant of demand. In such cases, rough estimates of likely demand can be obtained by combining the predicted age and sex structure of the inmigrant population with age and sex-specific data on visiting rates to or by doctors, health visitors or social workers. The latter can be obtained from local and health authorities.

In the case of education services, demand also clearly depends on the age structure of the inmigrant population, since provision must be made for all children between the ages of 5 and 16. However, there are complications, given that this provision can be made either by the state or the independent sector, and that some children below and above compulsory school age may also require school or college places. In the space available, it is not possible to look at each service area in detail. This section instead focuses on state school services as an example of how the demand for additional service provision might be estimated.

Predicting the demand for local school places which will fall on state schools requires separate estimates of:

- *The total number of children aged 0–18 years* brought into the impact zone by inmigrant employees (see §3.4.1).
- *The number of these children below compulsory school age (0–4 years), aged 5–16 and above school leaving age* (see §3.4.1).
- *The proportion of those children below compulsory school age likely to require nursery education in the impact zone.* Information on the proportion of this age group currently attending nursery schools, both nationally and in individual LEAs, can be obtained from the Department for Education's *Statistical bulletin.* The proportion attending nursery schools in the relevant LEA area could then be assumed to apply to the inmigrant children associated with the project. This assumes that there will be no changes in LEA policies on the provision of nursery education before the project gets under way.
- *The proportion of children aged 5–16 attending primary and secondary schools in the impact zone.* All children of school age can be assumed to attend LEA, grant maintained or independent schools in the impact zone.
- *The proportion of children above school leaving age (16–18 years) likely to remain in full or part time education in the impact zone.* Data on the proportion of this age group attending secondary schools or further education colleges are published in the *Statistical bulletin* and are updated annually. This information is available for individual LEAs, as well as

nationally. Again, these proportions could be assumed to apply to the children brought into the area by inmigrant employees. In practice, the likelihood of this age group remaining in education depends very much on the availability of alternatives, in the form of employment and training opportunities. It must therefore be assumed that there will be no changes in the relative attractiveness of these various alternatives.

* *The proportion of inmigrant pupils attending independent schools.* The *Statistical bulletin* shows the proportion of pupils of different ages attending independent schools, in England as a whole, regions and some subregions. For example, in 1990, the proportion of pupils in England attending independent schools was 5% for 5–10 year olds, 9% for 11–15 year olds and 19% for those aged 16 or over. These national proportions could be assumed to apply to the children brought into the area, again assuming no changes in the relative importance of the state and independent sectors before the project gets under way. The estimated number of pupils attending independent schools should then be subtracted from the total school place requirement to show the number of places required in LEA and grant maintained schools and colleges.

The demand for additional school places is unlikely to be evenly distributed throughout the impact zone. The extent to which demand is geographically concentrated or dispersed will determine the total number of schools affected and the likelihood of strains on educational provision in individual schools. The distribution of school place requirements will largely reflect the place of residence of inmigrant families. Unfortunately, the latter is difficult to predict in the absence of relevant monitoring data. §3.4.1 outlined a possible approach to prediction, but it may be desirable to present a series of estimates based on different assumptions about the concentration or dispersal of inmigrant families.

Assessing significance

An important indicator of the significance of local service impacts is the extent to which *capacity thresholds* are exceeded as a result of the demands arising from the inmigrant population. Let us consider the example of the demand for local school places. If the current accommodation capacity in a school is expected to be almost fully utilized in the absence of the project, and pupil:teacher ratios are already high, then even a small project-induced increase in pupil numbers may create a need for additional classrooms and/or extra teaching staff. In the absence of such additional provision, the result may be overcrowding and an unacceptable increase in class sizes. By contrast, a large increase in pupil numbers in a school with a considerable amount of underutilized capacity and low pupil:teacher ratios may be much less significant. Increases in pupil numbers in such schools may still exert an impact, even if they do not put the available capacity under pressure. Class sizes will be larger than would otherwise have been the case, and additional staff time may need to be devoted to individual assessments of incoming pupils. Assessment of significance therefore requires

43

information not only on the likely project-related increase in demand, but also the existing (and projected) utilization of service capacity.

In certain circumstances, additional service demands may be seen as beneficial. For example, an influx of pupils into a small rural primary school with declining rolls may safeguard the future of the school, either in the short term (during construction) or in the medium to long term (during operation). The nature and significance of local service impacts will change as the project progresses through its various stages. The inmigrant population, including children, will tend to become older, with the result that the type of services demanded will tend to change over time. For example, there will tend to be a shift away from nursery and primary school demand towards secondary school demand. This tendency will be counterbalanced to some extent by the turnover of employees (bringing new, younger, families into the area) and by births in the original inmigrant families.

3.4.4 Local authority finances

Estimating implications for local authority revenue
Major projects can affect the revenues received by their host local authorities in two main ways. First, inmigrant employees buying properties in the impact zone will become liable to pay council tax in the local authority area into which they have moved. The likely *increase in council tax receipts* can be roughly estimated by multiplying the predicted number of inmigrant employees purchasing properties locally by the existing average council tax payment in the LAD concerned. Methods to estimate the proportion of inmigrant employees buying properties were outlined in §3.4.2. Information on existing council tax payments should be obtained directly from the local authority (or authorities) in the impact zone. If project employees purchase houses mainly in higher than average price bands, this simple method will underestimate the actual increase in receipts.

The second way in which the project will affect local authority revenues is through the *population changes* brought about by the arrival of inmigrant employees and families. These changes will affect the *standard spending assessment* of the local authority concerned. The standard spending assessment (SSA) is central government's assessment of how much it would cost the local authority to provide a typical or standard level of service. SSAs are a key determinant of the distribution of *revenue support grant* from central government to individual authorities. The SSA consists of a basic amount per head of population, with various weightings to reflect particular local circumstances such as the number of primary and secondary school pupils, daily commuting flows into and out of the area and the extent of social deprivation. The population data used in the calculation of SSAs are the official mid-year estimates published by the OPCS/GRO. To the extent that inmigrant employees and their families are picked up in these official estimates, the SSA for the authority concerned should be adjusted upwards.

However, these adjustments will not take place immediately. There is usually about a two-year time lag between an actual increase in population and the resulting increase in revenues.

In theory, if the project results in a 5% increase in a local authority's population (with no changes in the structure of that population), then this should be reflected – after a time lag – in a 5% increase in the authority's SSA and – all other things remaining equal – a similar increase in revenue support grant. It should therefore be possible to roughly estimate the likely increase in SSA and revenues associated with the project-induced increase in population. In practice, things are rather more complicated.

First, not all inmigrant employees or families will be picked up by the mid-year population estimates, especially during the construction stage. Construction employees not bringing families into the area are unlikely to appear on the electoral register in the impact zone, re-register with a local doctor or appear on local property registers for council tax purposes. They are therefore unlikely to be picked up by any of the data sources used to arrive at the mid-year estimates. As a result, any increase in the authority's SSA is unlikely fully to match the actual percentage increase in population. Any estimates of increased SSAs as a result of the project must incorporate some allowance for this underrecording of the actual population increase, at least during construction.

A second problem is that the increase in the SSA as a result of the project will reflect not only the size of the project-induced population increase, but its precise structure (e.g. the number and ages of children); the latter is more difficult to estimate accurately. A final problem is that an increase in the SSA for an authority does not necessarily produce an equivalent percentage increase in revenue support grant. For example, if the project results in an increase in the number of council tax payers (as outlined above), the additional revenue received will be taken into account in determining the amount of grant distributed to the authority.

Local authorities do not benefit directly from the payment of *non-domestic rates* by the project developer during construction and operation. Receipts from non-domestic rates are pooled nationally and then redistributed to individual authorities on a per capita basis. However, to the extent that the project-related population increase is recorded in the official mid-year estimates, the authority should receive increased receipts from the non-domestic rate pool. The likely increase can be calculated roughly by multiplying the existing per capita receipts from the pool by the expected increase in local population. However, for the reasons given above, the recorded increase in population in official estimates may not fully reflect the actual increase (especially during construction), and this should be allowed for in any estimates.

Information on existing levels of revenue support grant, SSAs and receipts from the non-domestic rate pool can be obtained from the local authority or the Chartered Institute of Public Finance and Accountancy's (CIPFA's) Statistical Information Service (see Appendix B).

Estimating implications for local authority expenditure

Inmigrant employees and their families will place demands on a range of services provided by local authorities, health authorities and other public bodies. These service demands will entail additional expenditure for the authorities concerned. Let us take the example of the arrival of pupils into LEA schools within the impact zone. Methods to estimate the likely number of such pupils were outlined in §3.4.3. But how can the additional expenditure necessitated by the arrival of these pupils be estimated?

The simplest approach would be to multiply the expected number of pupils by the existing annual average cost per pupil in the LEA concerned. Data on average expenditure per pupil in each individual LEA are published annually by CIPFA in *Education statistics, Estimates* (for the current year) and *Actuals* (for the previous year). This average cost method has two main weaknesses. First, costs per pupil vary according to the age group involved – they are invariably higher for secondary school pupils than for primary school pupils. A more sophisticated approach would therefore involve combining estimates of the expected numbers of inmigrant pupils in particular age groups with the average cost per pupil for each of these age groups. The CIPFA publications noted above provide data on costs per pupil for each LEA, broken down into single-year age bands.

A second and more fundamental weakness of the average cost approach is that it fails to distinguish between fixed and variable costs in service provision. Fixed (or overhead) costs do not vary in response to changes in the number of pupils in individual schools. Examples include most of the costs associated with school buildings, maintenance, heating, cleaning, rates and central support and management functions. Variable costs are those which change in response to changes in pupil numbers. Examples include capitation allowances (which are based on the number and ages of children on the roll at the beginning of each year) and teachers' salaries (if the increase in pupil numbers results in additional staff being taken on). Existing average costs per pupil include both fixed and variable costs, and are therefore unlikely to be a reliable guide to the actual costs incurred as a result of an influx of pupils. In schools with considerable surplus capacity, in which the arrival of pupils does not create a need for additional staff or accommodation, the additional cost per pupil is likely to be considerably lower than existing average costs in the LEA as a whole. Given these complexities, accurate prediction is unlikely to be achieved. It is recommended that any estimates should incorporate various assumptions concerning the extent of fixed and variable costs.

Similar estimates should also be made of the additional expenditure incurred in other service areas, such as police, fire, recreation and personal social services. Expenditure in some of these service areas may be rather unresponsive to small changes in population, unless critical capacity thresholds are likely to be approached. Information on existing local authority expenditure per head of population in these service areas is again available from CIPFA's Statistical Information Service. The proposed development may also necessitate the provision of improved infrastructure by the local authority (or authorities) concerned. This will typically include the construction of new roads or the improvement of existing ones. The local authority will normally require the developer to fund the full capital cost of such provision.

Assessing significance

Predictions should be made of the future stream of project-induced revenues and expenditures, in each year of the construction and operational stages of the project life. Although these two streams may balance over the lifetime of the project, there are likely to be periods during which there are shortfalls or surpluses. Any significant shortfalls in revenues, and their duration, should be noted. For many projects, the build-up of revenues is likely to lag behind the need for additional expenditure. For example, additional population will create immediate demands on local services, but will be reflected in increased revenue support grant only after a time lag. The construction stage may also see little increase in revenues, with most inmigrant employees not buying properties locally and not being recorded in official population estimates.

3.4.5 Other social impacts

Other social impacts can be wide-ranging and may include:
- increased crime levels locally, particularly during the construction stage, associated with an influx of young male itinerant employees into the impact zone
- changes in the occupational and socio-economic mix of the population, and
- linked to the above, problems in the integration of incoming employees and families into the local community and community activities. There may be a clash of lifestyles or expectations between incomers and the existing host community.

Prediction of such impacts is difficult, but is likely to require at least a comparison of the predicted age, sex and occupational profile of inmigrants with that of the existing population in the impact zone. The latter can be determined largely by reference to 1991 census data. Monitoring studies may be helpful in indicating the likely scale of certain impacts (see, for example, Glasson et al. (1992), for an assessment of the likely impact of a major construction project on local crime levels).

3.5 Mitigation

Several approaches to the mitigation of *demographic impacts* are available. The most basic would be to encourage the maximum recruitment of labour from within daily commuting distance of the project site, thereby reducing the number of employees and families moving into the impact zone. Possible methods to encourage the use of local labour by developers and contractors were discussed in Chapter 2. In addition, developer policies on travel, accommodation and relocation allowances might be used to influence the relative attractiveness of daily and weekly commuting versus relocation. Such policies might lead to some reduction in the proportion of inmigrant employees relocating and bringing families into the area.

The mitigation of local *accommodation impacts* is likely to involve attempts either to provide additional accommodation for the workforce or to encourage the use of unoccupied or underoccupied accommodation in the impact zone. Encouragement of the sharing of accommodation would also be a useful mitigation measure, but it is uncertain how this could be carried out in practice. The provision of accommodation specifically for the workforce, in the form of purpose-built hostel or additional caravan accommodation, has already been discussed in §3.4.2. The success of such provision as a mitigation measure will depend on its attractiveness in relation to the alternatives available locally, in terms of location, facilities and cost. The release of unoccupied accommodation is rather more difficult to influence. During construction, one approach might involve the placing of advertisements in the local press requesting those willing to provide workforce accommodation to contact the developer. This may alert potential providers of accommodation to the opportunities presented by the project. In some circumstances, it may be considered desirable to encourage the use of local B&B and other tourist accommodation (e.g. to boost occupancy levels outside a short tourist season). This could be achieved by the compilation of a directory of local accommodation establishments by the developer, and its use by contractors and individuals seeking accommodation in the area.

Impacts on local services and local authority finance can be partially mitigated by the direct provision of certain facilities by the developer. Examples might include a medical centre and firefighting equipment and staff located on the project site, as well as recreational facilities for the workforce. Developer funding of additional local authority provision necessitated by the project is also likely to be requested. Funding of local community projects may also be offered as partial compensation for the adverse impacts of the project.

3.6 Monitoring

Existing monitoring of demographic and social impacts is limited. Ideally, such monitoring should consist of three key elements. The first of these is *the establishment of administrative systems which ensure a regular flow of information* on the total number of people directly employed on the project and the mix of local and inmigrant employees. During most construction projects, the developer is likely to request this type of information from the contractors on site as a routine part of project management, for example to monitor earnings levels, bonuses and allowances across the construction site. The provision of such information can be made a contractual requirement. Existing monitoring systems can therefore often be used with only minimal modifications. For most projects, information on the operational workforce should be directly available to the developer via its own personnel records. However, this will not be the case for certain developments, such as business parks or retail projects, where several employers occupy the floorspace provided by the developer. In such cases, the developer should establish data collection systems covering all occupants, with the submission of information being required on a regular basis.

The systems described above will, at best, only indicate the total number of employees moving into the impact zone. Information on the number of these employees bringing families, the characteristics of these families, the type and location of accommodation taken up and the use of local services, can be obtained only directly from the workforce itself. The second component of any monitoring system must therefore be *a regular (possibly annual) survey of the project workforce*. This is likely to involve interviewing a sample of the workforce, with care taken to ensure a representative coverage of all types of employees. Such surveys can also be used to obtain information on other issues, such as workforce expenditure and journey to work patterns.

The final element in any monitoring system should be *the monitoring of various social and economic trends within the impact zone*. This can range from regular monitoring of house prices or rent levels, the amount of housebuilding, occupancy levels in local B&B and other accommodation, school rolls, doctors' list sizes or crime levels. Such trends should be compared with those in the wider county or region, as well as those nationally. In addition, periodic surveys of local service providers (e.g. head-teachers or doctors) should be carried out.

CHAPTER 4

Noise

Riki Therivel

4.1 Introduction

Virtually all new developments have noise impacts. Noise during construction may be because of such activities as land clearance, piling, and the transport of materials to and from the site. During operation noise levels may decrease for some forms of developments such as science parks or new towns, but may remain high or even increase for developments such as mineral extraction works, new roads, or industrial processes. Demolition is a further cause of noise. As a result, the EIAs for most developments consider noise.

Noise is a major and growing form of pollution. It can interfere with communication, increase stress and annoyance, cause anger at the intrusion of privacy, and disturb sleep, leading to lack of concentration, irritability, and reduced efficiency. It can contribute to or aggravate stress-related health problems such as high blood pressure. Prolonged exposure to high noise levels can cause deafness or partial hearing loss. Noise can also affect property values and community atmosphere. In 1990, local environmental health officers in the UK received almost three complaints about noise per thousand people, more than double the number ten years earlier. By the early 1990s, about half of the dwellings in England and Wales were exposed to daytime noise levels higher than the levels desirable to prevent significant community annoyance (DoE 1992b).

Although most EIAs – and this chapter – are limited to the impact of noise on people, noise may also affect animals, and in certain cases EIAs will need to include specialist studies on these impacts. Bregman & Mackenthun (1992) summarize previous studies on animals' reactions to noise. Although noise is linked to vibration, this chapter deals only with noise; most EIAs do not cover vibration.

4.2 Definitions and concepts

4.2.1 Definitions

Noise is defined as unwanted sound. *Sound* in turn consists of pressure variations detectable by the human ear. These pressure variations have two characteristics, frequency and amplitude. Sound *frequency* refers to how quickly the air vibrates, or how close the sound waves are to each other (in cycles per second, or Hertz (Hz)). For example, the sound from a transformer has a wavelength of about 3.5 m, and hums at a frequency of 100 Hz; a television line emits waves of about 0.03 m, and whistles at about 10,000 Hz or 10 kHz. Frequency is subjectively felt as the *pitch* of the sound. Broadly, the lowest frequency audible to humans is 18 Hz, and the highest is 18,000 Hz. For convenience of analysis, the audible frequency spectrum is often divided into standard octave bands of 32, 63, 125, 250, 500, 1k, 2k, 4k and 8kHz.

Sound *amplitude* refers to the amount of pressure exerted by the air, which is often pictured as the height of the sound waves. Amplitude is described in units of pressure per unit area, micropascals (μPa). Often the amplitude is converted to sound *power*, in picowatts (10^{-12} watts), or sound *intensity* (in 10^{-12} watts/m^2). Sound intensity is subjectively felt as the *loudness* of sound. However, none of these measures are viable because of the vast range which they cover (see Table 4.1). As a result, a logarithmic scale of decibels (dB) is used. A sound *level* in decibels is given by

$$L = 10 \log_{10} (P/p)^2 \text{ dB},$$

where P is the amplitude of pressure fluctuations, and p is $20\,\mu$Pa, which is considered to be the lowest audible sound. The sound level can also be described as

$$L = 10 \log_{10} (I/i) \text{ dB},$$

where I is the sound intensity and i is 10^{-12} watts/m^2, or by

$$L = 10 \log_{10} (W/w) \text{ dB},$$

where W is the sound power, and w is 10^{-12} watts. The standard used should be explicitly stated in any noise assessment. The range of audible sound is generally from 0 dB to 140 dB, as is shown in Table 4.1.

Because of the logarithmic nature of the decibel scale, a doubling of the power or intensity of a sound, for instance the combination of two identical sounds, leads to an increase of 3 dB, not a doubling of the decibel rating. For example, two similar lorries sound as loud as one lorry plus 3 dB. Multiplying the sound power by ten (e.g. ten lorries) leads to an increase of 10 dB. Figure 4.1 shows how the dB increase can be calculated if one noise source is added to another. An increase of 3 dB is barely detectable by the human ear; an increase of 10 dB is broadly perceived as a doubling of loudness.

The physical level of noise does not clearly correspond to the level of *annoyance* it causes, yet it is the annoyance caused by noise that is important in EIA. For this reason it is important to distinguish between objective (physical) noise

Table 4.1   Sound pressure, intensity and level.

Sound pressure (μPa)	Sound power (10^{-12} watt), or sound intensity level (10^{-12} watt/m^2)	Sound level (dB)	Example
200,000, 000	100,000,000,000,000	140	threshold of pain
	10,000,000,000,000	130	riveting on steel plate
20,000,000	1,000,000,000,000	120	pneumatic drill
	100,000,000,000	110	loud car horn at 1 m
2,000,000	10,000,000,000	100	alarm clock at 1 m
	1,000,000,000	90	inside underground train
200,000	100,000,000	80	inside bus
	10,000,000	70	street-corner traffic
20,000	1,000,000	60	conversational speech
	100,000	50	business office
2000	10,000	40	living room
	1,000	30	bedroom at night
200	100	20	broadcasting studio
	10	10	normal breathing
20	1	0	threshold of hearing

levels and subjective (perceived) loudness. The human ear is more sensitive to some frequencies than to others: it is most sensitive to the 1 kHz, 2 kHz, and 4 kHz octaves, and much less sensitive at the lower audible frequencies. For instance, tests of human perception of noise have shown that a 70 dB sound at 4 kHz feels like about 75 dB, and at 63 Hz it feels like about 45 dB. Since most sound analyses, including those in EIA, are concerned with the loudness experienced by people rather than the actual physical magnitude of the sound, a so-called *A-weighting curve* is used to give a single figure index which takes account

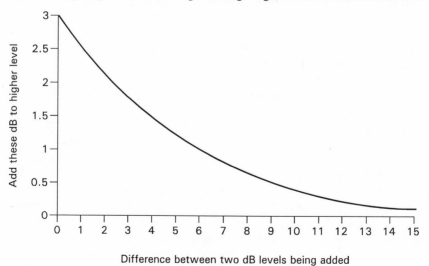

Difference between two dB levels being added

Figure 4.1   Adding two noise sources.

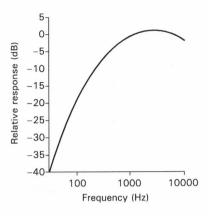

Figure 4.2   A-weighting curve.

of the varying sensitivity of the human ear (Fig. 4.2). Most sound measuring instruments incorporate circuits which carry out this weighting automatically, and all EIA results should be A-weighted (dB(A)). Other weightings exist, but are rarely used.

Noise levels are rarely steady: they rise and fall with the types of activity taking place in the area. Noise levels can be described in several ways. One is the LA_{10} level, the dB(A) level which is exceeded for 10% of the time; this indicates the noisier sounds and is generally used for road traffic assessment. Another is the LA_{90} level, the level exceeded for 90% of the time; this indicates the noise levels during quieter periods, or the *background noise*. LA_1, exceeded 1% of the time, is indicative of the loudest noise heard; alternatively the LA_{max} can be used. Finally, the equivalent continuous noise level LA_{eq} is a notional steady noise level which, over a given time, would provide the same energy as the intermittent noise. Figure 4.3 gives an example of these measures.

Many noise standards specify the length of time over which noise should be measured. For instance the Noise Insulation Regulations 1975 are based on measures of $dBLA_{10}$ (18 hour); the average of the L_{10} levels, in dB(A), measured in each hour between 6 a.m. and midnight. Mineral Planning Guidance Note 11 refers to $dBLA_{eq}$ (1 hour), the equivalent continuous noise level, in dB(A), during one hour of a weekday.

4.2.2 Factors influencing noise impacts

The level of noise from an activity (or source) heard by the receiver is influenced by a range of factors, which can be broken down into three categories: those related to (a) the components of the source (e.g. rock drill, dump truck), (b) the transmission paths from source to receiver (through direct sound, reflected sound, and groundborne vibrations), and (c) the receiver.

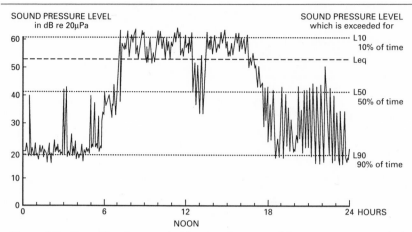

Figure 4.3   Sound levels exceeded for stated percent of the measurement period.

At the *source*, the sound level is the main determinant. The frequency of the sound is important, both because humans have different sensitivities to different frequencies and because higher frequencies are more easily attenuated over distance than are lower frequencies. Noise levels tend to be lowest in the early hours of the morning and highest during midday, with weekends quieter than weekdays; the same noise will be heard more clearly when background levels are low than when they are high (see Fig. 4.1). The duration of the sound, and whether it is intermittent or continuous will also influence its impact.

The directive pattern of the source has a great influence. Where a source is free to radiate sound in all directions (e.g. from an aeroplane), the sound energy received at any point is inversely proportional to the square of the distance from the source. If the distance from the source is doubled, the energy is reduced to a quarter. This means a reduction of 6dB for every doubling of the distance (halving the energy reduces the sound by 3dB, quartering the energy reduces it by 6dB). However, this rarely happens. Much more often the source is located on a reflecting plane such as the ground (e.g. from a car) or a wall (e.g. from a burglar alarm), so twice as much power is radiated to a given direction: 3dB are thus added to the original noise level before subtracting 6dB for each doubling of distance. Similarly, where a source is located on the ground *and* near a wall, four times as much power is radiated; 6dB are added. Seasonal variations can also affect noise at the source.

In the *pathway* from source to receiver the central determinants are distance, as was explained above, and barriers or screening, which reduce sound by reflection and absorption; screening is discussed further in §4.6. Other determinants include wind direction, topography, and atmospheric and meteorological conditions. Noise down wind from the source can be 20dB greater than up wind. Ground absorption can occur over long distance if soft ground surfaces such as grass can absorb sound. On the other hand, hard surfaces reflect sound: this reflection can either interfere with the direct sound waves, reducing them, or it can harmonize with them and accentuate them. Atmospheric absorption reduces sound over distance in function of the frequency of the source (higher frequencies are more reduced over distance than lower frequencies), the temperature, humidity and air pressure. The temperature gradient affects the propagation of sound because the velocity of sound increases with increasing temperature, which in turn varies with height; where temperature decreases with height (e.g. during the day), the sound is bent away from the ground, so that noise on the ground is less. Conversely, where temperature increases with height (e.g. at night), sound is bent towards the ground, so ground-level noise is greater.

Finally, characteristics of the *receivers* can affect the noise impact. Some receptors such as schools and hospitals are more sensitive than others, and sensitivities may vary with the time of day. Different noise levels are thus acceptable for different receivers. The receiver's perception of the attitude of the site operator can also affect sensitivity to noise.

4.3 Legislative background and interest groups

Noise is controlled in three ways: by controlling overall noise levels, setting limits on the emission of noise sources, and keeping people and noise apart. The type of regulation, standard or guideline applicable in an EIA will depend on the type of activity proposed. Table 4.2 lists UK noise regulations and standards for specific types of activity. The most important of these are summarized below; a longer discussion can be found in e.g. Garbutt (1992), Haigh (1990), Hughes (1992), Johnson & Corcelli (1989), or McManus (1992).

DoE/WO Circular 10/73, *Planning and noise* (DoE 1973) gives guidance to local planning authorities on how to assess the likely impact of noise from roads, aircraft and fixed installations on residential areas. The circular is under review, and emerging government advice is found in the draft PPG on planning and noise of 1992. The Scottish equivalent is Circular 23/73.

The Land Compensation Act 1973 allows people whose enjoyment of their property has been adversely affected by the execution of public works to be compensated. Under this act, regulations can be enacted to determine when compensation is due. To date, only the Noise Insulation Regulations 1975 have been introduced, which apply to new highways. They require highway authorities to provide noise insulation for residential properties if they are (a) within 300 m from a new or altered highway, (b) not subject to compulsory purchase, demolition or clearance, (c) not already receiving a grant for noise insulation works, (d) subject to 18-hour L_{10} noise levels over 67.5 dB(A), (e) subject to an increase of at least 1 dB(A) over the existing noise level, and (f) on a new road which contributes at least 1 dB(A) to the final noise level.

Under the Control of Pollution Act 1974, a local authority can control noise from construction sites. It can serve a Section 60 notice which specifies the plant and machinery which may or may not be used, working hours, and noise levels. A developer may apply for prior consent for construction works through a Section 61 consent. The local authority can also designate noise abatement zones in which specified types of development may not exceed specified noise levels.

Table 4.2   Noise regulations, standards, and guidelines.

Type of project	Relevant regulation, etc.
Road	DOT Design Manual 11
	DOT Calculation of Road Traffic Noise
	Memo on the Noise Insulation (Scotland) Regulations
	TRL Supplementary Report 425
	Land Compensation Act 1973
	Noise Insulation Regulations 1975 (SI 1975/1763)
	DoE/WO Circular 10/73, SO Circular 23/73
	Road Traffic Regulation Act 1984
Airport	DoE/WO Circular 10/73, SO Circular 23/73
	Civil Aviation Act 1982
Railway	DOT Railway Noise and the Insulation of Buildings
Industrial	DoE/WO Circular 10/73, SO Circular 23/73
	BS4142
	Control of Pollution Act 1974
	Environmental Protection Act 1990
Mineral workings	DoE/WO Circular 10/73, SO Circular 23/73
	Mineral Planning Guidance note 11
	BS5288
	Environmental Protection Act 1990
Construction and other open sites	Environmental Protection Act 1990
	BS5288
	Control of Pollution Act 1974
Occupational exposure	Noise at Work Regulations 1990
	Health and Safety at Work Act
	EU Directive 86/188

The Environmental Protection Act 1990 makes statutory nuisances, including noise from a premise which is prejudicial to health or a nuisance, subject to control by the local authority.

The 1988 DOT report *Calculation of road traffic noise* (DOT 1988) gives a procedure for predicting noise in areas where noise is dominated by traffic noise, at 1 m from the relevant facade, and at a stated height above ground; this can be extrapolated to distances of up to 300 m from the road. Calculations incorporate information about traffic volume, vehicle speeds, the percentage of heavy goods vehicles, the road gradient, road surface, and distance from source to receiver. This procedure must be used for calculations made under the Noise Insulation Regulations; in these cases the noise levels prior to a new road scheme are compared with the highest noise levels expected within 15 years of the opening of the scheme. The *Memorandum on the noise insulation (Scotland) Regulations 1975* is the Scottish equivalent of this report. The TRL supplementary report 425, *Rural traffic noise predictions – an approximation,* gives procedures for how changes in noise levels can be approximated for quiet rural locations.

The DOT's *Design manual for roads and bridges vol. 11: Environmental assessment* (DOT 1993), the successor to the *Manual of environmental appraisal* (DOT 1983), describes procedures for assessing the environmental impact of road schemes where traffic increases or decreases of 25% of more (about 1 dB(A)) are expected in the year the scheme opens.

BS4142, Rating industrial noise affecting mixed residential and industrial areas (BSI 1990), is a method for determining the increase in noise levels from new fixed buildings and plant, and the likelihood of this increase causing complaints. It does not apply to noise from open sites. *BS5288, Noise control on construction and open sites* (BSI 1984), covers noise from construction and other open sites. BS5288 gives tables of indicative noise levels from various types of mobile plant, and provides a method for calculating noise from vehicles on haul routes.

Mineral Planning Guidance Note 11, *The control of noise at surface mineral workings* (DOE 1993b) gives guidance for noise levels at extraction sites.

Other relevant legislation includes the Public Health Act 1961, Health and Safety at Work, etc. Act 1974, Motor Vehicles (Construction and Use) Regulations 1978, Road Traffic Regulation Act 1984, Civil Aviation Act 1982, Local Government (Miscellaneous Provisions) Act 1982, Town and Country Planning Act 1990, Town and Country Planning (Scotland) Act 1972, the DOT's 1991 publication *Railway noise and the insulation of dwellings*, BS8233 on sound insulation and noise reduction for buildings (BSI 1987), local authority by-laws, and building regulations that require houses and flats to be built to prescribed noise insulation standards. Various EU Directives control noise from vehicles, aircraft and construction plant. Individuals may resort to common law if they suffer annoyance from noise; this generally involves proving the existence of a private nuisance, namely an unlawful interference with their land, their use and enjoyment of their land, or some right enjoyed by them over the land or connected with it. Existing standards do not always directly apply to the development proposed. In such cases, reference may be made to conditions at existing developments, provided that they are similar in all relevant impacts.

Finally, the local authority environmental health officer has a duty under the Environmental Protection Act to ensure that noise nuisances do not occur in his/her area. The officer's view will be sought by the planning authority when an application is received, so the developer should discuss plans with him/her prior to submission. The local planning authority may require a Section 106 agreement concerning noise before granting planning permission.

4.4 Scoping and baseline studies

The details of an EIA's baseline or ambient noise survey are usually determined as a result of the relevant legislation and discussions between the developer and the local authority. The survey should (a) identify noise-sensitive locations in a relevant area, and (b) measure and describe the background noise in that area. The survey should record the quietest conditions that typically occur in an area (e.g. on a quiet Sunday morning in addition to or rather than a weekday afternoon); this is because the greatest increase in noise caused by a proposed development will be in comparison with these quiet conditions. If under particular conditions (e.g. a specific wind direction) higher background levels commonly occur, these should also be recorded. For projects that affect a large number of people (e.g. a road), locations may be identified that are representative of groups of affected people.

Noise is primarily a local impact, and it is unlikely that much information about the specific area in question can be obtained from desktop studies. Various maps may help to identify noise-sensitive locations (e.g. residences, hospitals, schools) in the area, but this should be confirmed by a site survey. Local authority environmental audits may include noise data, but are unlikely to be site-specific. Virtually all EIAs rely on noise measurements carried out at the site.

Noise-measuring equipment is portable and battery-powered, and usually consists of (a) a microphone which converts changes in ambient pressure into an electrical quantity (usually voltage), (b) a sound-level meter which amplifies the voltage signals, averages them, and converts them to dB, (c) an analyzer which records noise descriptors (e.g. L_{eq}, L_{10}) over a period of time, and (d) a reference sound source against which to calibrate the equipment. In some cases, several of these may be incorporated into the same piece of machinery. The sound-level meter may have different types of settings, corresponding to different ways of averaging voltage; slow (over 1 sec), fast (0.125 sec), and sometimes peak and impulse. A windshield may be needed in windy conditions.

The precise procedures for measuring noise (e.g. duration of measurement, location of equipment, and measurement levels) are specified in the relevant regulations or guidelines (see §4.3). Broadly, noise measurements involve:

- taking note of the equipment used, including manufacturer and type
- taking note of the date, weather conditions, wind speed, wind direction, and whether the equipment is attended or not
- calibrating the sound meter and microphone
- setting up the microphone at the appropriate site (check guidelines for criteria)
- noting the precise location where measurements are taken (e.g. on a map or using grid references)
- taking measurements using the criteria from the relevant guidelines (e.g. continuous for 24 hours, or for 1 hour; using fast weightings for traffic or slow for construction noise)

- noting start and finish times, and any factors (e.g. types of background noise, whether equipment was left unattended overnight) which could affect the measurements, and
- checking the calibrations.

Table 4.3 gives an example of baseline noise data. Generally an EIA includes such data and a description of how they were collected, along with a map showing the location of the measurement points. Where noise monitoring is carried out during construction and operation, the same measurement points will generally be used.

Table 4.3 Example of baseline noise data; location 3: Campus Close.

Date	Start of period	Noise levels, in dB(A)					Comments
		L_{90}	L_{50}	L_{10}	L_1	L_{eq}	
1 April	1500	56	57	60	62	58	Mostly traffic noise
	2200	46	49	53	55	50	Traffic, dog barking
2 April	0720	55	57	59	61	57	Traffic, birdsong

4.5 Impact prediction

The aim of noise prediction in EIA is to identify the noise levels which may occur, both in the short and long term, as a result of the development; the overall noise levels resulting from the accumulation of background noise, and noise from the proposed development and other concurrent developments; and the significance of these factors.

Predicting noise *levels* is a complex process that incorporates a wide range of variables, including:

- existing baseline noise levels
- the type of equipment, both mobile and fixed, used at the site (see BS5228 for indicative noise levels from mobile plant; Table 4.4 gives examples of typical sound levels from construction equipment)
- the duration of various stages of construction and operation
- the time of day when the equipment is used
- the attitude of the site operator
- the location of the receivers (e.g. residences, hospitals, footpaths), how they are used, and their sensitivity to noise
- the topography of the area, including the main forms of land use and any natural noise barriers, and
- meteorological conditions in the area.

These will affect the amount and type of sound coming from the site (e.g. type of equipment, duration of workings), how that sound travels (e.g. distance between source and receptor, topography, meteorology), and the response of the receivers (e.g. timing of workings, sensitivity to noise).

Table 4.4   Examples of typical sound levels from construction equipment (BS5228).

Type of equipment	Sound level, in dB(A), at 7 m
Unsilenced pile-driver	110
Unsilenced truck scraper, grader	94
Unsilenced pneumatic drill	90
Unsilenced compressor	85
Concrete breaker	85
Crane	85
Unsilenced generator	82
Sound reduced compressor	70

Procedures for predicting noise levels are specified in many of the regulations listed in §4.3. These procedures are too cumbersome and diverse to discuss in detail here; they are often set up as computer models. The reader is referred to the relevant regulations and standards for further information. It may often be necessary to carry out noise monitoring at a similar existing activity or development in order to predict the effects of a proposal. In cases where a project affects many receivers (e.g. a road or rail line), it may be necessary to predict noise levels for various types of "typical" receivers and to extrapolate these, with adjustments for local circumstances, to other similar receivers.

More consensus exists about the *significance* of changes in noise levels (the difference between predicted noise levels with and without the development). A change of 3 dB is barely detectable, a change of 5 dB is clearly detectable, and a change of 10 dB corresponds subjectively to a doubling or halving of loudness; Table 4.5 suggests possible significance criteria. Within this overall framework,

Table 4.5   Example of noise significance criteria (adapted from Arup Environmental 1993).

Criterion	Construction noise	Traffic noise
Severe adverse	Noise above traffic noise insulation thresholds for ≥8 weeks; insulation or permanent rehousing required	>15 dB increase
Major adverse	Noise above traffic noise insulation thresholds for <8 weeks; insulation or temporary rehousing required	10–15 dB increase
Moderate adverse	Noise above ambient levels for ≥8 weeks, but below traffic noise insulation thresholds	5–10 dB increase
Minor adverse	Noise above ambient levels for <8 weeks, but below traffic noise insulation thresholds	3–5 dB increase
None	Noise at or below ambient levels	<3 dB increase

however, variations exist. An increase in noise in an area already subjected to high noise levels may be more significant than a similar increase in an area with lower noise levels. The same level of noise at a noise-sensitive location will be more significant than that at a less sensitive location. Generally, to attain desirable limits for indoor comfort, outdoor levels of noise should not exceed 65 dBLA$_{eq}$ (daytime), and in new residential areas they should not exceed 55 dB. The World Health Organisation suggests that daytime outdoor noise levels should be below 50 dBLA$_{eq}$ to prevent significant community annoyance, but in cases where there are other reasons to be in an area, such as good schools, people may tolerate up to 55 dBLA$_{eq}$ (WHO 1980).

4.6 Mitigation

If the noise from the proposed development is likely to exceed the levels recommended in the relevant standards (see §4.3), mitigation will be necessary. Mitigation measures may be useful even if standards are met, to prevent annoyance and complaints, and as part of best practice procedures. Mitigation of noise is most effective, and potentially least expensive, when it is used at the source, before the noise has escaped. Less ideally, noise can be controlled at the receiver's end. Control on the transmission path is very difficult.

Control of noise *at the source* can take several forms. First, the equipment used or the modes of operation can be changed to produce less noise. For instance, rotating or impacting machines can be based on anti-vibration mountings. Internal combustion engines must be fitted with silencers. Aeroplanes can be throttled back after a certain point at take-off, to reduce their noise. Traffic can be managed to produce a smooth flow instead of a noisier stop-and-start flow. Well maintained equipment is generally quieter than poorly maintained equipment.

Secondly, the source can be sensitively located. It can be located (farther) away from the receivers, so that noise is reduced over distance. A buffer zone of undeveloped land can be left between a new road and a residential area. The source can be planned so that noisier components are shielded by quieter components; for instance housing can be shielded from a factory's noise by retail units. A noisy development can be sited near other sources of noise, since the combination of two noise sources leads to a lower increase in dB than having two separate sources (see Fig. 4.1). Natural or artificially constructed topography or landscaping can be used to screen the source.

The source can be enclosed to insulate or absorb the sound. Sound insulation reflects sound back inside an enclosure or barrier, so that sound outside the enclosure is reduced. However, merely enclosing the source is not the optimum solution, since the noise reverberates within the enclosure, and effectively increases the strength of the enclosed sound. Providing sound absorption within the enclosure avoids this happening. Sound absorption occurs where the enclosure or barrier absorbs the sound, converting it into heat. Sound absorbing walls need a method of cooling the heat generated; this in turn will allow some sound to escape because it provides a discontinuity within the enclosure. Most enclosures are constructed of both insulating and absorbing materials.

Details of requirements for noise enclosures and their effectiveness are very complex and require specialist knowledge. The reader is referred to the relevant standards and to textbooks on noise control (see for example Mulholland & Attenborough (1981), SRL (1991), or Lipscomb & Taylor (1978)). However, some general points can be made here. Methods of measuring sound insulation usually distinguish between airborne sound (noise) and structural sound (vibration), and any reference to insulation should distinguish between them. Broadly, the ability of a panel to resist the transmission of energy from one side of a panel to the other, or its *transmission loss*, will depend on (a) the mass of the panel (more mass = more transmission loss), (b) whether it is layered or not, with or without discontinuities between the layers, (c) whether it includes sound-absorbing material, and (d) whether it has any holes or apertures.

Acoustic fencing or other screens, either at the source or at the receiving end, can also reduce noise by up to 15 dB. The effectiveness of screens depends on their height and width (larger is better), their location with respect to the source or receiver (closer is better), their form (wrapped around the source or receiver is better), their transmission loss, their position with respect to other reflecting surfaces, the area's reflectivity, and whether they have any holes or apertures.

Noise screens can consist of topographical features or tree plantings, as well as of artificial materials. For instance, earth mounds (bunds) are often built alongside roads to absorb and reflect traffic noise away from nearby buildings. The primary acoustic benefit of plantings is perceived rather than actual: people appear to be less bothered by a development project's noise if they cannot see the project. Nevertheless, thick areas (30 m) of dense trees and underbrush may reduce noise by up to 3–4 dB at low frequencies and 10–12 dB at high frequencies. A mixture of deciduous and coniferous trees will give maximum noise reduction in the summer, and some reduction in the winter when the deciduous trees' leaves have fallen. Scattered decorative plantings are unlikely to be effective. It must be remembered that saplings take time to mature, and are unlikely to reduce noise for several years after planting. A recent development in noise screens is the "GreenWall": walls of woven dormant willow shoots 2 m apart are filled with soil and an inbuilt irrigation systems. When the willows root and grow, the wall combines the acoustic benefits of a soil bund with the aesthetic benefits of plantings.

Control of noise *at the receiver's end* is often similar to that at the source. Good site planning can minimize the impact of noise; for instance in a house by a busy road the more noise-sensitive rooms (e.g. bedroom, living room) can be shielded from the road noise by the less noise-sensitive rooms (e.g. kitchen, bathroom). A screen can be erected to reflect sound away from the receiver, for instance an acoustical screen between a highway and property. The equivalent of a noise enclosure can be achieved by soundproofing a house using double-glazed windows. The Land Compensation Act 1973 requires highway authorities to insulate houses affected by noise over a certain level.

4.7 Monitoring

Any conditions imposed as part of a project's planning permission are enforceable, including conditions related to noise. These can apply not only to noise levels (e.g. during construction, operation; during the day, night), but also to noise monitoring to be conducted by the developer (e.g. distance from the site boundary, frequency). If no planning conditions are set, local environmental health officers can still monitor noise from a site, for instance in response to local residents' complaints, to determine whether it is a statutory nuisance.

There are at present no requirements to compare any noise monitoring data with the noise predictions made in EIAs. A best practice EIA could propose not only noise-related planning conditions, but also a noise monitoring programme, and relate its findings to the EIA to improve future noise prediction methodology. The sites and noise measurement techniques used in carrying out baseline noise surveys should be such that comparable monitoring data can later be collected. However, given the current lack of legislative requirements for monitoring, this is unlikely to occur.

4.8 Conclusion

This has been only a brief introduction to a very technically complex topic. Noise predictions will require expert input, and possibly computer models. Readers are strongly urged to consult the relevant standards (see §4.3), as well as standard texts on acoustics and noise control.

CHAPTER 5
Traffic

Aidan Hughes

5.1 Introduction

It is implicit in the planning of many new developments that their success will depend on their attractiveness to car-borne visitors or users. Large retail parks, superstores and out-of-town business parks are examples. Industrial developments similarly are more attractive to the end user if they have high-quality access to motorways and trunk roads. Despite the growing awareness of the disbenefits of traffic congestion and the increasing interest in public transport, this trend seems likely to continue at least for the foreseeable future.

Assessment of the attraction of traffic to a development is not the same as assessment of the environmental impacts of new roads, and the latter are not considered in this chapter. However, much of the reference material is applicable to both, and a significant increase in traffic associated with a development may result in the need for additional roads. Traditionally, assessing the impact of the traffic generated by development has concentrated on the capacity of the road network to accommodate the increased traffic flow. Little or no interest has been shown in the wider environmental impacts. However, as with other EIA components, the need to identify and assess secondary impacts is increasingly recognized. This chapter is mainly concerned with primary impacts of a development on traffic; secondary impacts on components such as noise or landscape are touched on here, and are considered further in other relevant chapters.

5.2 Definitions and concepts

Traffic comprises (a) private cars and taxis, (b) vans, (c) goods vehicles, (d) buses, (e) motorcycles, and (f) pedal cycles. The mix of different vehicle types and their relationship to the immediate built environment will determine how changes in traffic flow might impact on the wider environment. It is possible to describe a stream of traffic on a length of road at a particular time with reference to:

- highway link capacity
- junction capacity
- driver delay
- speed
- number of accidents or accident rate
- proportion of heavy goods vehicles (HGVs)
- number of bus movements
- pedestrian/cycle flows crossing the road
- frequency of access
- turning movements
- location and type of on-street car parking, and
- the nature of frontage land uses.

5.3 Legislative background and interest groups

5.3.1 Legislative background

The need to assess the impact of a development on traffic as part of an EIA is identified in both European and national guidelines. *EC Directive 85/337* (CEC 1985) provides guidelines for the EIA of development and the impact on humans; within this could be classified impacts on traffic. The various UK regulations on EIA also refer to the issue. However, none of these provides information on the method of assessment.

A great deal of reliance has traditionally been put on the Department of Transport's *Manual of environmental appraisal* (DoT 1983) and its successor, the *Design manual for roads and bridges, volume 11: environmental assessment* (DoT 1993) when assessing the impacts of road traffic or, in the case of non-road developments, the secondary impacts of increases in traffic. The Scottish Office's publication *Scottish traffic and environmental appraisal manual* (SDD 1986) does the same for Scotland. However, the applicability of these publications to non-road schemes is limited, nor do they cover all of the impacts of a development on traffic.

With regard to the general consideration of traffic matters as part of a new development, several statutory procedures apply. A planning authority must by statute (the Town and Country Planning Act 1971 and the Local Government Act 1972 for England and Wales, the Town and Country Planning (Scotland) Act 1972 and the Local Government (Scotland) Act 1973 for Scotland) consult the relevant highway authority if the proposed development involves access to a highway, or will increase traffic on the highway network. For trunk roads or motorways the highway authority is the Secretary of State for Transport, and for other classified roads is the local highway authority (county council, metropolitan borough council, etc.). In Scotland, the Secretary of State for Scotland is responsible for trunk roads, special roads and motorways, and the local roads authority for other classified roads. It is becoming increasingly common for a highway authority to require the developer to carry out a formal traffic impact analysis to support the planning application. However, this is not a statutory requirement.

In those circumstances where the traffic impact assessment identifies a requirement to upgrade existing road infrastructure or provide new infrastructure as a result of the development, the works will normally be paid for in full by the developer. Alternatively, the relevant authority may require a contribution by the developer for the construction of a new road. At the time of writing the government was in the process of consultation concerning such contributions. The present legal framework for these arrangements is contained in Sections 106 of the Town and County Planning Act 1971 and Section 278 of the Highway Act 1980.

The Institution of Highways and Transportation's *Traffic impact assessment guidelines* (IHT 1994) propose methods for assessing a development project's impacts on traffic. The Institute of Environmental Assessment's *Guidelines for the environmental assessment of road traffic* (IEA 1993) focus on the environmental impacts of traffic from non-road developments, aiming to provide a "systematic, consistent and comprehensive coverage for the appraisal of traffic impacts for a wide range of development projects". Although these complementary publications are not official standards, they are useful and well accepted.

Table 5.1 provides a guide to more specific standards available for the assessment of various traffic impacts.

Table 5.1   Standards for forms of traffic impact.

Impact	Available standards
Driver delay and junction capacity	DOT programmes for the assessment of delay: ARCADY (roundabouts), PICADY (priority intersections), and OSCADY (traffic signal junctions)
Highway link capacity	Traffic Flows and Carriageway Width Assessment (DOT 1985)
Pedestrian delay	Design Manual 11 (DOT 1993)
Noise	Calculation of Road Traffic Noise (DOT 1988)
Air quality	Design Manual 11
Visual	Design Manual 11
Community effects (severance, etc.)	Design Manual 11
	The Appraisal of Community Severance (TRL 1991)
Accidents and safety	None, but refer to Guidelines for the Safety Audit of Highways (IHT 1990)
Vibration	Traffic Induced Vibrations in Buildings (TRL 1990)
	Damage to Structures from Ground-Borne Vibrations (BRE 1990)
Ecology	Design Manual 11

5.3.2 Interest groups and sources of information

The highway authority is the primary point of contact in establishing local priorities in applying these standards. The highway authority has statutory obligations with regard to maintenance, performance and control of development. They will have information on present traffic flows, trends in traffic flows, pedestrian and cycle flows, traffic growth forecasts and accident records, and possibly on likely future traffic flows based on traffic model predictions. Each highway authority produces an annual statement of its current and future programme of road construction, junction improvements and safety schemes.

The Department of Transport similarly hold information on traffic flows on the trunk road and motorway network. For larger road schemes the DoT will also hold information on predicted future traffic flows. A programme of current DoT road schemes is given in *Roads for prosperity* (DoT 1989b).

The relevant planning authority should also be contacted, since in many cases they will dictate the requirements for a wider assessment of traffic impacts than simply those related to driver delay, highway capacity, and road safety. Each local planning authority, depending upon its status, will have available copies of its future plans (e.g. unitary development plan, local plan). These provide information on land-use and transport planning at both a local and strategic level.

The planning authority will be able to advise on consultees such as the emergency services, resident groups, and local council members. However, an independent evaluation of those parties likely to be affected will provide a more comprehensive view.

5.4 Scoping and baseline studies

If the development is likely to increase traffic by 10%, or by 5% in congested or sensitive areas, this should trigger the production of a traffic impact assessment. The Institute of Environmental Assessment (IEA 1993) suggests two criteria for determining the area over which such an assessment should be carried out: (a) include highway traffic links where traffic flows, or the number of HGVs, will increase by more than 30%, and (b) include any other sensitive area where traffic flows will increase by 10% or more.

General information available on traffic conditions and trends is summarized in Table 5.2. The main components of a traffic baseline survey are (a) traffic conditions, namely the operational characteristics of the road network, (b) information on pedestrian and cycle movements, (c) information on public transport, (d) road safety statistics, and (e) secondary impacts such as noise and air quality measurements. Together with a full understanding of existing land uses and the proposed development, these will allow the baseline conditions to be described.

Table 5.2   Traffic information available.

Source of information	Information available
Local highway authority: data	– Traffic count data by hour of day, day of week, week of year and by mode – Trends in traffic growth from historical data – Accident data for whole road network – Pedestrian and cycle movement counts – Predicted traffic flows from traffic models – Car parking
Local highway authority: publications	– Transport Policies and Programmes, published annually, include current and future (5 year) road proposals, together with long term strategies, public transport proposals, cycle proposals, accident remedial schemes, traffic management proposals, funding and timing – Road Safety Plan, part of the TPP, identifies proposals for accident reduction through area wide and site specific safety measures
Department of Transport: data	– Automatic traffic count data collected nationally throughout the year at permanent counters on motorways and trunk roads – Predicted traffic flows for major road proposals from traffic models
Department of Transport: publications	– White Papers – Roads for Prosperity (DoT 1989b) – National Road Traffic Forecasts (DoT 1989c) – Calculation of Road Traffic Noise (DoT 1988) – Road Accidents Great Britain (DoT 1991) – Design Manual for Roads and Bridges Vol. 11 (DoT 1993)
Local planning authority	– Local plan – Structure plan – Unitary development plan – Other planning documents
Passenger Transport Authority: data	– Information on bus routes, frequency of services, proposed changes to services, etc. – Information on local rail and light rail services

5.4.1 Traffic conditions

Although it is useful to understand the trends in baseline traffic flows on the network and the daily flows, it is most important to obtain hourly traffic data. The DoT and local planning authority will hold information on trends in traffic growth and predicted traffic flows for the relevant area. The most critical period of the day is likely to be either that hour when the road network is carrying its heaviest volume of traffic or that hour when the development is generating its highest volume of traffic. Often these hourly peaks will coincide. When the peaks do not coincide, for example at a large wholesale market when many movements of HGVs take place during the night, the environmental impact is likely to be greater, since many HGV movements will be added to a low level of base flow.

For a comprehensive assessment, both turning movement flows at junctions, and link flows on the network, will be required. Consideration should also be given to variations in traffic flow that may occur as a result of seasonal factors (see DoT 1985).

From this information, an assessment can be made of the current operation of the highway network, based on the **capacity** of the individual links and junctions, together with an assessment of the baseline environmental factors. Figure 5.1 shows a simple example of such an assessment.

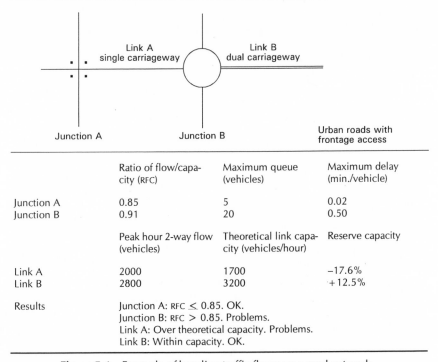

	Ratio of flow/capacity (RFC)	Maximum queue (vehicles)	Maximum delay (min./vehicle)
Junction A	0.85	5	0.02
Junction B	0.91	20	0.50
	Peak hour 2-way flow (vehicles)	Theoretical link capacity (vehicles/hour)	Reserve capacity
Link A	2000	1700	−17.6%
Link B	2800	3200	+12.5%
Results	Junction A: RFC ≤ 0.85. OK.		
	Junction B: RFC > 0.85. Problems.		
	Link A: Over theoretical capacity. Problems.		
	Link B: Within capacity. OK.		

Figure 5.1   Example of baseline traffic flows on a road network.

5.4.2 Pedestrians and cyclists

The most relevant information on pedestrian movements is that related to the location and use of pedestrian crossing facilities, and the location of major pedestrian generators such as schools, shops, offices or public buildings. Understanding the actual number of pedestrians in any one location is of only limited value, although the use of pedestrian facilities on the approach to junctions is taken into account in analyzing vehicular delay. Methods of measuring pedestrian delay are given in the *Design manual 11* (DoT 1993).

The location and use of cycle facilities in the vicinity of the proposed development provides useful background data for assessing impact on cyclists. However, even where cycle facilities are available, many, if not the majority, of cyclists will continue to use the road network. The number of cycle movements is often identified separately in traffic count data.

5.4.3 Public transport

The level of public transport accessibility to the development site can be critical in determining the volumes of traffic being generated by a development. Identifying the bus and rail routes, together with service frequency, will provide an important input to the assessment. Understanding the capacity of existing public transport services to take additional passengers would also be valuable. However, these data are not always available, because of their commercial sensitivity.

5.4.4 Road safety

In many cases it will only be necessary to understand the general location and nature of accidents on the affected road network. Typically, most attention should be paid to a cluster of accidents at a particular location, or to many accidents along a length of road. Local authorities will identify these accident clusters as problem sites and will include them in a programme of remedial measures. Different local authorities will use different criteria for identifying problem sites. In urban areas a problem site may be any location where there have been nine or more accidents in five years. In rural/suburban areas this rate would be lower. Changes in traffic flows, road conditions, or changes to the road infrastructure will affect road safety conditions (DoT 1991).

5.4.5 Projecting the baseline forwards

The most critical decision in assessing future baseline conditions without the development is deciding on the future year for the assessment. In terms of assessing the most significant environmental impact, the year of opening is most appropriate, since this is when the development traffic is largest in proportion to the total flow. However, it is common for the DoT and some highway authorities to insist on a baseline year 15 years after opening. This can be very misleading. Adding 15 years' traffic growth to present baseline conditions may increase baseline traffic flows by 50% or more. This increase can result in a breakdown in the operation of the network, including extensive delays and junctions over capacity. Taking this theoretical 15 year baseline case and using it to assess the impact of development traffic confuses the assessment to the point of losing any sense of robustness, since what one is measuring is the impact of general traffic growth rather than the impact of development traffic. As a rule of thumb it is thus advisable for the assessment first to be made for the year of opening. For a phased development, the opening year of each phase would provide an appropriate baseline.

The assessment of traffic growth between the baseline year and the future year should be based, where possible, on local trends. Where this is not available, the *National road traffic forecasts* (DOT 1989c) can be used. For two reasons, caution should be taken when using these forecasts. First, they include, as an element of growth, the growth that results from future development. Adding development traffic to the forecasts can therefore result in double counting. Secondly, the forecasts reflect growth trends as a result of increases in car ownership and annual vehicle mileage. Applying these trends to, for example, congested peak-hour periods is not necessarily appropriate. In assessing the future conditions, note should also be taken of other committed development that will result in changes to existing conditions, and of programmed changes to the highway infrastructure.

There are no comparable forecasting techniques for assessing growth in the use of public transport. Historical trends are therefore the best measure.

5.5 Impact prediction

Prediction of the traffic impacts of a development involves (a) describing the development, (b) assessing the number of trips in and out of the development, (c) determining the type of trips this entails, (d) determining the origins and destinations of the trips, (e) determining what routes the trips will be made on, and (f) assessing the impact of these changes in traffic on the local road network and environment.

5.5.1 Describing the development

Understanding the intended use of the proposed development is key to the assessment of its likely traffic impacts. The determining factors will be all, or a mix, of the following:
- the development size, including the mix of uses on the site
- the phasing of the development
- the period of construction and construction methods
- the number of employees or residents, and patterns of working use
- the number of visitors and hours of operation or opening
- access arrangements, including service access
- the number of parking spaces to be provided, and

- the proportion of employees and visitors using the different transport modes (e.g. bus, car, train).

Where the proposed development replaces all or part of an existing use, as much information as possible should be obtained about the existing use. The environmental impact of the new development will be the net difference between the impacts of the existing and proposed developments.

5.5.2 Assessing the volume of traffic attracted

Information on the likely number of trips in and out of the development may be available from the highway authority. However, it is common practice to supplement this information with data on similar developments from one of two databases: GENERATE, produced by the West Midlands Joint Data Team; and TRICS, produced by JMP Consultants Ltd (see Appendix B). Each database contains a wide range of land uses in a variety of geographical areas (e.g. city centre, edge of town, suburban), and the TRICS database covers land uses throughout the UK. The databases can be used to find a comparable existing development and apply the information on its trip attraction to the proposed development. The information available is generally in the form of a description of the development (by floor area, number of car parking spaces, number of employees, access by public transport, etc.) together with the daily traffic generation by hour of day. Caution must be exercised in applying the trip rates from these databases too rigorously. Each development is unique and it is important to understand the influence of other factors such as the rôle of public transport or other travel modes, location, and, if applicable, seasonal factors and competing developments.

Where there are no comparable data available for a particular development, assumptions can be made from first principles. The robustness of these assumptions depends upon gaining a full understanding of the proposed operation of the development. Both comparison with other development projects and the use of first principles are subjective to a certain extent, and any subsequent analysis should be subject to sensitivity testing.

5.5.3 Assessing the type of trips generated

In recent years the assessment of trip type and distribution has become increasingly complex. This is largely because of the abundance of large out-of-town retail developments which generate significant volumes of traffic. Traditionally it was assumed that all of the traffic generated by such developments was new to the network. Simply adding these trips to the existing traffic flow on the network would often result in predictions that links and junctions would become overloaded. The developer would then contribute to or pay for highway improvements to offset the traffic impact of the development.

However, it has become recognized that many of the trips to new developments are not really new, particularly in the case of retail developments. Most trips to new developments are already on the network, visiting other attractions such as shops, businesses, or schools in the vicinity. Recent research (MacIver & Dickinson 1991) has identified two main types of trip: (a) primary trips, which have the same origin prior to visiting the site as destination on leaving the site, and (b) linked trips, whose origin and destination are not the same, where the development is an intermediate stop on a route between the original origin and destination. Primary trips are further broken down into (a1) newly generated trips, which did not occur prior to the development opening, and (a2) redistributed trips that used to be made to an alternative development and now are made to the new development. Linked trips can be similarly further broken down into (b1) pass-by trips that visit the development without diverting from their existing route, and (b2) diverted linked trips that involve a significant diversion from the route that would originally have been followed. No clear methodologies exist to date for assessing trip types, and this assessment is usually based on experience and/or local knowledge.

For mixed development sites, there will also be an element of multi-purpose trips, e.g. a single trip combining a visit to a retail outlet and a visit to the cinema. There is little statistical information available on multi-purpose trips and therefore professional judgement must be exercised.

5.5.4 Determining the origin and destination of trips

Having established the number and types of trips taken, it will be necessary to determine their origin and destination. Several methods are available, ranging from relatively crude assessments which are little better than informed guesswork, to more complex methods including modelling techniques. One of the most frequently used methods is the gravity model. A simple gravity model to determine the distribution of home-based trips to a retail park would involve:

1. establishing the drive time in 5, 10 and 15 minute intervals from the retail park, and drawing the resulting drive time isochrones on a map
2. establishing the size of population within the area covered by the 15 minute isochrone
3. splitting the population into zones, probably based on census districts
4. determining a "centre of gravity" for each zone, namely the point in the zone which reasonably represents the centre of population
5. for each zone, establishing the population size P and travel time T to the retail park (or some more generalized estimate of cost), and
6. calculating the percentage of trips from any one direction. This will be a direct function of the percentage of the population in that direction, expressed as a function of P/T^2 (say).

"Gravity model" refers to the greater "gravity" exerted by nearer than by farther developments. Table 5.3 shows possible results from such a model.

Table 5.3 Possible results from a simple gravity model of traffic generation.

	Population P	Time T	P/T^2	Trips (%)
Zone 1 (North)	10,000	5	400	59
Zone 2 (South)	4,000	10	40	6
Zone 3 (East)	8,000	7	163	24
Zone 4 (West)	11,000	12	76	11
			679	100

5.5.5 Assigning trips

Although the assessment of distribution identifies the proportion of trips arriving from and going to a particular direction (north, south, east or west), it is the assignment of trips which identifies the particular routing of individual trips. As with distribution, several assignment techniques are available ranging from crude guesswork to sophisticated modelling techniques. In practice, the local highway authority may provide the best guidance based on their local knowledge.

The most robust form of trip assignment is carried out using a traffic model. In most cases the model will be built using data from roadside interviews and household interviews, which will give information on trip purpose and type, as well as traffic counts. The model replicates all of the origins and destinations of each trip in the network and assigns each trip to a particular route. If a new generator of traffic is added to the network it becomes a new zone. The trip assignment will then be carried out by the programme, assigning the new trips to the optimum route between the existing zones and the new zone.

5.5.6 Assessing significance

The assessment of significance should focus on (a) who will be affected, (b) to what degree, and (c) whether the change is significant. The Institute of Environmental Assessment (IEA 1993) suggest that the following *affected groups and special interests* should be addressed:

- people at home
- people in work places
- sensitive groups including children, elderly and disabled
- sensitive locations such as hospitals, places of worship, schools, and historical buildings
- people walking
- people cycling
- open spaces, recreational sites, shopping areas

- sites of ecological or nature conservation value, and
- sites of tourist or visitor attraction.

To this list one should add the general category of "road users".

The traffic impacts of the development will include not only the primary impact of changes in the traffic components listed in §5.2, but also secondary environmental impacts on:

- noise
- vibration
- landscape
- severance
- driver delay
- pedestrian delay
- pedestrian amenity
- accidents and safety
- hazardous loads
- air pollution
- dust and dirt
- ecology, and
- heritage and conservation (IEA 1993).

The *degree* to which these interest groups and environmental components are affected will vary depending upon the severity of exposure and time of exposure. For example, an increase of 50% in HGV movements during the day may not be significant for residential properties fronting a road, whereas a similar increase over one hour early in the morning may be.

The standards quoted in Table 5.1 provide a guide for the assessment of impact *significance*. However, with the exception of driver delay and junction and highway capacity, the guidance is poor in terms of using specific measures to assess specific impacts. A great deal of reliance has traditionally been placed on the DOT *Manual of environmental appraisal* (DOT 1983) and *Design manual 11* (DOT 1993) but, as discussed earlier, these do not specifically relate to non-road developments. Table 5.4 summarizes measures of significance that might be adopted, based on currently available guidance. More broadly, the *Manual of environmental appraisal* suggested that increases in traffic flows of 30% have a slight impact, 60% a moderate impact, and 90% a substantial impact. The assessment of both impact magnitude and impact significance is largely a matter of judgement, and needs to consider both local circumstances and local policies.

Table 5.4 Measures of significance for forms of traffic impact.

Impact	Measures and significance
Driver delay and junction capacity	– Has average driver delay per vehicle been increased? ($\geq 5\%$ significant) – Does there continue to be reserve capacity in the junction? – Is the capacity within the theoretical design capacity?
Highway link capacity	– Does there continue to be reserve capacity along the link? – Is the capacity within the theoretical design capacity?
Pedestrian delay	Depends on local factors
Noise	– Have noise levels changed? (increase or decrease of $<3\,dB(A)$ no impact, $3–5\,dB(A)$ minor impact, $5–10\,dB(A)$ moderate impact, $10–15\,dB(A)$ major impact, $>15\,dB(A)$ severe impact) – How noisy will the development proposal be? ($<55\ L_{Aeq}$ acceptable, $55–63\ L_{Aeq}$ to be taken into account, $64–72\ L_{Aeq}$ undesired, $>72\ L_{Aeq}$ unacceptable)
Air quality	– Does an 8-hour average concentration of $9\,ppm$ CO occur at least once per year? (if yes, significant) – Does a 1-hour peak concentration of $35\,ppm$ CO occur at least once per year? (if yes, significant) – See §8.2 and §8.4.4
Visual	Depends on local factors; see Chapter 6
Community effects (severance, etc.)	What are the changes in the traffic flow? (30% slight, 60% moderate, 90% substantial)
Accidents and safety	Depends on local factors
Vibration	No reliable data available to assess the significance of traffic induced vibrations
Ecology	Depends on local factors; see Chapters 11–13
Dust and dirt	– How does the proposed development compare with similar developments elsewhere? – Depends on local factors

5.6 Mitigation

The traffic impacts of a development are primarily related to conditions away from the site (i.e. the traffic network), and it is unlikely that mitigation measures will be included as part of the design of the development itself. It is, however, common for planning authorities to stipulate conditions relating to offsite works. Typically, such conditions will relate to making improvements to the road network, providing new or enhanced pedestrian facilities, or restricting the hours of operation for the movement of HGVs. Table 5.5 lists a range of typical mitigation measures which may be adopted to offset specific impacts. Mitigation measures for secondary impacts related to traffic, such as air pollution and noise, are discussed in other chapters.

Table 5.5   Mitigation measures for forms of traffic impact.

Impact	Typical mitigation measures
Driver delay and junction capacity	Junction improvement, road widening
Pedestrian delay	New or improved pedestrian facilities, e.g. central refuges, zebra crossings, pedestrian phase in traffic signals
Noise	Restriction on working hours, time restriction on movement of goods vehicles, specific routing for goods vehicles, provision of noise barriers or insulation of properties
Community effects	Traffic calming, speed reduction measures, improved pedestrian/ cycle facilities
Safety	Traffic calming, junction or link improvements, improved pedestrian/cycle facilities, prohibition of turning movements, prohibition of on-street parking, improved street lighting

CHAPTER 6

Landscape

Brian Goodey

6.1 Introduction

> Landscape is an important national resource . . . an outstanding natural
> and cultural inheritance which is widely appreciated for its aesthetic beauty
> and its important contribution to regional identity and sense of place.
> Although it is subject to evolution and change, the landscape is recognized
> as a resource of value to future generations. (DOT 1993)

The EC's EIA Directive and subsequent UK regulations require reference to the
landscape setting within which change is to occur, namely "a description of the
likely significant effects, direct and indirect, on the environment of the development" explained by reference to possible impacts on the landscape and the interaction of landscape and other impacts. Landscape impacts are possibly the most
subjective impacts addressed in EIA. The vast majority of landscape is cultural,
rather than natural heritage, and its national, regional or local identification
depends very much on the values and associations of residents and visitors. It is
appropriate, therefore, to consider "cultural heritage", both as formally designated, and in terms of popular recognition, as an integral element in landscape
evaluation and assessment. However, landscape is also linked to ecology, with
strong associations between landscape and ecological quality.

Landscape architects have been particularly involved in EIAs for special (and
especially road), waste disposal, extraction, and infrastructure projects (Fieldhouse 1993), and landscape assessments for coal and mineral extraction
schemes, wind farms and power stations have been of particularly high quality
(Mills 1992). However, other projects can also have significant landscape
impacts, if they are insensitively designed or sited in an area of high landscape
quality. Professional landscape input is essential to interpret these impacts, with
experience and judgement required at the various stages of public presentation
and determination.

6.2 Definitions and concepts

6.2.1 Overview

What does the term "landscape" mean? A picture on the wall, trees and green rolling hills, neither sea nor town? A prized, romantic, view from a remembered holiday visit, with partner, refreshment and the hour/day/year as key factors? Or perhaps a television news image of protesters astride advancing excavators, or a snapshot in the mind evoked by music, poetry or novel?

The following is an initial list of factors that contribute to landscape (CC 1993):

- physical: geology, landform, climate and microclimate, drainage, soil, ecology
- human: archaeology, landscape history, land use, buildings and settlements
- aesthetic:
  - visual, e.g. proportion, scale, enclosure, texture, colour, views
  - other senses, e.g. sounds, smells, tastes, touch
- associations:
  - historical, e.g. history of settlements, special events
  - cultural, e.g. well known personalities, literature, painting, music

Figures 6.1–6.4 offer examples of some of these factors, and of the visual and contextual complexity of landscape. The view experience and detached image (here in photographs) transfer the landscape to a mental file of place memories that incorporate weather, personal experience, accompaniment, and media information. Figures 6.1 and 6.2 both show urban scenes filtered through foliage. Figure 6.1 overlooks High Wycombe beyond a narrow hilltop of decayed trees, and Figure 6.2 is of Edinburgh's Waverley Station viewed from a public park adjacent to The Mound. Season, plant condition and the distance and function of the view beyond all contribute to the viewer's assessment of quality. Figure 6.3 shows a Black Country canal side, with buildings that may be read as urban dereliction, or alternatively, as significant remnants of industrial archaeology. This figure incorporates more "heritage" and, perhaps, cultural meaning, than does the severely modified recreational stream incorporated in the residential design of Milton Keynes (Fig. 6.4). Judged by cultural reference, Figures 6.2 and 6.3 generate more meaning; judged by the level of use, Figure 6.2 is seen most often, Figure 6.3 the least. Judged by the rusty concept of "nature", Figures 6.1 and 6.4 seem to fare best, although Figure 6.1 needs severe modification, and Figure 6.4 incorporates recent landscape design.

Most people know what they like when it comes to viewing the landscape, and they frequently compose views for a glance, for contemplation, or for a photograph. Thus, many of the basic components of landscape are already stored in their bank of experience. However, to structure and use this experience, if only to relate to the work of the landscape professional, some guidance is needed.

Figure 6.1 Contrasting landscapes: High Wycombe.

Figure 6.2 Contrasting landscapes: Edinburgh.

Figure 6.3 Contrasting landscapes: Walsall.

Figure 6.4 Contrasting landscapes: Milton Keynes.

6.2.2 Cultural values in landscape assessment

In EIA the view is not abstracted, except as a consequence of illustrated report writing, but many of the aesthetic and cultural values of the viewer remain. Contemporary popular meanings, media interpretations (Burgess 1990a) and perceptions of landscape (Goodey 1987, 1992) are always likely to exceed the constraints of professional methodology, but deserve attention when public response is under consideration, and must be recognized in EIA. As Kroh & Gimblett (1992) have shown:

> Landscape sites are not rated for their impact on a one by one basis but as links in a transitional interaction between person and place. Contextual information is acquired as people move through the landscape. This enables them to rate a scene not only for its unique qualities, but also as part of an area which affords immediate and potential opportunity . . . the use of traditional analysis and classification techniques used in visual assessment may not adequately represent all the elements experienced in the landscape.

Current research into the meaning of what are increasingly termed "cultural landscapes" – so as to clearly incorporate all landscape elements – has divided into two main streams. The first, advanced by Cosgrove (1984, 1985) and Cosgrove & Daniels (1992), is primarily concerned with the cultural meanings implicit in landscape and its representation (see also Rackham 1991). More practically, Burgess (1990b, 1993) and Burgess et al. (1988) have developed a qualitative methodology for determining popular attitudes towards local landscapes and the recreation opportunities they are intended to provide.

6.2.3 Balancing quantitative and qualitative methods

To give substance, then, to what is at least partly a qualitative, rather than quantitative, enquiry, the reader will also need to dip into the past literature (e.g. Penning-Rowsell 1974, Patmore 1975, Robinson et al. 1976, Relph 1984) in order to enlarge and structure the personal experiences of looking, examining and comparing, which are fundamental to landscape work. Amongst the most pertinent works are American papers edited by Zube et al. (1975), which focus on experimental assessment techniques in the resource context, Higuchi's (1988) study of the visual structure of Japanese landscapes, and Bourassa's (1991) theoretical framework.

Nicholls & Sclater (1993) suggest that appraisal and evaluation, although attaching value, can deal only with detached elements of the landscape, whereas assessment incorporates "all the ways of looking at, describing, analysing and evaluating landscape" (CC 1991a). In terms of both the spirit and the letter of EIA, it is this diversity of views that need to be incorporated in the process. Nicholls' research into local authority views suggests, however, that there is a preference for limited, landform-based studies; in short, a preference for evaluation over the complexities of assessment. For instance, Novell (1993) suggests that a landscape assessment should:

> make the clear distinction between the objective measurement of change in the landscape which a development is likely to bring about, and the more subjective assessment of its effect on character and visual amenity. Both aspects of study will require structured judgement on the impact of change in the landscape, both in terms of its intrinsic quality and of the effect on people's use and enjoyment. The techniques used by the experienced practitioner should be controlled, systematic, rigorous and capable of replication. Naturally landscape is defined as all external environments, both urban and rural.

The Countryside Commission (1990, 1992, 1993) has pressed hard to develop a pattern of assessment research that reflects the breadth of public responses to valued landscapes. Quantitative evaluations are, however, frequently sought by EIA clients, project co-ordinators and receiving authorities. Often the provision of such quantitative evaluations is then destroyed by a commentator who realizes a more personal recognition of landscape's diversity, or by an interested community which articulates landscape meanings in emotive, but compelling, terms.

6.2.4 Terminology and concepts

The lack of agreed precision in terminology is indicative of the problem of incorporating the diversity of personal and group values, and meanings of landscape, into an EIA process which seeks both the scientific precision (Powell 1989) offered by other EIA professionals, and an accurate representation of both existing and possible future situations. Some of the terminology and concepts developed for landscape assessment are summarized below, with sources for further reading.

The Institute of Environmental Assessment makes an initial distinction between landscape impacts – namely, "changes in the fabric, character and quality of the landscape" – and *visual impacts*, which relate to "changes in the appearance of the landscape and the effects of those changes on people (IEA 1994b). Thus, visual impacts can be seen as a subset of landscape impacts, which deal with impacts on views, viewers and **visual amenity** (see §6.2.1)

To describe the landscape, Fabos (1979) distinguished between *landscape as scenery*, and *landscape as environment*, neither of which need be limited to concepts of the "natural". Appleton (1975) has developed a qualitative "*prospect and refuge*" approach to landscape aesthetics, which he has latterly (1990) pursued into the arts. Tandy (1967) identified *areas of visual containment*, or landscape identity/character areas, as relatively self-contained areas whose landscape character changes little. In contrast, crossing a *visual barrier* such as a hilltop or a curve in a road, opens up a new vista. Martin (1993) differentiates between

83

generic and specific landscape types, and *character areas at different scales*. Maps illustrating *tranquil areas* (ASH 1993a, Rendel 1994), defined through a sieve of highway, railway, pylon and overflying variables, are an attempt to integrate visual and sonic disturbance at a macro scale, and further constructs involving additional sensory data to that of visual perception are to be expected.

In terms of project impacts, *intrusion* – "the quality of an element or factor which appears to stand out to the detriment of a design; a serious visual problem or conflict" (Lucas 1991) – is likely to be one of the most significant issues raised in EIAs. The literature is considerable and has developed from the belated recognition that developments generate visual (and psychological) as well as sonic impacts (Hopkinson 1974). Intrusion indices of various types have been used to quantify and graphically reflect the impact of development projects.

Although much of landscape and visual assessment is concerned with professional and public perceptions of the view – where a positive response to aesthetic attractiveness may be expected – users of the landscape are now also concerned with *personal security*. Burgess's studies (1988, 1990b, 1993) of perceived *opportunity and risk* in urban fringe woodlands are especially significant in illustrating the comparatively recent fears (Painter 1992) expressed by a wide section of the community with regard to woodland and open space areas. This is a significant issue when it comes to considering landscape mitigation later in the EIA process. Similarly, recognition of the clear overlap between the work of the ecologist and the landscape expert has caused the ASH consulting group (1993b) to develop the concept of *ecological identity areas*, which provides preliminary ecological assessment at the landscape scale.

6.3 Legislative background and interest groups

6.3.1 Regulations and designations

More than 20% of the area of England and Wales has been designated for landscape purposes as a result of such legislation as the National Parks and Access to the Countryside Act 1949, the Countryside Act 1968, and the Wildlife and Countryside Act 1981. The Countryside Commission (CC), Scottish Natural Heritage and Countryside Commission for Wales have the remit to protect and improve the landscape, and to provide new and improved opportunities for access to the countryside.

When initiating a landscape appraisal, one should always explore, through the local planning authorities involved, the possible designations applicable to, or adjacent to, the site under consideration. *National Parks* would seem to be the most significant British statement of landscape quality, although their origins represent a predictable complexity of aesthetic, ecological and recreation factors. The Norfolk and Suffolk Broads and the New Forest enjoy similar designations. Nationally designated *Areas of Outstanding Natural Beauty* such as the Quantock Hills or the Forest of Bowland again imply an aesthetic judgement as to quality. Other relevant non-statutory designations include *Heritage Coasts* and *National Trails* such as the Pennine Way and Ridgeway. The Midland New National Forest is a further national initiative. *Heritage landscapes* may benefit from tax relief under the terms of the Inheritance Tax Act 1984.

Through its policies and programmes, the CC has also encouraged a wide range of *regional and local initiatives* involving local authority, urban fringe and rural interests, including the recent designation of *Community Forests* (Forestry Commission 1992). An increasing number of county and district authorities have undertaken detailed landscape studies of their areas. Much of this work has been based on CC's 1987 guidance on landscape assessment (see also 1988, 1991a, 1991b), and has been expanded to embrace the needs of EIA (Stiles et al. 1991).

The "cultural heritage" extends the search into a range of *archaeological, building, townscape and structural designations*. Thus, a landscape assessment may need to consider Parks and Gardens of Special Historic Interest, National Trust properties and areas subject to Limestone Pavement Orders in England, and National Heritage Areas and National Scenic Areas in Scotland, the latter two being subject to statutory designation. Green Belts, local Areas of Great Landscape Value, or occasionally Special Landscape Areas, and County Parks (England and Wales) or Regional Parks (Scotland) will also require consideration, the parks having statutory designation. In Scotland there is an additional array of local landscape and scenic designations. Local Tree Preservation Orders and conservation area designations will be significant, and some sites, such as the Ironbridge Gorge industrial archaeology complex in Shropshire, enjoy international recognition by UNESCO as World Heritage Sites.

It is also essential to compare any landscape designation with *ecological and nature conservation zones*, which will seldom be co-extensive.

6.3.2 Guidance

Sclater & Nicholls's (1993) judgement that "there can be no simple textbook prescription for the format of landscape assessment" in EIA may need to be reappraised following the recent publication of the CC's manual of basic landscape assessment practice (CC 1993; see also Martin 1993). Part 1 of this manual outlines key terms and common stages in EIA, from planning, through desk study, field survey, analysis and presentation. Part 2 outlines the wide range of methods currently used in assessment from visual description and checklists to GIS approaches, and evaluates each in terms of effectiveness and appropriateness for various contexts. Part 3 is especially valuable for EIA, in that it not only provides methods for consolidating the assessment but also outlines the forms of assessment strategy which may be adopted in relation to the landscape's future use and image, illustrating conservation, enhancement, restoration and other strategies which might be employed in the mitigation stage of an EIA. Part 4 provides illustrated case studies of assessments for a county, district, Community Forest, Environmentally Sensitive Area, river catchment, and AONB, with details on the costs, time and skills involved.

The Institute of Environmental Assessment and the Landscape Institute have also prepared very useful guidelines for the assessment of landscape and visual impacts, specifically for EIA. Draft guidelines (IEA & Landscape Institute 1994) were produced in March 1994, and final guidelines are expected in late 1994.

The DOT's *Design manual for roads and bridges* (1993) gives a summary introduction to landscape assessment for road EIAs. This manual should perhaps be considered in association with three earlier American reports that chart the development of landscape and perceptual issues for roads (Appleyard et al. 1964, US Department of Commerce 1966, UAFHA 1968), but provides as concise a guide to the basics as one might need.

6.4 Baseline studies

The best safeguards to an effective landscape EIA are: (a) a clear statement of purpose, (b) initial consideration of the full range of landscape elements and meanings involved, (c) application of a comprehensive and tested methodology, and (d) clear communication of evidence in terms which can be understood and discussed by the community at large. The baseline study establishes the parameters and structure for the following investigation. It thus needs to be extensive and rigorous, establish a digestible account of the area and project concerned, and highlight specific details that will require later investigation. At this stage, the assessment requires a stated framework of analysis, but should not venture into aspects of design or enhancement (Nelson 1993). This section suggests (a) steps in baseline landscape appraisal, (b) methods for visually representing landscape, and (c) criteria for evaluating landscape quality.

6.4.1 Steps in baseline landscape appraisal

Kent County Council (1991) suggests that landscape appraisal "involves evaluating the views into and out of the site, and describing the type, extent and quality of these views. Open areas, contained spaces, important features, including historic landscapes need to be identified". Kent also notes the importance of establishing local "landscape character", derived from building materials, plant species and design features which might, then, be reflected in the development proposals. A suggested process for a baseline landscape study is outlined below. This indicative, rather than inclusive list, identifies key activities that provide evidence of the existing landscape.

1. **Landform** setting: from an initial analysis of topographic maps and plans, identify the proposed location and the **visual envelope** within which the project is to be located (e.g. topography, major ridge lines, direction of open views). This may be enhanced by field survey.
2. **Land-cover** setting: from available maps and photographs, establish major land-cover zones on and surrounding the proposed location.
3. Designated features: from consultation establish the location of any landscape, heritage and other significant designations for the area (see §6.3.1).
4. Cultural associations: from consultation, especially with libraries and museums, establish any significant cultural meanings for the area.
5. Field mapping of dominant elements: annotate maps with located photographs to show dominant and/or significant natural or cultural features (e.g. industrial chimneys, wooded hilltops) in their setting.
6. Landscape character areas: annotate maps to show main areas of common landscape character (e.g. areas of visual containment, ecological identity areas; see §6.2.4).
7. Intervisibility survey: build on the map of landform setting (1) with field observations, sketches and photography to show the depth and character of views into, and from the proposed site.
8. Evident landscape processes: carefully observe plants (especially tree groups), drainage and erosion patterns, agricultural practice and evidence of development in order to identify continuing processes of change.
9. Patterns and levels of community interest: identify existing patterns of landscape use and interest in the area; interview those most affected; sample broader zone and consider leisure and cultural interests.

6.4.2 Visual representation of baseline landscape

Much of the evidence derived from the above process can be synthesized into a series of visual representations that illustrate indicative and significant views and evidence of special features. Methods for visual representation include (Mills 1992):

- plans or maps of the site, land uses, statutory designations, viewpoints, landscape quality, listed buildings, and/or tranquil areas (see §6.2.4)
- photos from key viewpoints into and out of the proposed site
- diagrams of landscape features such as topography or visual barriers
- aerial photos, and
- videos.

The recent reports by the CC (1993) and IEA (1994b) illustrate and evaluate these techniques, and discuss the value of overlay mapping and computer-aided landscape classification methods. In many cases, several aspects of visual impacts can be incorporated on the same map or figure, perhaps linked to a visual impact schedule enumerating viewpoints and key locations shown on the map.

6.4.3 Assessment of baseline landscape quality

The criteria for evaluating landscapes for designation as national AONBs, or the more local Areas of Great Landscape Value, can also serve as an initial checklist in assessing any landscape (CC 1993):

- *Landscape as a resource:* the landscape should be a resource of at least national (regional, county, local) importance for reasons of rarity or representativeness.
- *Scenic quality:* it should be of high scenic quality, with pleasing patterns and combinations of landscape features, and important aesthetic or intangible factors.
- *Unspoilt character:* the landscape within the area generally should be unspoilt by large-scale, visually intrusive industry, mineral extraction or other inharmonious development.
- *Sense of place:* it should have a distinctive and common character, including topographic and visual unity and a clear sense of place.
- *Conservation interests:* in addition to its scenic qualities, it should include other notable conservation interests, such as features of historical, wildlife or architectural interest.
- *Consensus:* there should be a consensus of both professional and public opinion as to its importance, for example as reflected through writings and paintings about the landscape.

Although economic assessment of landscape as a component of amenity and leisure environments and opportunities is a significant research growth area, a typical project is likely to generate a wide range of perceptions, concerns and involvements that do not fit neatly into any quantitative or graphical framework. At the baseline stage, it will be sufficient to identify the local, regional, national (and possibly international) constituencies and special interest groups that might be involved, and to note aspects of the cultural heritage that deserve special consideration.

6.5 Impact prediction

6.5.1 Overview

Effectively communicated predictions of the nature, likelihood and significance of changes that may occur, over various subareas and periods as a result of the proposed development, is at the heart of successful landscape assessment. In brief, such prediction requires that a landscape's likely evolution without the development is described in both written and graphic terms as noted above, and that similar modes of description are then applied to the landscape expected as a result of the development. A structured comparison between these two descriptions will highlight the landscape impact. Its significance in affecting image and activity can then be determined.

It would now be useful to outline an effective, quantifiable, method by which these predictions are to be achieved. But those landscape qualities most prized by observers reflect a well digested combination of preferences, attitudes, glanced snapshots and media reinforcement, whereas the predictor may only have vegetation cover and slope maps to hand.

All prediction is imprecise. The site's future evolution without the development may be affected by vegetation change, crop selection (e.g. rape and linseed), or the discovery of a buried hoard. Seasons will continually alter the aesthetics of most views. Similarly, the function of a new development within a rapidly shifting political and legal framework, or where new scientific evidence or technological development may change the form or operation of the site, makes prediction difficult. That landscape change is usually registered by the observer in foot-, or car-borne sequence of three-dimensional views, in subtle colour variation, and including visible movement as well as sound and textures, means that landscape description techniques seldom capture their subject. The legal requirement to consider indirect and cumulative impacts widens the scope for error still further, and although discussion of these impacts may provide the opportunity for exploration, few developers will tolerate the expense involved in what might be seen as an academic exercise.

Despite these limitations, impact prediction is feasible. It should consider the development's impacts on the features listed in steps 1–9 of §6.4.1, in terms of (a) magnitude and (b) significance.

6.5.2 Impact magnitude

The project's location, dimensions (especially vertical), materials, colour, reflectivity, visible emissions, access routes, traffic volumes and construction programme will all affect the magnitude of its impacts on the landscape. The magnitude of visual impacts will be related not only to the development's *zone of visual intrusion*, but also to the number and types of viewers affected, their distance from the development, and the degree of intrusion to which they will be subject (IEA 1994b). Methods for describing these impacts include:

- plans or maps showing the area over which the development can be seen (the zone of visual intrusion or influence), phasing of the development, and the degree of visual impact experienced by affected properties
- sketches or artists' impressions of the development (perhaps from previously identified viewpoints), mitigation measures, or alternative options
- cross-sections, perhaps with viewlines
- photos from key viewpoints into and out of the proposed site
- photographs with artists' impression overlays
- photomontages of the proposed development, alternative options, and/or

the site before, during and after the development
- GIS methods, and
- virtual environments.

Parallel illustrations of the present landscape's predicted evolution without the featured development are especially valuable. The Essex Development Control Forum (1992) notes that "the scale of the proposal can sometimes be illustrated by reference to some existing feature in the landscape. The impact of new large areas of lighting (from roads, storage areas, car parks, sport complexes, etc.) at night may need to be considered". The DoT (1993) suggests that two assessments should be prepared, covering different time spans and seasonal conditions.

Many EIAs still use the simpler methods listed above. However, the basic "drawing" has increasingly become part of a photomontage (Branson et al. 1993) and computer-aided process which incorporates initial survey, impact analysis, presentation of alternative forms of mitigation, and preferred solutions. For example, RPS Clouston (1993) explain how their computerized *photomontage* for the proposed A5225 Wigan Hidley and Westhoughton bypass was prepared:

An aerial photograph of the site was scanned; the new road layout and ground modelling were plotted by computer onto a transparent overlay, to match the final print size supplied for the artwork.

The landscape consultants had prepared a 1:2500 layout showing the main structure planting. A new overlay was prepared, adjusting the computer plot to the reality of the photography where necessary, eliminating confusing computer detail, incorporating the significant landscape elements and including diverted footpaths and road lighting positions. This simplified drawing was then traced onto the photographic print, with traffic signals, road markings, lighting columns and other incidental detail to be added later.

The road surfaces were painted in first, using masking film to achieve definition. Once this was done the engineering concept could be clearly seen. Prior to this it appeared as a multitude of indecipherable lines on the complex photographic background. This was followed by the retaining walls and bridge abutments.

Basic landscape tones were then added, eliminating all redundant roads and structures. Unless the photograph has been taken with a low sun angle, or there are great differences in ground levels, variations in ground form will seldom be visible from the air.

The critical point in photomontage is the addition of detail and the blending of new elements with old. It is on the colour matching and interpretation of detail that the final realistic effect will depend. The correct representation of tree planting at, say, 15 years may be much more important to the landscape architect in the team than to the consulting engineer, who might well give a higher priority to concrete finishes and road marking. The ultimate client may be just as much concerned with limiting the cost of the exercise, and providing a minimum accuracy level which will not give a misleading impression of the proposals to the general public.

Steinitz (1990) illustrates the potential for user participation in the selection of alternative landscape strategies using an integrated sequence of contemporary methods, including GIS (see Appendix D). A further application of GIS methods to landscape change proposals is reported by Priestnall et al. (1993) in work undertaken for the Forestry Commission.

The current development of *virtual environments* for landscape assessment and prediction – far advanced from their beginning in flight simulation – is likely to be in widespread in practical operation by the end of the century (see Orland 1992, Cassettari 1993). To the increasing ability of computers to simulate spaces is added the potential for interaction with simulation through touch and sound, permitting participatory generation and evaluation of alternative future images. Current research programmes (e.g. Pittman 1992) place strong emphasis on environmental and landscape topics. There are two major obstacles to the uptake of these developments. The most obvious is cost, which is likely to restrict such technology to laboratories and high investment entertainment for the immediate future. The second is the scepticism of professionals and the public, suspecting that important value issues may be lost in the supposed technological sophistication.

6.5.3 Impact significance

The CC's criteria for assessing landscapes (see §6.4.3) can also be used to assess a development's impacts on the landscape, by comparing how well the criteria are fulfilled with and without the proposed development. An assessment of visual impacts involves determining more (a) how many people – usually classified into groups such as residents, footpath users, etc. – are affected, and (b) how much. The DoT (1993) suggests a four-point scale for describing the latter:

- substantial adverse or beneficial impact: a significant deterioration or improvement in the existing view
- moderate impact: a noticeable deterioration or improvement
- slight impact: a barely perceptible deterioration or improvement, and
- no change: no discernible deterioration or improvement.

Time and mental predisposition affect the level of visual intrusion identified by various viewers. One cannot assume a distance decay of interest or concern as the viewer recedes from a site. Age, gender, activity patterns, family grouping, education, holiday preferences and inherited traditions with regard to places unseen (such as images of an unvisited "home" area) will require consideration. However, although it should never be presumed that "people will get used to it", major features such as power-station cooling towers can be recognized positively as place markers, evidence of new technology, or as attractive design by some, while remaining offensive intrusions to others. The impact of visual intrusions will often depend on the viewer's commitment to, and time spent within, a given landscape.

6.6 Mitigation and enhancement

The concept of mitigation is, perhaps, an inappropriate starting point when it comes to landscape. Frequently the surveyor and assessor is also part of a landscape design team that will recognize, in the process of mitigation, an opportunity for creative environmental enhancement. A lack of regard for mitigation, and for complementary and comprehensive landscape design, has been one of the criticisms levelled at EIAs by local authorities (Fieldhouse 1993), and has led to local authorities taking the initiative in proposing mitigation measures.

The most common areas of mitigation are likely to involve (a) sensitive building design and siting (e.g. building height, colour and shape; siting in relation to local topography and microclimate); (b) sensitive choice of site level; (c) mounding, which will in turn affect surface-water drainage and may itself offer visual intrusion; (d) planting both on and off the site, and retention of existing vegetation; (e) ecological conservation; and (f) other hard and soft landscaping (e.g. street furniture, design of access roads, lighting). As mentioned earlier, mitigation may also need to incorporate measures for public safety. Manipulation of views through the use of form or colour, or through the introduction of new landmark features may serve to deflect the eye, or imply cultural benefit. Landscape design also has a part to play in other areas of mitigation aside from direct landscape modification. For example, it may need to be integrated with technical advice on water holding facilities, acoustic fencing or bunding. The mitigation process may also require the generation of alternative future scenarios which focus on, or stress, particular characteristics of the modified site.

Four characteristics are evident in landscape mitigation measures. First, the landscape consequent upon development is largely the *visible manifestation of all the physical, and some human, changes* achieved by the development. Although mitigation measures and environmental gain can aim to enhance views and improve habitat, their success is predicated on their integration with predicted modifications to landform, drainage, surface and subsurface condition, local climate and access patterns. Landscape mitigation cannot be viewed as the "green icing" on the newly baked cake. Ecological and visual suitability can be achieved only through an integrated, team, approach.

Secondly, *time* is the major factor in the success of landscape mitigation measures. Even a suburban retail park planted densely with semi-mature trees and decorative shrubs takes time to settle, both into the surrounding townscape, and in the eyes and minds of the users. Many interested parties tend to propose, claim or require that a development is hidden from view, or at least "integrated" with its surroundings in the shortest possible time. Rapid integration demands high investment in preparation, stock and management, and may not best serve longer-term ecological or land-use intentions for the area.

Thirdly, there is a very strong tendency towards *heritage reference*. People's response to a new development is to search for historic landscape elements which can either be re-stated or contributed anew. Contemporary landscape values point strongly to this reinforcement of tradition, which has become enmeshed with a presumed "ecological" approach to landscape design, with the consequent assumption that nothing that can be added to the landscape is as good or appropriate as what already exists. For example, the advice offered by the Essex Development Control Forum (1992) is that "There is a presumption in favour of conservation *in situ*, i.e. reinforcement of existing planting, management of areas, fencing to control grazing and overuse, etc. Only as a last resort should consideration be given to relocating seed beds or redistribution of wild flower seeds, etc.". This last point refers to the translocation of ecological material from development sites, for instance the transfer of 20 acres of unique magnesium limestone grassland at Thrislington Quarry in County Durham after 1982. Such proposals, which have been enthusiastically supported by some development interests, do, however, have possibly adverse aesthetic and symbolic meanings, as well as ecological implications.

There is no reason, however, why a new development without such sensitivities should not be mitigated through the development of an essential novel landscape experience, while still retaining ecological fit with the local environment. It is possible, and often desirable, to contribute further qualities to the visual environment. New landscapes with new images and meanings may become the much-loved views of future generations.

The fourth characteristic of landscape mitigation is the importance of *process* in facilitating the desired landscape outcome. Many fundamental characteristics of the emerging landscape can be established through careful design of the development process itself, including the initial retention of local plant and surface materials, use of exposed substrata for visual effect, diversion of watercourses, and the pattern of vehicular use on site during construction. Mitigation starts before the development begins, not after.

6.7 Monitoring

Monitoring and evaluation of a professionally directed process of change is only slowly being recognized as an essential element in environmental management and EIA. Unfortunately, the planning consent target, to which the EIA submission is linked in the British procedure, militates against any client investment or professional interest following this short-term objective. Given that initial predictions of landscape impacts tend to have relatively high levels of uncertainty, immediate change is easily recognized by the broad community, and mitigation requires time to be effective, it is important that landscape monitoring is undertaken.

Establishing a regular programme of specific, comparable, observations (and measures where appropriate) achieves three significant advantages: (a) an early-warning system with regard to likely impacts of the development, permitting changes in the construction and/or mitigation procedures; (b) a learning experience which may feed directly into other projects; and (c) regular evidence as a basis for discussion with authorities or the public. Being able to illustrate a process of change is particularly significant should public debate be focused on a specific event, or newsworthy opinion.

6.8 Concluding issues and further reading

A series of characteristics unique to landscape within the EIA process will shape the future development of methods in this area. In summary these are:

- Landscape change and expectation of change is often the most accessible area of involvement and concern for the majority of the community involved. Symbolic of a very wide range of emotions and relationships, the physical change can be represented and communicated in an immediate and provocative form.
- Given the inevitability of qualitative evidence, or quantitative summaries of qualitative observation, there must always be an admitted subjective element in the communication of baseline data, impact analysis, and mitigation measures. It is very unlikely that hard landscape evidence will ever be sufficient for full integration with other aspects of EIA.
- Rapid development is to be expected in the area of visualization, both for professional analysis and as a medium through which a public response may be sought. The complex values evident in earlier debates on landscape assessment may well be obscured or forgotten in the drive to achieve technologically sophisticated solutions with attractive and superficially convincing outcomes.
- Landscape mitigation embodies elements of the design process, and thus generates options for aesthetic change which will inevitably extend beyond the intention to meet, or exceed, standards. The modified landscape must incorporate all other mitigation measures, within an evolving scene which carries current visual values.
- Landscape planning is likely to become an increasingly important element in the statutory planning process, encouraged by the scale of development and land-use issues, the increasing conflicts over land use and appearance, and public interest in the look of the land. Within the past few years,

landscape methods have advanced rapidly into the daily activity of planning authorities.

- Although regional and national landscape values might seem to be established and enduring, they are currently being reappraised by national agencies, with likely reformulations which, through policy and investment, will have a considerable impact on the landscape characteristics that we are encouraged to value.

- In the medium and longer term it is inevitable that European and individual European state landscape policies will influence and modify British policies and practice. Although technological developments, rather than legal instruments, are likely to advance homogenization of landscape image, shared attitudes towards nature, sustainable tourism, personal security and peripheral development will begin to be registered in individual landscape attitudes.

As with many areas of EIA, the available literature specific to landscape has hardly kept pace with the subject's development. Fabos (1979) provides an account of early literature in the field. Landscape planning texts by Hackett (1971), Turner (1987) and Preece (1991) provide accessible introductory accounts, and two recent manuals, both derived from the work of the Forestry Commission, are essential reference. In *Elements of visual design in the landscape*, Bell (1993) provides a clear account of the basic elements of landscape design, with frequent illustration appropriate to EIA. There are urban, as well as rural, examples and clear line sketches link with photographs which provide substance to frequently observed patterns and characteristics, including time, light and order. In *The design of forest landscapes*, Lucas (1991) concentrates on the application of design and management methods in the shaping of forest landscapes. Lucas covers most possible forest contexts, including road and watersides, field systems and woodland, making this a good guide to the design of most British landscapes. Frequent colour photographs, line drawings, indicative plans and elevations, together with a textbook structure, make this an easy source of analytical and design concepts. The reader is also directed to recent EISs as available, for professional journals seldom carry detailed accounts of consultancy reports.

CHAPTER 7

Archaeological and other material and cultural assets

Nicola Bourdillon, Rosemary Braithwaite, David Hopkins, Roger France

7.1 Introduction

Europe has known some 500,000 years of human activity and settlement, from the earliest hunter gatherers to the present day. As a result, almost all sites on mainland Britain have had previous human occupation and are therefore of potential historical interest. The study of archaeological and other historical resources is important to:

- fulfil an innate curiosity about the past, since the origins and development, lifestyles, economy and industry of previous generations can be traced and understood through archaeological remains
- contribute to a sense of tradition and culture, and
- promote a sense of national identity.

Archaeology is a vital component of recreation, since many people enjoy visiting archaeological sites and studying archaeological remains. It contributes to education; archaeological study is used as a basis for integrating the teaching of several other subjects, and can promote an understanding of the rôle of the past and its relevance to today's society. Britain's historic heritage is also important to the tourism industry. It attracts visitors from all over the world and, if well interpreted and presented, it can be an important financial asset. However, archaeological and other historical remains are a fragile and finite resource that should be carefully managed and conserved, and are therefore one of the many elements that need to be addressed in any EIA. On most sites these remains are not important enough to affect development, but a site's historical and cultural interest is always monitored by planning authorities, and EIAs should show that it has been considered.

7.2 Definitions and concepts

7.2.1 Overview

The DOE recommends that a site's "[a]rchitectural and historic heritage, archaeological sites and features, and other material assets" should be described in an EIA, as well as the "[e]ffects of the development on buildings, the architectural and historic heritage, archaeological features, and other human artefacts" (DOE 1989). However, what precisely *material and cultural assets* are is open to interpretation. The general requirement to consider such aspects ensures that an EIA is comprehensive and that issues of strong local feeling or wider social and cultural heritage are considered. Some EIAs have interpreted material assets very widely, including, for instance, agriculture, forestry plantations, recreation and amenity, utilities and other services, communications, rights of way, and potential future resources. However, here material and cultural assets together are taken to be:

- archaeological remains, both above and below ground
- historic buildings and sites (including listed buildings, cemeteries and burial grounds, parks, village greens, bridges and canals), and
- other structures of architectural merit.

In practice, there is no precise distinction between *archaeology* and *other aspects of the historical environment*. For instance, English Heritage (EH), which is responsible for the major archaeological sites in England, is also known as the Historic Buildings and Monuments Commission. Academic historians are concerned with the past on the basis of written evidence (in the UK, effectively from Roman times); archaeologists use a much wider range of evidence and therefore may go back to the earliest human occupation (in the UK, from perhaps 500,000 years ago to World War II). The legislation covering the historical environment is a patchwork of regulations and guidance that draws an arbitrary distinction between archaeology and ancient monuments on the one hand, and other aspects of the built environment such as listed buildings and conservation areas on the other. This chapter broadly follows the legislative distinction, so each section discusses first archaeology and then historical buildings and sites. However, it is recognized that this distinction is not clear and may not be applicable to all EIAs. Other architectural issues are addressed in Chapter 6 on landscape.

7.2.2 Archaeology

The range of archaeological evidence reflects the diversity of human experience; the need for water, food and shelter, the use of changing technologies, and the religious, cultural and political needs of society. The physical remains of human activity and endeavour are known as the *archaeological resource*. These remains range in size and complexity from individual objects, used and discarded, to settlements. They include many details in the landscape which itself is the product of human use and adaptation of the natural environment. The physical evidence may survive as earthworks such as burial mounds, hillforts, field banks and lynchets. It can also survive as structures such as buildings, canals, bridges and roads. However, the majority of the archaeological resource is smaller and often hidden below ground, surviving as features such as pits, postholes, gullies and ditches cut into the subsoil. Very often the evidence is in the form of artefacts, like coins, pottery sherds, stone tools and metal objects. Archaeological remains lie below many of the buildings and streets of British cities and towns. Over 600,000 archaeological sites are presently known in the UK, or about 200 per parish. The *archaeological record* is the sum of present archaeological knowledge, i.e. that part of the archaeological resource which has been identified to date. Table 7.1 summarizes the principal archaeological periods and likely remains from these periods

Table 7.1   Principal archaeological periods and likely remains (based on DoT 1993).

Period	Dates	Likely remains
Prehistoric	Earliest Palaeolithic (~500,000 BC to AD 43)	From early rock shelters and stone artefacts to the circles, barrows, Celtic field patterns, farmsteads, villages and hillforts of the Late Iron Age
Roman	AD 43 to 410	Native and immigrant farms, Roman towns and cities, military forts, roads
Medieval	5th–16th centuries	Origins of most modern towns (e.g. postholes from wooden buildings, masonry), Norman castles, deserted villages, ridge and furrow agriculture
Post-medieval	Late 16th to early 18th centuries	Civil War constructions, beginnings of industrial-scale extraction and manufacture, country houses and their parks and gardens
Industrial	Mid 18th century onwards	Buildings and infrastructure linked with industrialization, industrial relics
Post-Industrial	World wars	Defences, e.g. pillboxes

The rich pattern of archaeological remains that can be seen today is the result of the impact of successive generations on the remains left by previous generations. This process involves a degree of damage and destruction that is an inevitable part of the evolution of the archaeological record. However, the current threat to the archaeological resource is more significant than in the past, as a result of the technological changes and the rapid increase in development that have occurred particularly since World War II. Today the archaeological record is more likely to be deleted than altered or added to. Land that has been marginal since prehistory has, with the use of modern machinery and chemicals, become viable for arable farming, with the resultant damage by ploughing and soil erosion. The increase in road building, housing and industrial developments, and the need for materials for their construction, continually deplete the archaeological resource. Already in some areas, gravel extraction since 1945 has been so extensive that our ability to understand the evolution of the landscape has been badly reduced. As a result of high land prices in towns, there is now a prevalence of deep basementing, below-ground carparks, and substantial foundations to support high buildings. In some historic centres only a small proportion of the archaeological resource remains intact.

The significance of archaeological finds is derived both from the nature of the finds themselves in their contexts, and from the interpretation archaeologists are able to put on them given contemporary understanding. Although the ability to learn about the past is based on the investigation and interpretation of archaeological remains, this investigation often results in the destruction of the archaeological resource being studied. Archaeological excavation aims to dismantle remains to their constituent parts in order to understand the processes by which they were formed (see §7.4.1). This work is closely documented, with all the elements drawn and photographed, and the objects that are found removed and conserved. Although this enables future study and reinterpretation of the results, the site cannot actually be reconstructed. However, archaeology is an evolving study and is constantly harnessing new technologies, techniques, procedures and theories. Preserving archaeological remains for future study is therefore important. Just as archaeologists today can learn substantially more from the archaeological resource than their counterparts of yesterday, so preserving a site *in situ* for future archaeologists will allow even more information to be gained. In addition, the more visible sites that are used for tourism, recreation and education need to be preserved and conserved. Although the preservation of all remains would be impractical, and would lead to the stagnation of archaeology, the case for the preservation of the archaeological resource must always be carefully considered.

7.2.3 Historical buildings and sites

Listed buildings form the most visible and tangible of all aspects of the historical environment. They are a finite resource and they cannot undergo change without cultural loss. The careful appraisal of their history and condition, together with their protection through effective policies and careful professional practice, can lead to improved decisions concerning their conservation. Three main sources of judgement apply to changes to the character of buildings, deriving from the disciplines of archaeology, architecture and architectural history. In the face of change, decisions have most commonly been made on visual criteria. Since the legislation only implies preservation, the retention of inherited and historically interesting building fabric has been seen as the less important aspect of conservation activity. The encouragement of the conservation of "architectural character" has given rise to imitative architectural styles, often of mediocre quality, and to the encouragement of facadism. Little or no understanding of the concept of intervention has yet appeared in architectural practice in the UK; nor has the concept of stewardship on the part of building owners and local authorities been thoroughly adopted. However, listed buildings form a distinctive and finite stock of the built cultural heritage, and development that may affect this national resource needs careful management, including EIA.

Wider areas of townscape value are analogous to listed buildings; as areas of special architectural or historic interest they too are an important aspect of the built heritage, and in the UK account for some 4% of the built environment. As with listed buildings, an assessment of an area's interest and value is needed at the outset.

7.3 Legislative background and interest groups

7.3.1 Archaeology

The archaeological resource is protected by the Ancient Monuments and Archaeological Areas Act 1979, and the Town and Country Planning Act 1990. Important further guidance is provided in Planning Policy Guidance Note 16 (PPG 16). In Northern Ireland the Historic Monuments Act (Northern Ireland) 1971 is expected shortly to be replaced by the Historic Monuments and Archaeological Objects (NI) Order.

The Ancient Monuments and Archaeological Areas Act 1979 provides legislative protection to a selection of archaeological sites or monuments that have been identified as being of national importance and are included within a schedule maintained by the Secretary of State for National Heritage. These are consequently referred to as Scheduled Ancient Monuments. Some ancient monuments of national importance are not yet scheduled, and EH is currently undertaking a review of the schedule, the Monuments Protection Programme, which is considerably increasing the number of scheduled sites. Any works to or within a Scheduled Ancient Monument and likely to damage that monument require the prior consent of the Secretary of State; this consent is referred to as Scheduled Monuments Consent. Where consent is issued, it is frequently subject to conditions to prevent damage or to limit damage to agreed levels and with appropriate archaeological recording. Unauthorized works that damage a Scheduled Ancient Monument are a criminal offence, and significant penalties exist. The act also protects the setting of such monuments. The Secretaries of State are advised by EH, Cadw and Historic Scotland on Scheduled Ancient Monuments and other archaeological matters.

The 1979 Act also enables the designation of Areas of Archaeological Importance (AAIs). Five pilot AAIs were designated in 1984: York, Chester, Hereford, Exeter and Canterbury. Once an area is designated, developers are required to give six weeks "operations notice" to the planning authority of any proposals to disturb the ground, tip on it or flood it. A designated "investigating authority" then has the power to enter the site and, if necessary, undertake archaeological excavations for up to four months and two weeks. After that time the investigating authority must cease excavation but can continue to enter the site to record and inspect the works. This legislation did not address key areas such as preservation or funding, and has subsequently been overtaken by the procedures in PPG 16. No further AAIs have been designated.

The Town and Country Planning Act 1990 enables local authorities to protect a wide range of archaeological remains. Where development threatens to destroy remains, the authority can require appropriate investigation through a planning condition or legal agreement. In certain circumstances it can also secure the positive long-term management of sites. These provisions are usually expressed in the policies relating to archaeology within development plans (EH 1992a).

The impact of development on archaeology has for some time been recognized as a material consideration within the planning system. In 1990 the DoE issued Planning Policy Guidance Note 16 on Archaeology and Planning (DoE 1990a), which describes how archaeological matters are to be dealt with in the planning system. PPG 16 is an extremely useful reference document and should be carefully considered when preparing an EIA. Broadly, PPG 16 requires local planning authorities to acquire sufficient information to enable the full impact of a development to be considered. These powers had already been frequently used for the archaeological resource, but were formalized by the EIA regulations. Accordingly, the manner in which archaeological considerations are already dealt with in the planning system is closely akin to the requirements of EIAs. PPG 16's Welsh equivalent was published in 1991 (WO 1991). Its Scottish equivalent, published in January 1994, is in two parts: National Planning Policy Guidance Note 5, *Archaeology and planning*, and Planning Advice Note 42, *Archaeology – the planning process and scheduled monument procedures*.

7.3.2 Historic buildings and sites

Several interrelated regulations apply to listed buildings and conservation areas, in particular the Planning (Listed Buildings and Conservation Areas) Act 1990, and Circular 8/87 on principles of selection for listed buildings for England (whose appendix is notably and unusually detailed); Welsh Office Circular 81/81 for Wales; the Town and Country Planning (Scotland) Act 1972 and the Scottish Development Department's *Memorandum of guidance on listed buildings and conservation areas 1988* for Scotland; and the Planning (Northern Ireland) Order 1991 for Northern Ireland. Circular 8/87 is expected to be replaced by a new DoE/Department of National Heritage PPG 15 on historic buildings and conservation areas. This section reviews the principal requirements of the current regulations, but this should be updated with the most current official publications.

Listed buildings

A listed building is one that has been included in a list compiled by central government as being of "special architectural or historic interest". A developer cannot demolish, alter or extend any listed building in a way that affects its architectural or historic character unless listed building consent has been obtained from the local planning authority, and listed buildings must be taken into account when local planning authorities undertake land-use planning decisions. A small team of specialist investigators from EH identifies buildings to be listed: this method echoes that earlier employed for compiling schedules under the Ancient Monuments Acts 1882. Listed buildings account for some 2% of the building stock. EH's proposals are closely scrutinized by the Secretary of State before confirmation.

Central to architectural conservation – the conservation of whole buildings – is the definition of building character, for it is against this definition that judgements are made about the nature and extent of permitted changes. The criteria used to identify buildings of "special architectural or historic interest" are as follows:

- all buildings built before 1700 which survive in anything like their original condition
- most buildings between 1700 and 1840, although selection is necessary
- between 1840 and 1914 only buildings of definite quality and character, including the principal works of the principal architects
- between 1914 and 1939, selected buildings of high quality, and
- after 1939, a few outstanding buildings.

In choosing buildings for the list, particular attention is paid to qualities other than the chronological, including: (a) special value within certain types, either for architectural or planning reasons or as illustrating social and economic history (e.g. industrial buildings, railway stations, schools, hospitals, theatres, exchanges, almshouses, mills); (b) technological innovation or virtuosity (e.g. cast iron, prefabrication, or the early use of concrete); and (c) group value, especially as examples of town planning (e.g. squares, terraces or model villages). Theoretically, legislation protects architectural or historical character by embracing everything within the property's grounds, although a detailed evaluation of each property is needed to make a judgement about those features that are of worthwhile architectural or historical significance.

Listed buildings are graded to indicate their relative importance: Grade I buildings are of exceptional or outstanding interest, Grade II* are particularly important and of more than special interest, and Grade II are of special interest and warrant every effort being made to preserve them. Grade III is a non-statutory and now obsolete grade. Grade III buildings are those which, although not qualifying for the statutory list, were nevertheless considered to be of some importance. Many of these buildings are now considered to be of special interest by current standards, particularly where they possess "group value", and are being added to the statutory lists as these are revised. Slightly different grades apply to Scotland and Northern Ireland.

Local planning authorities define the way in which listed buildings are to be protected. The special character of these buildings is to be "taken into account" when development and land-use planning decisions are made. Accordingly, EIA studies need to consider the declared policies of county and district councils to determine the strength and quality of the protection that is offered. The characteristics of protection are variable over the nation as a whole, being a mixed array of legal provisions and aesthetic judgements.

If a local authority considers a non-listed building to be of special architectural or historic interest and in danger of demolition or significant alteration, it can serve a Building Preservation Notice, which effectively lists the building for six months; this allows the Secretary of State to determine whether the building should be included in the statutory list or not. The National Trust for England and Wales, and the National Trust for Scotland, have the power to declare their properties inalienable so that they cannot be removed from the Trusts' ownership without their consent except by special Parliamentary procedure.

Conservation areas

According to the Planning (Listed Buildings and Conservation Areas) Act 1990, conservation areas are sections of land or buildings designated by local authorities as being "of special architectural or historic interest, the character or appearance of which it is desirable to preserve or enhance". Local authorities must have regard to conservation areas when exercising their planning functions, and conservation area consent must be obtained from the local authority before a building within a conservation area can be demolished.

Conservation areas have proved to be a popular and positive element of town planning, although it was not until the Civil Amenities Act of 1967 that the notion of architectural and historic interest was extended to protect townscapes. The act highlighted the need for standards of amenity for the outdoor environment, for the enhancement and highlighting of special townscape qualities, and for the protection of the identity and character of the *genius loci*. Central government guidance allows liberal interpretation for local authorities to declare conservation areas. They

> . . . will naturally be of many different kinds. They may be large or small, from whole town centres to squares, terraces and smaller groups of buildings. They will often be centred on listed buildings, but not always. Pleasant groups of other buildings, open spaces, trees, an historic street pattern, a village green or features of historic or archaeological interest may also contribute to the special character of an area. (DOE 1987a)

The act encourages the involvement of local communities through conservation area advisory committees. The concept of conservation areas has found wide-spread support with the public as a whole, in spite of early reluctance by many councillors and continuing fears from the design disciplines.

Other legislation

There is no specific national legislation addressing the World Heritage Sites promoted by the UNESCO Convention for the Protection of the World Cultural and Natural Heritage; their protection lies in the importance given to them within the planning process and through policies relating to the development plans.

EH maintains a Register of Parks and Gardens of Special Historic Interest in England, namely sites that are regarded as an essential part of the nation's heritage. The register grades parks and gardens from Grade I of exceptional interest, to Grade II of special interest. These sites are not afforded statutory protection, but are protected by recognition of their importance through the planning system, and policies relating to them in development plans. Historic Scotland and Scottish Natural Heritage compile a similar Inventory of Gardens and Designed Landscapes in Scotland (SO Circular 6/92).

7.3.3 Interest groups and sources of information

The *county archaeologists* should be involved early in the EIA process. They advise on the care of archaeological sites, maintain the Sites and Monuments Record for their area (see §7.4.1), screen planning applications for archaeological impacts, and make recommendations to the planning committee. They will be able to make a rapid initial assessment (see §7.4.1) and suggest professional contacts (e.g. members of the Institute of Field Archaeologists with local knowledge and experience) if further specialist knowledge is required. Each county has county archaeologists, and some district councils also have district archaeologists. In London the rôle of the county archaeologists is fulfilled by EH. Annex 2 of PPG 16 contains addresses of county archaeologists. In many counties the *Museum Service* works closely with the county archaeologists.

English Heritage is a body under appointed commissioners set up to advise the Secretary of State on archaeological, historical and heritage matters. It administers the most important sites. In Wales, *Cadw* has a similar function and is part of the Welsh Office; in Scotland, the equivalent is *Historic Scotland*, part of the Scottish Office. Historic Scotland also fulfils some of the functions of the county archaeologists. In Northern Ireland these rôles are performed within the DOE's Environment Service, which also fulfils many of the rôles of the county archaeologists.

In many areas *local history or amenity societies* have detailed local knowledge and take active interest in anything that affects their area. Local planning authorities must consult the national amenity societies when the demolition of a listed building is proposed. In practice, the societies are also consulted when more ordinary changes are proposed, as their expertise is substantial and unique. The advisory societies are the Ancient Monuments Society, the Society for the Protection of Ancient Buildings, the Georgian Group, the Victorian Society, the Council for British Archaeology, and the Royal Commission on the Historical Monuments of England; their addresses are given in Appendix B. Local amenity societies are more ephemeral; planning authorities maintain lists of societies in their localities which they consult over changes to listed buildings. The archaeologists' professional body is the *Institute of Field Archaeologists*; they publish lists of their members and their specializations.

7.4 Scoping and baseline studies

7.4.1 Archaeology

The aim of a baseline study is to identify and describe the nature, location and extent, period(s) and importance of the archaeological resources likely to be affected by the development. The resulting report should include:

- a summary of the archaeological context;
- an inventory of archaeological assets found both at the site and in the wider area likely to be affected by the development;
- an evaluation of these assets;
- an informed estimate of potential assets to be found in further investigation or likely to be at risk from development; past construction activities which might have already destroyed archaeological resources should be noted;
- a map of the project area showing the location of these assets;
- a note of any inherent difficulties which may limit the study's usefulness (e.g. problems of access).

Several sequential stages of data gathering can be identified. However, not all stages would be necessary for every EIA.

Rapid appraisal

Rapid appraisal of the archaeological resource involves the collation and review of existing and easily accessible data. This will certainly include a review of the Sites and Monuments Record (SMR) and consultation with the county archaeologist. It may also include a site visit. This appraisal will enable a preliminary view of the likely nature and scale of the archaeological constraint. It may in itself be sufficient to meet the aims of the EIA, or may identify the need for subsequent stages of data gathering.

The main source of archaeological information is the SMR. The SMR is an archaeological database containing information about the known archaeological sites and finds in each county. The SMR has a statutory locus, in that it is referred to within the General Development Order 1990; certain types of permitted development, such as mineral extraction, require permission where they affect an archaeological site registered on the SMR. The SMR information is gathered from several sources and in a variety of ways, from detailed surveys to chance finds. As a result there is considerable variance in the reliability of the data and the interpretation that can be placed upon it. It is often not very intelligible to non-archaeologists (and is not in fact a public document) and it may need professional interpretation to assess the significance or potential of archaeological sites. The county archaeologists will usually be familiar with the nature and shortfalls of the data being considered, and will be able to advise on the appropriate interpretation of the archaeological data. It is important to note that the interpretation of archaeological data is rarely straightforward.

Although the SMR is a comprehensive statement of the archaeological resource as currently known, it is not a definitive statement: new archaeological information becomes available all the time. Therefore, the sites on the record represent only a part of the actual archaeological resource and many archaeological sites remain as yet unlocated. This has two major implications for compiling an EIA. First, as the SMR only reflects current knowledge, there many be other important archaeological remains as yet unlocated that may be affected by a proposal. Secondly, if considerable time elapses between when the SMR is consulted and when that information is used, additional evidence may have become available in the meantime. These unknown sites are nonetheless a material consideration and therefore should be addressed when considering a development proposal. This is recognized in PPG 16:

> Where early discussions with local planning authorities or the developer's own research indicate that important archaeological remains may exist, it is reasonable for the planning authority to request the prospective developer to arrange for an archaeological field evaluation to be carried out before any decision on the planing application is taken. This sort of evaluation is quite distinct from full archaeological excavations. It is normally a rapid and inexpensive operation, involving ground survey and small-scale trial trenching, but it should be carried out by a professionally qualified archaeological organisation or archaeologist . . . Evaluations of this kind help to define the character and extent of the archaeological remains that exist in the area of a proposed development, and thus indicate the weight which ought to be attached to their preservation. They also provide information useful for identifying potential options for minimizing or avoiding damage. On this basis, an informed and reasonable planning decision can be taken.

Local planning authorities can expect developers to provide the results of such assessments and evaluations as part of their application for sites where there is good reason to believe there are remains of archaeological importance. If developers are not prepared to do so voluntarily, the planning authority may wish to consider whether it would be appropriate to direct the application to supply further information under the provisions of Regulation 4 of the Town and Country Planning (Applications) Regulations 1988 and if necessary authorities will need to consider refusing permission for proposals which are inadequately documented. In some circumstances a formal Environmental Assessment may be necessary. (DoE 1990a)

As mentioned in §7.3.3, the *county archaeologists* both maintain the SMR and are therefore a source of initial data, and advise the local planning authority and are therefore an initial source of advice. They will also be able to advise on the scope and content of the archaeological elements of the EIA. They are extremely knowledgeable about the archaeological potential of sites in their areas, and are also usually very realistic about development pressures. The county archaeologists will be anxious to ensure that the archaeological content of an EIA has been properly addressed, and will generally be happy to supply both data and advice. A charge may be made to cover the costs incurred in supplying data (ACAO 1992). As the county archaeologists usually advise the local planning authority regarding the acceptability of these elements, it is important to be aware of their opinions at an early stage. Where failure to consult results in additional archaeological concerns being raised, there is the potential for uncertainty, delay and additional costs which will negate the benefits of having carried out the EIA.

Where a development is likely to affect a Scheduled Ancient Monument or its setting, EH should be consulted and Scheduled Monument Consent may be required. The need to obtain this consent is independent of the planning process, and unless identified early could introduce substantial delay or even compromise the development altogether. Furthermore, where a development is likely to affect a monument of national importance which, although not scheduled, may be considered for scheduling in due course, it is advisable to seek the advice and opinion of EH or equivalent.

Desk-based assessment

A desk-based assessment should identify and collate as much existing information as possible and frequently requires some original research. Information may be retrieved from several sources, but the SMR is usually the most useful starting point.

Aerial photographs are an important source of data. Earthworks are often more easily recognized and interpreted from the air than from the ground. Buried archaeological remains can also be traced from the air in certain circumstances. The buried remains can affect the growing crop. For instance, a buried wall or road surface may retard crop growth, or in a dry year create a parch mark. A buried pit or ditch may promote crop growth. The patterns that result can be interpreted as archaeological features or sites. Different soil colours may also reveal archaeological sites. Aerial photographs may be found in national, local authority, and possibly private collections. The record office may contain historic maps or plans and other documents relating to the land, and it may be possible to find other data not yet assimilated into the SMR. The Victoria County Histories and Local Archaeological Societies may have additional information.

Desk assessment is usually undertaken at an early stage in project planning, when there may be an issue of commercial sensitivity. If so, it may be reasonable to use the county archaeologist and the Victoria County Histories, but not to approach the voluntary societies until later.

Field survey

A wide range of field survey techniques is available, including geophysical techniques, field-walking, augering, test-pitting, machine trench digging and earthwork surveys. These are described below. Not all of these techniques will be applicable in all circumstances. Some can act as useful preliminaries to other techniques. A phased approach to field survey is often the most sensible and cost effective, so it is common to use a suite of techniques as the proposal develops: perhaps starting with a rapid appraisal and then a desk assessment in the earliest stages, then field-walking before the actual site is proposed, and machine trenching afterwards. When considering the appropriateness of the various techniques, consultation with the county archaeologists may be valuable.

The county archaeologists usually produce a brief or specification for the work when a field survey is being undertaken through the planning process. A brief is an initial statement regarding the aims and scope of the archaeological work required, identifying certain working standards. It would form the basis of any specification produced, which should be referred back to the county archaeologists to ensure that all matters in the brief have been properly addressed. Alternatively the county archaeologists may issue a full specification that sets out in detail the works required in the field survey and would be sufficient to enable the project to be implemented and progress to be monitored.

The county archaeologists may also wish to make arrangements for monitoring the field survey to ensure that works are carried out to professional standards and to any specification that has been issued. This has benefits both for the archaeological resource and for the developer, who may have no independent means to monitor the value of the work being undertaken. It also enables the county archaeologists to keep up to date with any archaeological sites discovered during the fieldwork. Some county archaeologists charge for monitoring.

Geophysical techniques The ground has characteristics and properties that can be altered by previous land uses, some of which can be investigated and interpreted by geophysical techniques. The principal techniques used are resistivity and magnetometer surveys, although others are also available. Resistivity surveys measure the ground's resistance to the progress of an electrical current. Measuring increases and decreases in the resistance can indicate the nature and location of buried features. Magnetometer surveys measure the magnetic properties of the soil and can be used to identify locations of past human activity, particularly those that involved burning or heating.

Geophysical techniques can only be applied in suitable site conditions, and an experienced geophysical operator should visit the site to assess their feasibility. Where they are appropriate, geophysical techniques have an advantage over many of other field techniques in that they do not damage the archaeological resource. Because of this they are particularly appropriate for Scheduled Ancient Monuments, although Scheduled Monument Consent or a licence may still have to be obtained before surveys can be undertaken.

Although the results of geophysical techniques can sometimes be ambiguous, these techniques often successfully identify the location and extent of archaeological sites and can give some idea of their nature. The results can therefore help to focus subsequent stages of field survey to maximize data recovery. However, geophysical techniques are unlikely to provide sufficient information on their own, are not universally applicable, and are often expensive.

Field-walking   Field-walking, also known as *surface artefact collection,* is confined to ploughed fields. A plough breaks and turns over the surface soil. In ploughed fields there is a tendency for buried material to be brought to the surface, and where the plough intrudes into a buried archaeological site this will include archaeological artefacts. Rigorous collection and plotting of this material will enable the location, date, and extent of certain types of archaeological site to be described. The archaeological material collected can be anything that reflects human activity, such as pottery sherds, worked stone, coins, building material and even stone that is not local to the area and may have been imported.

The county archaeologist will be able to suggest a field-walking strategy that ensures that the data gathered will be comparable to other field-walking data already on the SMR. The area being studied is divided up by a grid, usually based on the national grid. Artefacts are then collected from along the lines of one axis of the grid, usually the north–south axis, and stored and recorded according to where on the grid they were recovered. The size of the grid thus determines the size of the collection units, and the precision of the results. A survey on a large grid will be rapid, but will represent a small sample of the available artefacts. A survey on a small grid will be more time-consuming, but the results will be based on a larger sample. The size of the grid is usually determined with reference to the sorts of archaeological sites that are anticipated. For instance a smaller grid would be required to locate small Mesolithic camps than a Roman villa. In general, grid spacing is about 20 m or 25 m.

Where a site has already been located, intensive field-walking (called *total collection*) can be used to determine spatial distributions across the site. Total collection involves laying out a small grid across the site, perhaps 5 m × 5 m, and collecting all the artefacts within each grid square.

Field-walking is a relatively rapid and inexpensive technique that can be applied over large areas. However, the results can be ambiguous or misleading. Where a site is located by field-walking, it is by definition being damaged. It is hard to judge from field-walking results alone how intact the site is, or whether it solely survives as artefacts trapped in the plough soil. A site surviving intact below the plough soil will not be represented on the surface. Certain periods do not produce artefacts that are likely to survive the ploughing action. The results of field-walking therefore need to be qualified by some understanding of the relationship between the depth of ploughing and the depth of the archaeology.

Augering   Augering is most frequently used in river valleys where **alluvial**, **colluvial** or **peat** deposits have masked the original land surface and where slightly higher ground in a wet environment may have acted as a focus for human activity. By recording the soil sequence from auger holes located over a wide area, the underlying and hidden subsurface topography can be mapped and the archaeological potential of the area inferred. Augering alone is unlikely to confirm the presence or absence of archaeological deposits, but it can clarify the archaeological potential and so focus subsequent stages of survey. It can also be used to clarify the nature of features located by geophysical techniques, and in certain areas to assess the potential for the preservation of palaeoenvironmental data.

Test pitting   Test pitting involves the hand excavation of an array of small pits of a predetermined size. It provides a clear picture of the nature of the soil structure and the upper layers of the underlying geology.

As with field-walking, the spacing and array of test pits usually reflect assumptions about the expected archaeological resource. Test pits can be varied in size and array in order to meet the requirements of the survey. They are usually 1 m × 1 m, or 1 m × 0.5 m for ease of excavation. The soil from test pits is often sieved through a wire mesh of a set size to ensure consistent artefact recovery, enabling a rigorous statement to be made regarding the number, type and depth of artefacts. Analysis of the different artefact recovery rates over an area gives an indication of the date, location and extent of archaeological sites. Test pitting is often used instead of field-walking where the land is pasture rather than arable, and in woodland where machine trenching may not be possible.

Machine trenching   Machine trenching uses trenches, usually cut with a toothless ditching bucket, laid out in a pattern across the site. The trench pattern will attempt to maximize information retrieval, possibly on the basis of existing data such as aerial photographs, field-walking or geophysical results. The extent of trenching required is usually an agreed sample of the land. The size of the sample is currently the subject of considerable debate, but is commonly around 2%, depending on local circumstances. When archaeological deposits are encountered excavation continues by hand. The excavation is controlled by a supervising archaeologist at all times. Machine trenching quickly locates features cut into the subsoil but, where large amounts of earth are rapidly removed, there is limited opportunity to collect artefacts, and the rate of artefact retrieval is low. Higher rates of retrieval can be achieved by hand-digging parts of the trench, equivalent to a test pit, and the use of metal detectors.

Trenching is very disruptive and it intervenes directly into the archaeological levels. This has the advantage of producing unambiguous information, but is potentially damaging to archaeological remains one might otherwise wish to protect. It is also not always possible to get a machine onto a site.

Earthwork surveys   Archaeological sites may be visible as earthworks such as banks, ditches, burial mounds, and sites of deserted or shrunken settlements. Sites that survive as earthworks are generally more intact than other sites. Ploughing can degrade earthworks, and the success of earthwork surveys is limited in fields that have been arable for a long time; generally, such land is more productively scanned from aerial photographs. Pasture can have visible earthworks surviving. When they are obviously visible, they will often have been recorded by the Ordnance Survey or the SMR. They can also be identified through aerial photographs. Woodland, particularly ancient woodland, holds the greatest potential for producing previously unrecorded earthworks. The sites will often be obscured from the air by trees and on the ground by undergrowth, so it is best to undertake the survey during the winter or early spring.

The nature of the earthwork survey will depend on the aims of the evaluation. The survey can vary from sketch plotting the earthworks onto an OS map, through two-dimensional surveys such as plane table surveys, to a three dimensional survey producing an accurate contour or hachure plan.

Finds   Some of the artefacts recovered may be subject to the laws of treasure trove; specifically, all discoveries of gold or silver should be reported to the coroner, who will consider whether the items were hidden with a view to being retrieved at a later date. If this is concluded to be the case, the State may retain any of these items, paying the landowner the market value. In all other situations the artefacts are the property of the landowner. It is usually recommended that they are donated to a local authority museum, so that they can be stored in appropriate conditions and made available for future study. All finds of human bone, from any period, have to be reported to the coroner.

The developer's responsibilities arising from the destruction of the archaeological resource often continue beyond excavation. If finds are donated to the appropriate local authority museum, it is likely that the planning authority will consider the developer to have met these responsibilities. If the developer wishes to make alternative arrangements, they may need to demonstrate that this alternative is appropriate. Some museums make a charge for accepting the long-term responsibility of storing archaeological material.

Some problems with field surveys   *Access to the site* will not be a problem where the developer already owns the land, although there may be problems where the project has offsite implications, e.g. as a result of dewatering. For projects such as road schemes, a field survey may not be possible until the route is finally selected and the land acquired. This is undesirably late because it does not allow a route to be chosen which would preserve important remains *in situ*.

The *project timetable* may constrain the fieldwork options. Field-walking is not possible in a standing crop, and can only be done after the fields are ploughed. Conversely, crop patterns show best in a well grown crop and should be photographed just before the harvest.

The *cost* of archaeological surveys depends upon the extent and nature of the survey and the techniques employed. Surveys are frequently labour intensive and some elements can be expensive. Where the developer is liable to pay compensation to the landowner for damage arising from the evaluation, the scale of compensation will depend upon the techniques used. However, the costs should be seen against the background of the cost resulting from unexpected delay to the progress of the planning application or indeed the progress of the development if significant archaeological deposits are located at a late stage in the process.

7.4.2 Historic buildings and sites

Although listed buildings account for only some 2% of the UK's building stock, they are a fragile and valuable resource. Only a full assessment of a listed building's inherited character at the outset will allow well informed judgements to be made about the significance of a proposed development's impacts. Owners, developers and local authorities all have their respective rôles to play in such assessments.

An initial review of the listed building register will identify any listed buildings likely to be affected by a proposed development. If such buildings are identified, a baseline survey will be necessary, involving an audit of the buildings' special architectural and historic interest. Such a survey consists of a detailed archival search of local history libraries and other property record depositories. The written product should be supported by plans, sections and elevations, together with a photographic survey and diagrammatic analysis of the individual buildings' evolution over time. This information is evaluated in terms of the relative importance of the building's component parts. This survey involves specialized work and is best undertaken only by those with a qualification in historical or architectural conservation.

Baseline studies for conservation areas have a wider remit than those for listed buildings. An initial survey will identify characteristics of significance, including archaeological features of interest (whether buried remains or standing structures), all listed buildings with an indication of their property curtilages, building age, and geological, topographical or landscape features. Those townscape features that constitute the area's special architectural and historic interest then need to be appraised, including vernacular characteristics, indigenous building materials, spatial characteristics, sections of group coherence or special townscape value, and long-distance views within, outside or across the conservation area that are of importance in the perception of its inherited character. The problems and policies that affect the present or future wellbeing of the area also need to be appraised. This consists of a statement of problems that adversely affect the physical amenity of the area (e.g. traffic intrusion, noise, visual intrusion, architectural disfigurements, fumes, waste-disposal deposits, etc.), the position with respect to present and future district-wide policies for preservation and enhancement, evaluations of specific problem sites, and opportunities for area-wide enhancements and improvements, including vehicular and pedestrian movement.

7.5 Impact prediction

7.5.1 Archaeology

Prediction of archaeological impacts involves three unknowns: what the archaeological remains are (discussed in §7.4.1), what the proposed development's impacts would be, and how significant the impacts would be. *Identification of impacts* must include both direct and indirect impacts. The direct impacts are often clear, and usually involve the removal of archaeological materials. Some of the direct impacts may not be immediately obvious, when they result from secondary operations such as drainage and landscaping works associated with the development. A development's indirect impacts are often more difficult to define. For example, dewatering associated with a development may lead to the destruction of some types of archaeological deposits on adjacent undisturbed sites which had previously survived as a result of waterlogging. A residential development may increase recreational pressure on a nearby earthwork or affect the visual setting of an adjacent archaeological site. Positive impacts are often indirect, e.g. when a road scheme relieves congestion in a historic town centre.

The *significance* of a development's impacts depends on factors linked to the interpretation archaeologists are able to put on finds given contemporary understanding. When assessing whether an ancient monument is of national importance, and thus whether it should be scheduled, the Secretary of State for National Heritage makes reference to eight "scheduling criteria":

- period: the degree to which a monument characterizes a particular period
- rarity: the scarcity or otherwise of surviving examples of the monument
- documentation: the significance of the monument may be enhanced by records, either of previous investigations or contemporary to the remains
- group value: the significance of the monument may be enhanced by its association with related contemporary or non-contemporary monuments
- condition: the condition or survival of the monument's archaeological potential
- fragility: the resilience or otherwise of a monument to unsympathetic treatment

113

- diversity: the combination and quality of features related to the monument
- potential: where the nature of the monument cannot be specified but where its existence and importance are likely.

These criteria are further described in Annex 4 of PPG 16 (DoE 1990a). They can be used to help establish the importance of not only ancient monuments but also other archaeological remains.

Lambrick (1993) suggests that cultural impacts can be evaluated in terms of who is affected. He lists the resources: archaeological remains, palaeoenvironmental deposits, historic buildings and structures, historic landscape and townscape elements, sites of historical events or with historical associations, and the overall historical integrity of the landscape. He then gives a list of human receptors who may be affected by impacts on these resources: owners and occupiers of historic properties and monuments, visitors to sites and buildings specifically open to the public, local communities, the general public as regards general enjoyment of historic places through informal public access, and individuals or groups with a special interest in the historic environment, including academic archaeologists. He then suggests:

Perhaps the best means of considering [significance] is to say that an effect is significant if it makes an appreciable difference to the present or future opportunity for people [receptors, as defined above] to understand and appreciate the historic environment [resources] of the area and its wider context (Lambrick 1993).

Impact significance may also be considered in geographic terms. The DoT suggests four categories of importance for archaeological remains, namely (a) sites of national importance, usually Scheduled Ancient Monuments or monuments in the process of being scheduled as such; (b) sites of regional or county importance; (c) sites of district or local importance; and (d) sites which are too badly damaged to justify their inclusion in another category (DoT 1993).

7.5.2 Historic buildings and sites

A proposed development action can directly affect a listed building in a variety of ways, ranging from the minor to the extensive:
- repairs of minor elements using replacement materials
- changes to the interiors of buildings, where decorations or other architectural features may enrich the understanding of the building's interest
- modifications to individual elements of the building which form a significant part of its character
- new extensions
- partial demolitions
- complete demolitions, and
- severance of part of a property from other parts (for instance, a house from its gardens or outbuildings).

Indirect impacts to listed buildings include noise and disturbance from nearby developments leading to a loss of amenity, and air pollution which can lead to deterioration of buildings and damage to garden and park vegetation. Nearby developments can cause visual intrusion and change the building's original landscape setting.

Direct impacts on conservation areas from the private sector are most commonly related to proposals for development, whether new-build or refurbishment. Extensive damage can also be created by permitted development for which special directions under Article IV of the General Development Order are needed (DoE 1987a). Public sector developments such as those by highway authorities or utility companies can affect conservation areas without reference to conservation area policies; these may be brought under the control of the Town and Country Planning Acts by specific directions under Article IV of the General Development Order. A conservation area can be directly affected through the loss of buildings, through cumulative impacts resulting in a general deterioration in the setting of the buildings, or through severance. Development can also result in the neglect of a building or site, resulting in its deterioration or destruction. More generally, development can alter or destroy open spaces and change the character of historic districts.

Any proposed development constitutes a potential intrusion into an acknowledged heritage object. Individual building owners, as much as public agencies and professional advisors, play a curatorial rôle in a building's conservation and should be involved in predicting the impacts of a proposed development. As such, impact prediction is best undertaken as a dialogue between the owner or developer and the local authority, which respectively represent the private and the public aspects of curatorial influence. The developer determines the extent of change that is anticipated, and thus the utilization of the property and its financial value. The local authority makes a judgement about the extent of architectural and historic change that can be allowed, taking into account national and local policies and standards. The outcome may take the form of agreement, compromise or disagreement. This evaluation constitutes a special negotiation over and above that needed for normal building refurbishment or property development. The local planning authority classes such a dialogue as an exploratory meeting. Agreement between the two parties at this stage can constitute an agreement for the later stages of design.

The *significance* of these impacts will depend on the significance of the building or site affected, as well as on the magnitude of the impact. Assessing a development's impacts on a listed building involves judgements that are architectural and aesthetic by nature, although it is possible to amplify these quantitatively according to the type of impact involved. §7.3.2 summarized the grading systems for listed buildings and parks and gardens, which provide an initial indication of relative importance. However, no such gradings exist for conservation areas; criteria for area-wide character and standards of amenity are needed for effective protection, and to allow judgements to be made about project impacts.

Applications for listed building consent should be made for any change that would affect the character of a listed building, and for planning permission to undertake development of the land. An application is often a way of confirming the earlier evaluation, and for determining the full structural and condition surveys to evaluate the building's structural condition and the implication of any changes to the building fabric. These surveys should be undertaken only by architects, building surveyors or structural engineers who are qualified in historical or architectural conservation. Most old buildings do not meet regulatory requirements governing modern building construction, but this does not necessarily make them unsafe. It takes training and experience to make judgements about their conditions which avoid the destruction of the building's character. Detailed application for full planning permission and listed building consent can be made with confidence only once the initial surveys and evaluations have been successfully concluded.

7.6 Mitigation and enhancement

7.6.1 Archaeology

Having identified the nature of the archaeological resource and considered the development's impact upon it, mitigation strategies may be recommended. For the majority of development proposals, *no further archaeological activity* is required because no archaeological resource has been identified, or there is no significant impact on any archaeological resource, or the scale or nature of the impact or the nature of the resource does not warrant further action.

An *archaeological watching brief* may be carried out during the relevant stages of development. These stages are likely to be earth moving, topsoil stripping, and the digging of foundations and services. The watching brief should enable any archaeological evidence encountered to be recorded, and removed if appropriate. It may be accepted that this will not cause unreasonable delay to the progress of the development; if some delay is considered likely, the circumstances which would warrant a delay should be described and agreed upon in advance.

In some circumstances the need for development may override the case for preserving an archaeological site. In this case the site should not be thoughtlessly destroyed, and the local authority may satisfy itself that appropriate provision has been made (DoE 1990a). This will involve the *archaeological excavation* of the site prior to the development. The developer's responsibilities also include post-excavation (e.g. the long-term storage of the excavated material and the appropriate dissemination of the results). Depending on the nature and extent of the remains, excavation, post-excavation and publication can be expensive and time-consuming.

Preservation in situ means leaving the archaeological site undisturbed. This is the only mitigation measure which wholly meets the EIA Directive's principle of preventing environmental harm at source. It is supported by PPG 16, which states that preservation *in situ* is the preferred action. The local planning authority may require preservation *in situ* if the archaeological remains are important, or the developer may choose to preserve *in situ* if mitigation requirements are too expensive. Preservation *in situ* can be achieved in several ways. The development can be avoided altogether, or, if the archaeological constraint has been identified sufficiently early, by site or option selection. A common solution is to preserve the site within the design of the development, for example as an area of open or recreational space. The local authority may attach a fencing condition to the planning permission to prevent inadvertent damage during construction work. This secures the erection of a fence around a stipulated area and prohibits work within that area. Provision may also be made for positive management of the archaeology to secure its long-term future from any indirect impact of the development. Preservation *in situ* can be achieved within the construction of a development. For instance, the less structurally demanding elements of a development, such as car parking, can be built on raised levels or rafted foundations above the archaeological deposits. Although these options are feasible, they can cause technical or engineering problems, such as shrinkage of buried material as it dries out. It may be possible to preserve the majority of an archaeological site by agreeing an acceptable level of destruction. For instance a low-density pile foundation may be acceptable where the pile has been designed to avoid the most significant deposits. Ultimately, preservation *in situ* may need to be achieved by abandoning elements of the development or indeed abandoning the development entirely. Where the importance of an archaeological site merits it, the local planning authority can refuse an application on archaeological grounds.

7.6.2 Historical buildings and sites

Mitigation measures in EIA should include policies to highlight and strengthen the historic building's or site's inherited and intrinsic qualities and special interest, as well as to preserve them. Preservation starts with the declaration of a listed building or conservation area: all subsequent actions should strengthen and reinforce architectural characteristics and retain historical interest. Without such intent, the intrinsic qualities of a listed building or a conservation area can be diluted and destroyed.

For conservation areas, unlike listed buildings, the legislation specifically allows their preservation to be accompanied by enhancement measures. Proposals for area-wide preservation and enhancement may consist of programmes of building maintenance and repair, and their implementation; programmes of building restorations involving the rectification of disfigurements and their implementation; programmes of face-lift enhancements; strategies for the enhancement of floorscape treatments and their integration into the design of public and private domains; strategies for building materials; and new infill building developments within clearly established building envelopes.

Proposals for the enhancement of conservation areas should be drawn up by local planning authorities and discussed at public meetings in the localities concerned. Such proposals may be compiled by local citizen groups with the advice and support of professionals qualified in architectural conservation or urban design, provided that the meetings at which proposals are presented are genuinely open to all local interests and involve elected representatives of the local authority. Where citizen groups take such initiatives, it still remains the province of the local authority to make formally adopt the proposals presented.

7.7 Monitoring

The prediction of archaeological impacts is not an exact science, and unexpected problems can arise. The chances of this happening are considerably reduced by a thorough evaluation, but some contingency should still be made for the un-expected. The planning authority has the power to revoke planning permission where an unexpected and overriding archaeological constraint warrants it. In this circumstance, compensation would have to be paid. This can prove to be an expensive option and it is one reason why local authorities are empowered to ensure, by field survey if necessary, that the full archaeological implications of the development have been properly identified prior to the determination of the application.

If unexpected archaeological remains are located, additional discussion between the developer and the county archaeologist will be needed. Where agreement cannot be reached, EH may be able to arbitrate between the two par-ties. Where these unexpected remains warrant it, the Secretary of State for National Heritage may schedule them, and the developer would then need con-sent to continue work. Developers can insure themselves against the risk of loss from encountering unexpected archaeological remains.

7.8 Conclusions

The historical environment is a specialist discipline, covering many different periods and types of remains. Further reading on archaeological impacts in-cludes DoE (1990a), EH (1991a,b, 1992b), Lambrick (1992), Morgan Evans (1985), Ralston & Thomas (1993), RICS (1982), and DoT (1992b). Few publica-tions exist on listed buildings and conservation areas. The most generally read-able treatment is by Ross (1991), who gives wide coverage to the rationale and evaluation of historic conservation in the UK. Suddards (1988) appraises the legislative provisions that underpin the aims of building conservation. DoE Circular 8/87 (DoE 1987b) provides the most detailed official guidance. Feilden's (1982) *Conservation of historic buildings* is a substantial reference volume, and several other publications on techniques of repair are provided by EH and the national amenity societies.

In EIA, it is important to contact the county archaeologists or their equivalent as early as possible, since they are valuable sources of archaeological data and advice. Where consultation is left to a later stage, unexpected problems and delays are more likely to occur. The EIA should be carried out by specialists trained not only in survey, excavation and post-excavation techniques, but also in interpreting the data for the relevant period and type of remains. Specialist knowledge will be needed to interpret the relative importance of these results and suggest appropriate mitigation strategies. Using specialists in archaeology or historical or architectural conservation, from the earliest stages of the EIA when the data gathering programme is first being considered, will ensure that the correct type and amount of data are obtained. The result of using inappropriately qualified staff may be that, after the EIA is completed, additional historical constraints may be identified, or additional information required, potentially introducing delay and reducing the benefits of carrying out the EIA.

Problems may arise where the developer gathers inadequate or inappropriate data for use in EIA. This frequently occurs as a result of cost-cutting on the data-gathering strategy. This can be a short-sighted saving when compared to the cost of delay to the progress of the application, or delay to the progress of the development.

CHAPTER 8
Air and climate
Derek M. Elsom

8.1 Introduction: definitions and concepts

8.1.1 Air and climate changes

A proposed development that will add pollutants to the atmosphere or alter the **weather** and **climate** may result in adverse effects on people, plants, animals, materials and buildings (Elsom 1992). These effects can occur at the local, regional or even global scale.

Major developments, such as power stations, oil refineries, waste incinerators, chemical processing plants and roads, pose obvious potential pollution problems, but even developments that emit little or no pollutants when completed and operating can create a local dust nuisance during the earth-moving and materials-handling operations of the construction stage, especially during dry weather conditions. Developments may give rise to both routine and non-routine pollutant emissions, if they intend to use one type of fuel for most of the time, but on a few occasions have to switch to an alternative fuel. In the UK this can occur when an industrial plant intends to use an "interruptible" natural gas supply. This type of supply allows British Gas the right to cease supplying gas during peak periods of national demand, during which the plant has to switch to a standby fuel such as heavy fuel oil for up to 30 days a year. Whereas natural gas produces no emissions of sulphur dioxide (SO_2), fuel oil emits significant amounts depending upon its sulphur content. Another example of non-routine emissions to consider is the possibility of an accident at a proposed development that intends to store or process toxic chemicals or nuclear fuels.

8.1.2 Effects of air pollutants

Pollutants can affect the health of a person during inhalation and exhalation as the pollutants inflame, sensitize and even scar the airways and lungs. On reach-

120

ing deep inside the lungs, they may enter the bloodstream, thus affecting organs other than the lung, and they can take up permanent residence in the body. In addition, some pollutants affect health through contact with the skin and through ingestion of contaminated foods and drinks. Pollutants affect health in varying degrees of severity, ranging from minor irritation through serious illness to premature death in extreme cases. They may produce immediate (acute) symptoms as well as longer-term (chronic) effects. Health effects depend upon the type and amount of pollutants present, the duration of exposure, and the state of health, age and level of activity of the person exposed.

Pollution damage to plants and animals is caused by a combination of physical and chemical stresses that may affect the receptor's physiology. Pollutants can affect crops by causing leaf discoloration, reducing plant growth and yields, or by contaminating a crop, so making it unsafe to eat. Effects on terrestrial and aquatic ecosystems can occur locally or even regionally in the case of pollutants that contribute to **acid deposition** (acid rain), especially in areas where the soils and lakes lack substances to neutralize or buffer the acidic inputs (see Ch. 9). Pollution problems for buildings can be short term and reversible, such as soiling by smoke (which can be removed by cleaning), but the effects of acid deposition can be cumulative and irreversible by causing erosion and crumbling of the stone.

8.1.3 Effects of climate changes

Weather and climate changes can occur locally when a development changes the characteristics of the area in terms of its radiation balance, surface friction and roughness, and moisture balance. Adverse microclimate changes include:
- alterations to the airflow around large structures such as office blocks, multi-storey car parks and shopping arcades, causing wind turbulence that affects the comfort and sometimes the safety of pedestrians
- the addition of moisture from industrial cooling towers and large reservoirs, causing an increased frequency of fog or even icing on nearby roads
- the reduction in sunlight for greenhouse crops lying beneath a persistent industrial pollution plume, and
- the ponding of cold air behind physical barriers such as road and railway embankments, so increasing the incidence of frost which can damage agricultural and horticultural crops in those areas.

Macroclimatic changes can result from emissions of greenhouse gases (gases that are strong absorbers of outgoing terrestrial infrared radiation) such as carbon dioxide (CO_2), methane (CH_4) and nitrous oxide (N_2O). These gases contribute to global warming, which may cause changes in the position and intensity of weather systems and so lead to changes in regional wind, temperature and precipitation patterns. Some regional climate changes may bring benefits, but others are likely to bring adverse impacts. Global warming will also cause global sea levels to rise, because of thermal expansion of the sea water and because of melting of mountain glaciers and polar ice sheets (Elsom 1992).

8.2 Legislative background and interest groups

8.2.1 World Health Organisation air quality guidelines

After reviewing pollution–health studies from around the world, the World Health Organisation (WHO) specified air quality guidelines based on the lowest level at which a pollutant was shown to produce adverse health effects, or the level at which no observed health effect was demonstrated, plus a margin of protection to safeguard sensitive groups within the population. Some pollutants, notably carcinogenic pollutants (e.g. arsenic, benzene, chromium, polycyclic aromatic hydrocarbons and vinyl chloride) have no safe threshold limit. Sensitive groups include asthmatics, those with pre-existing heart and lung disease, the elderly, infants and pregnant women and their unborn babies. Such groups form one-fifth of the population in the UK.

People are at risk when air quality exceeds the guidelines advocated by the WHO to "provide a basis for protecting public health from adverse effects of **air pollution** and for eliminating, or reducing to a minimum, those contaminants of air that are known or likely to be hazardous to human health and wellbeing". The WHO set guidelines for 28 air pollutants in 1987 (WHO 1987). They are not mandatory, but instead provide a guide to governments in setting their own standards (Table 8.1). WHO guidelines are based on health considerations alone and do not consider the economic and technological implications of attainment. Given the costs and problems involved in attainment, this explains why some national **air quality standards** are not as strict as the WHO guidelines (Murley 1991).

8.2.2 European Union air quality standards

The EU has introduced air quality standards in the form of mandatory health-based "limit values" and more stringent non-mandatory "guide values" to protect the environment. EU limit values have been set for lead, nitrogen dioxide (NO_2), ozone (O_3), sulphur dioxide (SO_2), and suspended particulates. Guide values are intended to be long-term objectives which, when met, will protect vegetation as well as aesthetic aspects of the environment such as long-range visibility and soiling of buildings (see also Box 8.1).

Some countries have set national air quality standards in addition to the EU standards. The UK's Expert Panel on Air Quality Standards has recommended national standards for benzene (running annual average of 5 ppb) and ozone (8h standard of 50 ppb) and is likely to recommend such standards for 1,3-butadiene

Table 8.1 EU and WHO air quality standards and guidelines, and UK public information air quality bands.

(a) Sulphur dioxide (SO_2)

Organization and standard	Concentration (μgm^{-3})[a]	Concentration (ppb)
WHO		
Annual mean[b]	50	19
24hr mean	125	47
1hr mean	350	130
10min mean	500	188
EU *limit values*		
Annual mean	80 if black smoke >40	30 if black smoke >34
	120 if black smoke <40	45 if black smoke <34
Winter mean	130 if black smoke >60	49 if black smoke >23
	180 if black smoke <60	68 if black smoke <23
98th percentile[c]	250 if black smoke >150	94 if black smoke >128
	350 if black smoke <150	132 if black smoke <128
EU *guide values*		
Annual mean	40–60	15–23
24hr average	100–150	38–56
WHO – *vegetation*		
Annual mean	30	11
24hr mean	100	38

a. $1\mu gm^{-3} = 0.376$ppb
b. Annual average limit values expressed as median values, annual average guide values as arithmetic means
c. Daily mean measurements in one year

(b) Suspended particulate matter

Organization and standard	Concentration (μgm^{-3})
WHO	
Annual mean, black smoke	50
24hr average, black smoke	125
Annual mean, total suspended particulates	60–90
24hr mean, total suspended particulates	120
24hr mean, thoracic particles	70
EU *limit values*	
Annual mean, black smoke	80 (68)[a]
Winter mean, black smoke	130 (111)
98th percentile, black smoke	250 (213)
Annual mean, gravimetric	150
95th percentile, gravimetric	300
EU *guide values*	
Annual mean, black smoke	40–60 (34–51)
24hr mean, black smoke	100–150 (85–128)

a. Numbers in parentheses refer to British Standard Smoke (BS = 0.85 EU value).

(c) Carbon monoxide (CO)

Organization and standard	Concentration (mgm^{-3})[a]	Concentration (ppm)
WHO		
8 hr mean	10	9
1 hr mean	30	26
30 min mean	60	52
15 min mean	100	86

a. $1 mgm^{-3} = 0.859 ppm$

(d) Nitrogen dioxide (NO$_2$)

Organization and standard	Concentration (μgm^{-3})[a]	Concentration (ppb)
WHO		
24 hr mean	150	78
1 hr mean	400	209
Monthly 1 h maximum[b]	190–320	99–167
EU limit value		
98th percentile	200	105
EU guide value		
Annual mean (median)	50	26
98th percentile (1 yr)	135	71
WHO – vegetation		
Annual mean	30	16
4 hr mean	95	50

a. $1 \mu gm^{-3} = 0.523 ppb$.
b. Not to be exceeded more than once per month.

and carbon monoxide (CO) in the near future. In 1990, the UK introduced broad non-mandatory air quality bands for NO$_2$, O$_3$ and SO$_2$ to be used for informing the public when "poor" air quality is forecast (Table 8.1g). Predicting whether a proposed development is likely to increase the frequency of occasions when "poor" air quality occurs would be another way of assessing significance in an EIA. Some urban areas in Britain, such as Bristol and Sheffield, have adopted air quality bands stricter than the national bands.

8.2.3 Emission standards

Air quality standards refer to the levels of air pollution to which people are exposed. Another type of legislated standard is the **emission standard** which specifies the maximum amount or concentration of a pollutant allowed to be emitted from a given source. Emission standards are usually derived from consideration of the cost and effectiveness of the control technology available. The

(e) Ozone (O_3)

Organization and standard	Concentration ($\mu g m^{-3}$)[a]	Concentration (ppb)
WHO		
8hr mean	100–120	50–60
1hr mean	150–200	76–100
EU limit values		
8hr mean	110	55
1hr public information	180	90
1hr public warning	360	180
WHO – vegetation		
Growing season mean	60	30
24hr mean	65	33
1hr mean	200	100
EU – vegetation		
24hr mean	65	33
1hr mean	200	100

a. $1 \mu g m^{-3} = 0.5 ppb$.

(f) Lead

Organization and standard	Concentration ($\mu g m^{-3}$)
WHO annual mean	0.5–1.0
EU annual mean	2.0

(g) UK public information air quality bands (concentrations in ppb).

Pollutant	Very good	Good	Poor	Very poor
O_3	0–49	50–89	90–179	>180
SO_2	0–59	60–124	125–399	>400
NO_2	0–49	50–99	100–299	>300

UK Environmental Protection Act 1990 introduced the system of Integrated Pollution Control which requires that pollution sources adopt the best available techniques not entailing excessive costs (BATNEEC) in order to minimize pollution (DoE 1990b). The act also established two groups of industrial plants and processes for regulation purposes. Major developments such as power stations belong to the Schedule A group and they require authorization by Her Majesty's Inspectorate of Pollution (HMIP); the less polluting Schedule B group are controlled by local authorities. For specific types of pollution sources, the existence of an emission standard implies the type of operating process or pollution control equipment that should be employed (Process Guidance Notes have been issued by HMIP).

Box 8.1 The nature of EU air quality standards

An example of an air quality guideline is the WHO 1-hour NO_2 guideline of 209 ppb ($400 \mu g/m^3$). This represents a maximum value that should not be exceeded at any time. It has been recommended on the basis that a concentration of 293 ppb ($560 \mu g/m^3$) was judged as the lowest at which adverse effects were found in asthmatics and that the guideline allowed for a further margin of protection. Rather than adopt the WHO guideline, the EU employed a 1-hour standard of 105 ppb ($200 \mu g/m^3$), which refers to the 98th percentile and not to the maximum value as used by the WHO. What this means is the 98th percentile of 1-hour average concentrations over one year (air pollution concentrations usually follow a log-normal distribution) should not exceed 105 ppb for more than 2% of the time, that is 175 hours. Expressed in another way, the 176th highest of a series of 8,760 hourly average concentrations should not exceed 105 ppb. In other words, a location attains the mandatory standard, even if up to 175 hourly concentrations during the year exceed levels recognized to pose a health risk to some people.

A similar point applies to, say, EU SO_2 air quality standards for daily (24-hour) values expressed in terms of the 98th percentile of all daily mean values taken throughout the year. In this case, SO_2 levels can exceed the standard – with no upper level indicated – for up to seven days during the year without failing to meet the standard. As a general but not infallible guide, the 98th percentile value is 2–3 times greater than the annual mean value; the maximum value is typically around twice the 98th percentile value.

One important consideration when selecting pollution control devices, and in effect any mitigation measure, is the requirement that the best practicable environmental option (BPEO) is adopted. For example, use of a mitigation measure which removed gaseous pollutants from an industrial stack by converting them to a sludge would not be permitted if disposal of the sludge created an even worse environmental problem in the form of landfill pollution.

Emission limits for pollutants can apply nationally, as in the case of SO_2, nitrogen oxides (NO_x) and suspended particulates. In 1988 the UK committed itself under the EU Large Combustion Plants Directive to reduce emissions of SO_2, NO_x and suspended particulates from existing installations with a capacity greater than 50 MW (e.g. coal-fired power stations) by 40%, 30% and 40% respectively by 1998 (taking 1980 emissions as the baseline). Further reductions of 60% for SO_2 have been agreed for 2003, but it is likely that EU pressure may lead to this target being raised to 70% or even 80%.

8.2.4 Regulations for hazardous chemicals

In the case of a proposed development involving materials that could be harmful to people in the event of an accident, the EIA should include an indication of the preventive measures to be adopted, so that such an occurrence is not likely to have a significant effect (DoE 1989). Details are included in the Control of Industrial Major Accident Hazards (CIMAH) regulations of 1984, which were introduced by the EU as a consequence of chemical accidents at Flixborough (UK) in

1974 and Seveso (Italy) in 1976. CIMAH regulations set limits to the quantities and combinations of chemicals that can be stored at a site and they require onsite and offsite plans in the case of an emergency to be drawn up to an approved standard. County Council Emergency Planning Officers can provide the latest update of the CIMAH regulations.

8.2.5 Climate standards and regulations

There are few legislated standards with regard to climate. In pristine areas, such as national parks and wilderness areas in the United States, regulations exist to ensure that visibility is protected. Persistent and coherent pollution plumes from industrial stacks during daylight hours are considered intrusive and objectionable, and mitigation measures to minimize or eliminate the plume are required. Similar regulations have not been introduced in the EU or UK, but nevertheless "plume blight" may be considered in an EIA.

At the global scale there are regulations concerning pollutants that contribute to global warming and those that cause stratospheric ozone depletion. In 1992 the UK and other nations, co-ordinated by the United Nations Environment Programme (UNEP), agreed to stabilize emissions of CO_2 – the major greenhouse gas – at 1990 levels by the year 2000. Consequently, any proposed development that is a significant source of CO_2 needs to consider this national CO_2 emission limit. Pollutants that damage the ozone layer, such as chlorofluorocarbons (CFCs; they contribute to global warming too), methyl chloroform, carbon tetrachloride and hydrochlorofluorocarbons (HCFCs), are subject to international agreements to reduce and eventually ban these pollutants.

8.3 Scoping and baseline studies

Before the impact of a proposed development can be predicted, it is necessary to establish the current baseline conditions concerning air pollution and climate and to establish whether they are likely to change in the future, irrespective of the planned development. Knowledge of baseline pollution conditions is essential because, even when a development is likely to add only small amounts of pollution to the area, it could lead to air quality standards being exceeded if air quality in the area is already high or may become high in the future. This requires obtaining measurements of the ambient levels of the pollutants of concern at one or more locations in the study area, so as to assess the amount of pollution present.

127

8.3.1 Selecting air quality and climate indicators

Aspects of air and climate that need to be addressed in carrying out an EIA are summarized by the UK guidelines (DoE 1989) as (a) level and concentration of chemical emissions and their environmental effects, (b) particulate matter, (c) offensive odours, and (d) any other climatic effects. Depending upon the development project, there is a wide range of atmospheric pollutants with which an EIA may need to be concerned; these are summarized in Table 8.2. Pollutants commonly examined for their potential impacts include SO_2, suspended particulates (dust, smoke), CO and NO_2. Ozone and volatile organic compounds (e.g. benzene) are receiving increasing attention, as are the toxic organic micropollutants (TOMPS) including polycyclic aromatic hydrocarbons (PAHs), polychlorinated biphenyls (PCBs), dioxins and furans. Toxic chemicals (ammonia, fluoride, chlorine) and toxic metals (cadmium, lead) are of concern for specific industrial plant proposals, as is ionizing radiation (radionuclides) in the case of nuclear power plants. Offensive odours could be a problem around proposed sewage treatment works, chemical plants, paint works, food-processing factories and brick works. Odours often generate great annoyance when residents are subjected to them in their gardens and homes, and they may adversely affect health (e.g. ranging from discomfort, nausea and headaches through to severe respiratory illness).

Climate indicators include temperature, relative humidity, solar radiation, precipitation, wind speed and wind direction. All developments are likely to modify the microclimate to some extent, but in most cases the changes to local temperature, amount of sunlight and shade, and airflow, are minor and not considered in EIA unless there are special reasons for doing so. Significant effects on sensitive environmental receptors could arise as a result of local changes in the frequency of weather extremes such as fog, frost, ice, precipitation and wind gusts.

8.3.2 Pollution data availability

Using information from existing pollution monitors is the simplest and least expensive approach. Various national monitoring networks collect pollution data, and many local authorities, universities and other organizations undertake short-term or long-term monitoring of various pollutants. Information from national survey sites (Table 8.3) can be obtained from the National Environmental Technology Centre (NETCEN), which publishes annual pollution data summaries. Local authority Environmental Health Officers can provide information concerning their own pollution monitors. Many produce annual reports summarizing the pollution data collected and assessing its significance in relation to air quality standards.

Table 8.2 Key air pollutants and their anthropogenic sources.

Pollutant	Anthropogenic sources
Sulphur dioxide (SO_2)	Coal- and oil-fired power stations, industrial boilers, waste incinerators, domestic heating, diesel vehicles, metal smelters, paper manufacturing
Particulates (dust, smoke, PM_{10})	Coal- and oil-fired power stations, industrial boilers, waste incinerators, domestic heating, many industrial plants, diesel vehicles, construction, mining, quarrying, cement manufacturing
Nitrogen oxides (NOx, e.g. NO, NO_2)	Coal-, oil- and gas-fired power stations, industrial boilers, waste incinerators, motor vehicles
Carbon monoxide (CO)	Motor vehicles, fuel combustion
Volatile organic compounds (VOCs), e.g. benzene	Petrol-engine vehicle exhausts, leakage at petrol stations, paint manufacturing
Toxic organic micropollutants (TOMPS), e.g. PAHs, PCBs, dioxins	Waste incinerators, coke production, coal combustion
Toxic metals, e.g. lead, cadmium	Vehicle exhausts (leaded petrol), metal processing, waste incinerators, oil and coal combustion, battery manufacturing, cement and fertiliser production
Toxic chemicals, e.g. chlorine, ammonia, fluoride	Chemical plants, metal processing, fertilizer manufacturing
Greenhouse gases, e.g. carbon dioxide (CO_2), methane (CH_4)	CO_2: fuel combustion, especially power stations; CH_4: coal mining, gas leakage, landfill sites
Ozone (O_3)	Secondary pollutant formed from VOCs and NO_x
Ionizing radiation (radionuclides)	Nuclear reactors, nuclear waste storage
Odours	Sewage treatment works, landfill sites, chemical plants, oil refineries, food processing, paintworks, brickworks, plastics manufacturing

Although it is unlikely that an existing pollution monitor will be found at the precise location of the planned development, there may be one or two monitors not too distant in a similar type of land use (e.g. city centre, residential, commercial, industrial). Expert opinion obtained from university or environmental consultants can be used to assess whether these recorded pollution levels are likely to be similar to those in the area under study. Alternatively the data can be modified to reflect the location of interest by using established relationships. For example, data collected by a kerbside monitor can be converted to an "urban background" site, 40–50 m away from the road, by employing accepted rates of change in pollution levels with distance (QUARG 1993).

Table 8.3   Air quality information available from UK monitoring networks.

Pollutant	Network	Type of monitor	No. of sites
Particulates (smoke)	EU Directive[a]	Filter/smoke shade	160
	Local authorities	Filter/smoke shade	240
Particulates (PM_{10})	Enhanced Urban Network (EUN)	Tapered element Oscillatory microbalance	13
SO_2	EU Directive	bubbler/net acidity	160
	Local authorities	bubbler/net acidity	240
	EUN	UV fluorescence	13
CO	EUN	infrared absorption	13
	Local authorities	infrared absorption	6
NO_2	EU Directive	chemiluminescence	7
	EUN	chemiluminescence	13
	Local authorities	diffusion tubes	1000
VOCs, e.g. benzene	Hydrocarbon	gas chromatography	7
	London Boroughs	diffusion tubes	60
O_3	EUN	UV photometric	13
	Rural network	UV photometric	17
PCBs, PAHs, dioxins, furans	TOMPS	polyurethane filter	4
Acid aerosols (e.g. sulphates)	Acid deposition	gas chromatography	35
Lead	EU Directive	filter/spectroscopy	11
Various metals	Multi-elements survey	filter/spectroscopy	5

a. The EU requires member states to undertake monitoring, especially in zones where the air quality limit value set by an EU Directive may be exceeded currently or in the future.

8.3.3 Onsite pollution monitoring

If pollution data are not available or are insufficient then onsite monitoring will be required and should be planned and initiated during the scoping exercise of an EIA (Harrop 1994). A baseline monitoring programme needs to consider (a) what pollutants to monitor, (b) what type of monitor to employ, (c) the number and location of sampling sites, (d) the duration of the survey, and (e) the time resolution of sampling.

Selecting the equipment to measure air pollution concentrations depends upon (a) the intended use of the data, (b) the budget allocated to purchase or hire the equipment, and (c) the expertise of personnel available to set up and maintain the equipment and, in some cases, to undertake laboratory analyses of collected samples. The cost of one of the fixed automatic stations forming part of the UK Enhanced Urban Network is £100,000, together with £10,000 annual operating costs. However, similar sophisticated equipment producing hourly concentrations of various pollutants can be hired for short-term sampling at a fraction of that cost from many companies (a list of addresses and details of equipment and consultant expertise is given in the Members Handbook available from the National Society for Clean Air and Environmental Protection). It is important that the equipment selected for monitoring is accredited nationally, so that the data collected can be compared with national air quality standards.

Local authorities faced with the need to monitor pollution, in order to assess whether air quality standards are being attained, are turning to relatively simple and inexpensive equipment such as passive diffusion tubes. Passive diffusion tubes absorb the pollutant onto a metal gauze placed at the bottom of a short cylinder open at the other end to the atmosphere. After exposure the tubes are sent for laboratory analyses. They can provide useful information for pollutants such as ammonia, benzene, hydrogen sulphide, NO_2, O_3 and SO_2 (one for CO is being developed currently). In areas of high pollution concentrations they can produce results for daily or even three-hourly exposures, although in areas with low concentrations they are usually exposed for two weeks at a time. QUARG (1993) give details of how monthly exposure readings of NO_2 from these tubes can be used to provide estimates of the annual mean and 98th percentile concentrations. Pollution bio-indicators such as lichen, which are sensitive to SO_2, may be used to reveal long-term pollution changes if earlier surveys have been undertaken in the area. Soil and vegetation analyses can also provide long-term levels of pollutants such as metals.

When siting any type of monitor it is necessary to consider (a) the need to protect against vandalism, (b) access to the site, (c) the avoidance of pollution from indoor and localized sources which may make the data unrepresentative of the wider area, and (d) the availability of a power supply (if needed).

8.3.4 Meteorological monitoring

Meteorological data are readily available from hundreds of sites throughout the UK maintained by the Meteorological Office (MO), local authorities, universities, schools and individual weather enthusiasts. Some pollution monitoring sites also monitor meteorological conditions. The MO can supply hourly, daily, monthly, annual and long-term averages of temperature, relative humidity, air pressure, precipitation (including fog), wind speed and wind direction for any of its stations at a small cost. Although the site for which data are available may be some distance away from the study site, the MO and other meteorological consultancies can provide expert advice concerning how local factors such as altitude, topography and nearness to the coast may lead to differences between the two locations.

8.3.5 Projecting the baseline forwards

Having established current baseline pollution levels, it is then necessary to consider how these levels are expected to change in the future, irrespective of the possible effects of the proposed development. If emission sources and strengths, as well as climate conditions, in the area are not expected to change in the future, then future pollution baseline emissions can be assumed to be the same as current levels. However, changes in population and activity patterns, new industrial developments or closures, changes in fuels (e.g. decline of coal in favour of gas, more cars using unleaded petrol) and stricter emission standards (e.g. increasing number of vehicles fitted with catalytic converters) can affect emission rates. Weather conditions that favour a build-up of pollutants (e.g. periods of calm or light winds, higher temperatures promoting increased evaporative emissions) may alter too but, in practice, these are not usually considered. Changes to emission rates and patterns are much more important and their implications for changes to future pollution concentrations need to be assessed. Local, district and county authorities can usually supply details of likely population and land-use changes, as well as information on new developments under construction. These will then have to be translated into changes in emission rates, as will the effects of current national and EU pollution-control legislation. The process of predicting future baseline conditions is then similar to the method of predicting the pollution consequences of the proposed development, in that a numerical prediction model is required for which the emission rates are the inputs. This process is explained in §8.4.

Future climate baseline levels are not usually predicted for the purposes of an EIA, given the major limitations of current models in predicting regional changes, let alone local changes, attributed to, say, global warming as a result of the increase in atmospheric concentrations of greenhouse gases. Improved models may alter this situation in the future (DoE 1991d).

8.4 Impact prediction

8.4.1 Physical models and expert opinion

Several types of models are available to predict air pollution concentrations. Physical (scale) models using wind tunnels or computer graphics are employed occasionally, in situations involving complex hilly terrain or where numerical models suggest uncertainty concerning the possible effects of nearby buildings on dispersion of pollution emissions.

Predictive methods include the use of expert opinion, providing it is backed up with reasons and justification which support that opinion, such as comparison with similar existing developments or planned projects for which prediction has already been undertaken. The use of expert opinion can be justified readily on cost when several similar projects are being proposed in different locations.

8.4.2 The Gaussian model

Overview

The type of model used most frequently in predicting air pollution is the numerical dispersion model, and the most commonly employed version is the Gaussian model (Zannetti 1990, Hunt et al. 1991, Cernuschi & Giugliano 1992). The model, usually in the form of a computer program run on a personal computer with a large memory, calculates how specified emission rates are transformed by the atmospheric processes of dilution and dispersion (and sometimes chemical and photochemical processes) into ground-level pollution concentrations. The model is first applied to the current situation, so that pollution monitoring can assess the accuracy of the model in calculating baseline levels. If it is of acceptable accuracy, it can then be applied with confidence to future conditions with and without the planned development.

The most appropriate model outputs that should be incorporated in an EIA are predictions of short-term pollution impact (e.g. highest or "worst-case" hourly average concentration) and long-term impact (e.g. annual average concentration). Such outputs can then be compared with the appropriate air quality standards and guidelines in order to identify any locations that approach or exceed the standards. Annual average concentrations are usually shown spatially as an isoline map compiled from concentrations predicted by the model for a grid spacing of say, 1km (Fig. 8.1). Hourly maxima may be shown as a plot of concentration versus downwind distance for a range of specified meteorological conditions, including those conditions that give rise to the highest concentration (Fig. 8.2). The "worst-case" situation usually corresponds to low wind speeds.

The Gaussian model assumes that the pollutant emissions spread outwards from a source in an expanding plume aligned to the wind direction, in such a way that the distribution of pollution concentration decreases away from the plume axis in horizontal and vertical planes, according to a specific mathematical (Gaussian) equation. Although a plume may appear irregular at any one moment, its natural tendency to meander results in a smooth Gaussian curve after ten minutes of averaging time. The horizontal axis of the plume does not normally coincide with the height of the stack or point of emission, as the density and momentum of the emissions quickly carry the plume to a higher elevation, known as the "effective release height" (sometimes two to even ten times higher than the stack or point of emission). Generally, the higher the effective release height and the stronger the winds, the greater is the dispersion of the pollutants and the lower the resulting ground-level pollution concentrations. The maximum ground-level concentration experienced from a pollution plume is where the plume first touches the ground.

Gaussian models are used frequently in an EIA to predict pollution concentrations from a single source such as an industrial stack. Various improvements to the basic model take into account nearby buildings, type of terrain and dry deposition of the pollutants. Models are available not only for individual stacks but

133

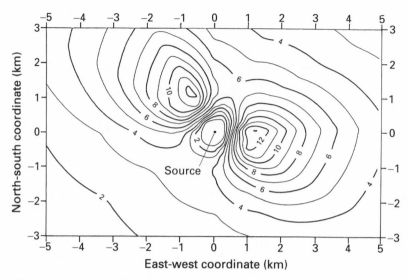

Figure 8.1 Predicted distribution of annual averaged ground-level concentrations of SO_2 ($\mu g/m^3$) as a result of emissions from a 50-metre high stack using the USEPA Industrial Source Complex model. As is often the case with UK climate data (in this example, data from Aughton, near Liverpool), the result is a distribution with two distinct peaks (to the northwest and east of the source).

for area sources (e.g. construction sites, urban areas, regions) and line sources (e.g. roads). Computer software and manuals for user-friendly numerical dispersion models can be obtained from the United States Environmental Protection Agency (USEPA), which has developed various models for regulatory purposes. The USEPA recommend the models of the UNAMAP series available as an ASCII disk containing test data for 31 air quality simulation models. These well established models can be obtained directly for the cost of downloading via a telephone line and the small cost of the manuals. Alternatively, variations of these models with user-friendly input and output routines can be purchased from specialist software companies. One USEPA model commonly used is the Industrial Source Complex (ISC) model, with its short-term (ISCST) and long-term (ISCLT) variations, producing results illustrated in Figures 8.1 and 8.2 (USEPA 1987).

UK models include the R91 model developed originally by the nuclear industry and the ALMANAC model developed by the former Central Electricity Generating Board (now National Power) for its power stations, but neither is available commercially. A new, but relatively expensive, UK commercial model called UK-ADMS (Atmospheric Dispersion Modelling System) was introduced in April 1993. It was developed by Cambridge Environmental Research Consultants on behalf of a consortium including National Power, the Meteorological Office, British Nuclear Fuels and HMIP, and so is likely to become the UK standard. It can even predict concentrations over averaging times of a few seconds, as is needed in the case of odours. Another commercially available model is the INDIC

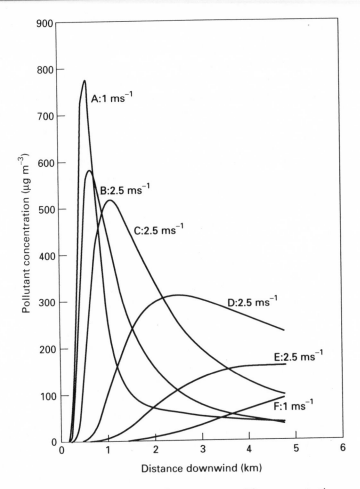

Figure 8.2 Predicted one-hour average SO$_2$ concentrations (μg/m^3) as a result of emissions from a 50-metre high stack using the USEPA Industrial Source Complex model for the "worst-case" wind speed in each Pasquill atmospheric stability class.

AIRVIRO model, developed in Sweden and used by Sheffield local authority and by the London Air Quality Network. The INDIC AIRVIRO model employed by the latter network, co-ordinating the work undertaken by the London Boroughs Association and the 33 London local authorities, incorporates the Warren Spring Laboratory long-period dispersion model (Munday et al. 1989).

Meteorological data inputs to the model
Gaussian models assume that the rate of dispersion of the plume, and consequently the pollution concentrations experienced at any location at the surface,

are a function of wind speed, wind direction and atmospheric stability (Barrow-cliffe 1993, Harrop 1986). Estimates of atmospheric stability for the simpler versions of the model can be obtained using a table or nomogram involving solar radiation, cloud cover and mean wind speed, and expressed in the form of seven Pasquill stability categories. Stability categories range from class A (very unstable), occurring during hot, sunny conditions with light winds, to class G (very stable), during cold, still nights with clear skies. For the purposes of the model, it is assumed that each stability class is characterized by a specified depth of boundary layer into which the pollutants are mixed. Summaries of Pasquill stability classes can be obtained from the Meteorological Office Air Pollution Consultancy Group for the nearest meteorological station. These tables indicate the annual percentage frequencies of each stability class by 30° wind direction sectors in six wind-speed bands, averaged over several years of data.

Some attempt may be made to modify the meteorological data employed, if the monitoring site is several tens of kilometres distant from the study area, or is located inland while the study site lies near the coast (being affected by sea breezes). Newer models, such as the UK-ADMS, using surface heat flux and boundary layer depth instead of Pasquill stability categories, include options to take into account whether the prediction applies to an inland or coastal site.

Emission data inputs to the model

The simplest application of a Gaussian numerical dispersion model would be to predict concentrations of a primary pollutant, say SO_2, produced by emissions from the single stack of an oil-fired power station in an area of flat terrain, with no other existing or planned pollution sources in the vicinity. Input for the model would be details of the emission rate of SO_2, the stack height, internal diameter of the stack, flue-gas temperature, flue-gas exit velocity (or volumetric flow rate) and meteorological information. Emission rates of SO_2 can be calculated from the amount of fuel used and the sulphur content of the fuel (typically 1.5%). If the power station used coal instead of oil, then some of the sulphur would be retained in the ash (typically 10%) and not released as SO_2, and so the estimate of emissions would have to be reduced by the proportion retained. Emissions of SO_2 (E) in tonnes per year can be calculated from:

$$E = 2 \times C \times S \times (1-A)$$

where

C = mass of fuel consumed (e.g. tonnes per year)
S = fractional sulphur content (e.g. 0.015)
A = fraction of sulphur retained in the ash (e.g. coal: 0.10; liquid fuels: 0)
2 = factor to convert sulphur to SO_2.

If the stack were equipped with a flue-gas desulphurization system (now required by legislation for new power stations if the output exceeds 50 MW), the efficiency of removal of SO_2 would need calculating (typically 90%).

Coal, oil and gas power stations also emit NO_x. There is no simple equation to estimate emissions of NO_x, as they depend upon the nitrogenous content of the fuel and the temperature of the combustion (high-temperature combustion oxidizes nitrogen in the air). In such situations, and for most other emission sources and pollutants, it is necessary to rely on emission factors. Emission factors are available from national pollution regulatory agencies such as NETCEN, USEPA and the Transport Research Laboratory (TRL) from emission measurements made on samples of each source such as vehicles, incinerators and boilers. Calculating emissions of NO_x (E) from, say, boilers of differing size and design requires the use of emission factors derived from detailed measurements of a sample of those boilers and then weighted using fuel consumption figures:

$$E = M \times F$$

where

M = measure of the level of activity (e.g. amount of fuel consumed)

F = emission factor (e.g quantity of pollutant emitted per unit of activity such as grams of NO_x per cubic metre of fuel consumed).

8.4.3 Compiling an emissions inventory

In an area with many existing sources of pollution, it may be necessary to compile a current **emissions inventory** and to indicate how the pattern of emission rates is expected to alter in the future in relation to projected changes in population and activity patterns, etc. (as explained in §8.3.5). The current inventory can be used as input into the numerical dispersion model to assess its accuracy in calculating current pollution concentrations, and the inventory calculated for the future (e.g. five years ahead when the proposed development is operating) enables future baseline pollution concentrations to be predicted. In practice, future baseline pollution prediction is undertaken only when emissions are likely to increase significantly in the area and when pollution levels are currently quite high in relation to air quality standards.

An emissions inventory should include emission rates from individual large-emitter point sources such as industrial plants and power stations, from area sources containing multiple smaller sources such as a housing estate, and from line sources, such as motor vehicles along a road and trains on a railway track (Timmis & Walker 1988). Occasionally, emissions inventories for some pollutants such as SO_2 are available from local authorities, which can save much time and effort.

Industrial plants keep detailed records of stack emissions of the major pollutants, because they normally have to demonstrate they are complying with legislated emission limits. Copies are received routinely by local authorities and/or HMIP as part of the process of authorizing plant operating licences. Residential (domestic) emissions are usually estimated as an area emission over, say, 1 km grid squares. Emissions are determined from records of heating systems within the area, weighted by emission factors for each type of equipment and details of household fuel consumption or fuel deliveries.

Calculating vehicle emissions at a particular location or along a stretch of road requires traffic census data on the number and different classes of vehicles. Emissions from motor vehicles are then calculated by summing up the number of vehicles of each class being driven in different speed bands and weighted by the percentage of vehicle class fuelled by a specific fuel (e.g. unleaded petrol, diesel) and by emission factors per vehicle type and speed. The emission factors are available from annual NETCEN and TRL laboratory measurements of exhaust emissions from samples of vehicles.

8.4.4 Predicting pollution concentrations near roads

Several models have been developed specifically to predict pollution concentrations in and around roads. The simplest is the graphical screening model which can be used to indicate those areas, if any, where air pollution is likely to cause concern (DOT 1983, 1993, Waterfield & Hickman 1982). Should this approach indicate that, say, CO concentrations in some locations may approach or exceed the WHO guideline, then further investigation is recommended using one of the more sophisticated models (Harrop & Carpenter 1992). Numerical dispersion models use known traffic flow (mean vehicle speed, flow rate) and meteorological conditions (e.g. wind speed and direction). Examples include the widely used PREDCO model of the TRL, the more advanced CALINE4 model of the Californian Department of Transportation, the CAR model developed in the Netherlands and the HIWAY model of the USEPA. The PREDCO model treats a road as a line source which is divided into a series of sections, each represented by a single point source, to which the Gaussian plume dispersion model is applied (Hickman & Colwill 1982; Hickman & Waterfield 1984). The model is confined to predicting CO, but the concentrations of other pollutants can be derived from empirical equations relating each pollutant to CO. The impact of a road at a specified location is then the sum of the impacts from all the points along the road under different traffic and wind conditions. The results of the prediction can be conveniently represented in terms of the expected number of days per year when the WHO 8-hour guideline of 9ppm ($10mg/m^3$) will be exceeded at a specified location; the WHO advice is that this level should not be exceeded more than once per year. Additionally, the results can be presented in map form, either showing the isoline enclosing the area where this criterion will be exceeded (Simms 1991) or showing isolines of various pollution concentrations.

8.4.5 Model limitations

All predictions have an element of uncertainty and it is important to acknowledge this and not treat the model as a "black box" by concentrating only on the results produced. Models are simplifications of reality, and their limitations, accuracy and confidence levels should be recognized and explained (Benarie 1987). Confidence in the accuracy of the model employed is gained by assessing its ability to predict the current baseline conditions in the study area, since the results can be verified using monitored pollution data.

8.4.6 Assessing significance

The significance of the likely pollution impacts of a proposed development is assessed by comparing the predicted changes in the area to air quality standards or guidelines, and determining whether these are likely to be exceeded at any locations, after taking into account the existing and predicted baseline pollution levels. If the planned development is predicted to increase pollution levels in excess or close to the air quality standard, then mitigation measures need to be proposed. If the changes are well below the standard, it is useful to express the increase in ground-level pollution concentrations in a meaningful way. For example, an EIA may conclude that a proposed coal-fired power station is expected to increase the annual average SO_2 concentration at the worst-affected location 10 km downwind by only 3% and that this increase is well within the year-to-year variability of annual average concentration produced by meteorological fluctuations. Even when a development is likely to add only small amounts of pollution to the area, it is important that an EIA makes specific assessment of what effect (perhaps negligible) this will have on any sensitive receptors and locations, such as residential areas, schools, nature reserves, SSSIs and historical buildings.

Not all assessments of significance are straightforward. An example is the case of "plume blight", since this involves a subjective assessment (see §8.2.5). The use of expert opinion involving comparisons with similar developments becomes important in such cases. Although a model may predict the percentage of time that a persistent plume exceeding 1 km in length would be generated by the proposed development, it would be useful to state how this compares with any existing plumes in the area and whether or not they have been the subject of public complaints.

Determining the level of significance of climate changes can be difficult in some cases. A local increase in temperature, wind turbulence, fog or frost may affect people, plants and wildlife directly or indirectly (e.g. fog causing road accidents), but the level of significance of the changes may require the use of expert opinion.

8.5 Mitigation

8.5.1 The need for mitigation measures

Mitigation measures should be advocated to avoid, reduce and, if possible remedy any significant adverse effects that a proposed development is predicted to produce. At one extreme, the prediction and evaluation of likely impacts may indicate such extreme adverse effects that abandonment or complete redesign of the proposed development is the only effective mitigating measure. More likely, modifications to the development can be suggested in order to avoid or reduce potential impacts (Wood 1989a, 1990). Some mitigation measures may be required by law for new – although not for existing – developments, such as flue-gas desulphurization systems for large coal-fired power stations, but the use of others depends upon the significance of the predicted impacts.

Various mitigation measures may be suggested to solve a potential problem. The likely effectiveness of each measure should be assessed in terms of the extent to which the problem will be reduced, as well as the costs of implementation. Whatever mitigation measures are proposed, it is important to ensure that they do not create problems of their own. Mitigation feeds back into design, so mitigation measures proposed to minimize adverse impacts of the project can be incorporated as alternatives in the project description. Subsequent proposed developments can make use of the information contained in a previous EIA in order to incorporate appropriate mitigation measures at the outset, rather than wait for its own EIA to identify potential problems.

8.5.2 Mitigating adverse pollution impacts

If the pollution impact from an industrial stack is predicted to approach or exceed air quality standards, this impact can be reduced by encouraging greater atmospheric dispersion and dilution of emissions by (a) raising the stack height, (b) reheating the flue gases to higher temperatures, and (c) emitting them at greater velocity. If a planned development is likely to exceed, say, maximum hourly pollution standards only during periods of poor atmospheric dispersion, then one possible mitigation measure would to keep a cleaner standby fuel for use during those forecasted occasions. Improved fuel combustion designs can reduce pollutant emissions, for instance by using low NO_x burners in furnaces. In many cases, the type and amount of pollutants emitted are a function of the fuel being burned, so alternative fuels can be proposed, such as fuel oil with a low sulphur content ($<1\%$) or natural gas. Traffic-generated pollutants decrease rapidly away from roads, and this process can be enhanced by roadway trenching, embankments, walls and trees, to reduce the pollution concentrations in nearby residential areas.

The construction stage of most projects has the potential to cause localized wind-blown dust problems, either when excavation is taking place or when materials are being transported and stored in stockpiles. Careful design of construction operations, including the selection of haulage routes into the site and the location of stock piles, can help to minimize dust problems in nearby residential areas. Mitigation measures can include: (a) frequent spraying of stockpiles and haulage roads with water, (b) regular sweeping of access roads, (c) covering of lorries carrying materials, (d) enclosing conveyor-belt delivery systems, and (e) early planting of peripheral tree screens where they are part of the planned development.

The need for mitigation measures may not always be clear. For example, should action be taken to ensure that odours from a food-processing plant are not experienced by residents of a few isolated houses on several days each year when the wind blows in their direction? In such a situation, consultation with the local planning authority will be needed to agree whether the impacts are sufficiently adverse to justify the cost of mitigation measures. Alternatively the local authority may suggest that the developer offers compensation to the affected residents, or offers to purchase the affected properties in order to create a buffer zone around the plant. If potential odour problems are to be tackled at source, solutions include taller stacks to encourage greater dispersion of the emissions, or removal of the pollutant completely by absorption, adsorption, oxidation or chemical conversion.

8.5.3 Mitigating adverse microclimate impacts

Adverse microclimate changes, such as increased wind turbulence around a proposed shopping precinct, can be minimized by the widening of narrow gaps between buildings, roofing of open spaces and changing the height and layout of buildings (Oke 1987). Unwelcome high air temperatures in open shopping precincts during summer can be reduced by the choice of building materials, consideration of building layout in relation to areas of sun and shade, and the planting of trees. Frost pockets affecting agricultural and horticultural crops can be prevented by landscaping and creating openings through road or railway embankments, which allow for the passage of cold air. The frequency of icing of roads can be reduced by landscaping and choice of road surface materials. The frequency of fog forming on cold clear nights along proposed motorways can be lessened by (a) eliminating any nearby areas of standing water, (b) reducing air pollution (suspended particulates) in the vicinity, (c) raising the road onto pillars above the fog-shrouded valley floor, and (d) planting tree belts which help reduce cold air drainage and scavenge fog droplets. Water vapour plumes from power station cooling towers, which have the potential to increase fog and icing of nearby roads, can be designed so that the banks of towers are orientated along the direction of the prevailing wind, such that the merging of individual plumes enhances buoyancy and reduces the number of occasions when plumes are brought to the ground.

8.6 Monitoring

Numerical prediction models contain uncertainties, so monitoring should be continued after the development has been completed in order to confirm the predictions and provide credibility to the process of EIA. This is particularly appropriate if similar projects are likely to be proposed in the future for other locations. Continued monitoring is also necessary to assess the effectiveness of any mitigation measures proposed in an EIA and to ensure that any potential air and climate problems identified have been minimized or eliminated.

8.7 Further reading

Additional sources of information concerning impacts on air and climate include Canter (1977), Samuelsen (1980), Wood (1989b), Lee & Lewis (1991), Watkins (1991), and Graber (1992).

CHAPTER 9
Soils and geology
Martin J. Hodson

9.1 Introduction

Much has been written on the effects of geology and soils on developments, but considerably less is available about the reverse. The Department of the Environment (1989) include soil, agricultural quality, geology and geomorphology as topics in their checklist that should be included in an EIA. Soil is here defined as the top layer of the land surface of the Earth, composed of small rock particles, humus (organic matter), water and air; geology concerns the rocks beneath the soil. Some types of development have effects on the underlying geology, and almost all have an effect on the soil. Soil is a major factor affecting plants, including agricultural crops, and plants provide the food and habitats for animals. Thus, avoiding major impacts of a development on the soil can go a long way towards preventing the degradation of a whole ecosystem.

When surveying a site for an EIA, it is common to include a section on the underlying geology, although there is generally less emphasis on geology than on soils. This chapter therefore summarizes methods for geological EIA, before dealing in greater depth with impacts on soils.

9.2 Geology

The DOE (1989) suggest that an EIA on geology should consider "Loss of, and damage to, geological, palaeontological and physiographic features". The first requirement is a map of the relevant area. Fortunately, nearly all of the British Isles have been surveyed and mapped by the Institute of Geological Sciences. These maps are of two types: (a) solid, showing only pre-Quaternary rocks (those formed over one million years ago), and (b) drift, which also show superficial and Quaternary deposits (which were formed within the past one million years, and principally since the last ice age). Drift maps show all of the information that might be needed, but are often not in the best format for EIA, as the rocks are often classified by age rather than by type. Nevertheless, a Geological Survey map can be quite informative about a site. The mineralogical and chemical composition of the rocks often have a direct effect on the soil type and chemistry, and hence an indirect effect on the plants.

One major development type with a considerable geological impact is quarrying and mining. In this case, the entire geological resource may be removed, and the geological impact is totally negative. In most cases, however, secondary impacts on the local environment are equally or more important. These include impacts on soils, hydrology (Ch. 10) and ecosystems (Chs 11–14). Quarrying for aggregate, its environmental impacts and its sustainability are discussed by Plowden (1992). The draft revision of Mineral Planning Guidance Note 6 (DoE 1993a) forecasts the need for 20 "superquarries" to meet the needs for aggregate from the South of England. These will have considerable impacts on both the geology and the local environment. After quarrying is completed, the sites are often used for landfills, and these have environmental impacts of their own, which are beyond the scope of this chapter.

Three other geological factors should be taken into account in EIA:

- Geological SSSIs are sites with unusual formations or particularly good fossils. These should be avoided where possible.
- Seismic risk is not usually a great problem in the UK, although there are occasional small earthquakes. In some parts of Europe (e.g. Italy) this may be a more serious problem. Even in the UK it is necessary to consider whether developments such as pipelines (liable to fracture) or nuclear power stations could be affected by an earthquake. With a long timescale, even a small risk may be a problem. Volcanic risk in the UK is negligible, but in some parts of the world a section of the EIA should be devoted to this topic.
- Subsidence caused by previous mine workings may be a problem, and it is also important to consider the potential impacts of any new mine workings on subsidence in the future.

9.3 Soils: definitions and concepts

An appreciation of a development's impacts on soils requires an understanding of basic soil features. The coverage of soil science here is, of necessity brief, and the reader is referred to Bridges (1978), Avery (1990), Brady (1990), Miller & Donahue (1990) and Rowell (1994) for further information.

9.3.1 Soil composition

There are two major types of soil: organic and mineral. Organic soils, or peats, largely consist of decayed plant material. They are the major soil types in some parts of the world, but they cover a relatively minor fraction of the land surface of the UK (only 3% of England and Wales, but rather more in Scotland and Ireland). Typically, mineral soils have four major components: mineral matter, usually derived from the parent rock (about 45% of the volume), organic matter (humus) derived from plant and animal remains (about 5%), water (about 25%), and air (about 25%). Thus, solid material accounts for only about half of the soil volume. The water and air fractions vary according to climatic conditions and land-phase hydrology, and are quite critical. If there is too little water, the soil dries out, causing a soil drought. If there is too much, the soil becomes water-logged.

The soil mineral matter consists of particles of different sizes. There are several classifications of these particles, all based on size. The following is a simplified version from the British Standards Institution:

	Particle size (mm)
gravel	> 2.0
sand	0.06–2.0
silt	0.002–0.06
clay	< 0.002

These categories are known as separates. The percentages of sand, silt and clay can be plotted on a textural triangle diagram (see Fig. 9.1) to give a textural class name.

The three major classes of soil are (a) sands, which have at least 70% sand separate and less than 15% clay separate; (b) clays, which usually have no less than 40% clay separate; and (c) loams, the most complex group, which are a mixture of sand, silt and clay separates. The textural properties of a soil are of great practical significance. Sandy soils are often termed "light", do not hold much water, and are thus prone to drought. They are, however, easy to till and cultivate. Clay soils are "heavy" and are likely to be difficult to cultivate. They frequently show poor infiltration (Ch. 10) and are susceptible to waterlogging. Most agricultural soils are loams. Knowledge of soil texture will greatly assist in determining soil properties (e.g. how freely it will drain, how liable it is to compaction and erosion, its potential as a source of mineral nutrients for plants).

9.3.2 Soil structure

In most soils, the soil particles mentioned above are organized into aggregates. These structures are called peds, which vary in size from less than a centimetre to several centimetres. Sandy soils frequently have little structure because sand particles have low cohesiveness. Soil structure is important in providing pores within the soil (between the peds) which allow aeration for roots. Many soils contain pans, which are dense layers that interfere with water and root penetration. A typical example is the iron pan which is frequently found in the podzol soil type (see §9.3.4). A very common type of pan found in agricultural soils is the plowpan, which is as a result of agricultural machinery causing soil compaction.

9.3.3 Soil fertility

Soil fertility is a vast topic; the reader is referred to Brady (1990) and Cresser et al. (1993) for more details. Essentially, the two major soil chemistry problems that are of importance in an EIA are low soil fertility, and toxicity, both of which will lead to poor plant growth. Low soil fertility is attributable either to low levels of nutrients (e.g. nitrogen, phosphorus, potassium) in the soil, or their being made unavailable for plant uptake in some way. Soil toxicity is caused by high levels of elements or compounds being present in the soil. Some elements that are essential for plant growth, and are beneficial at low concentrations, can be

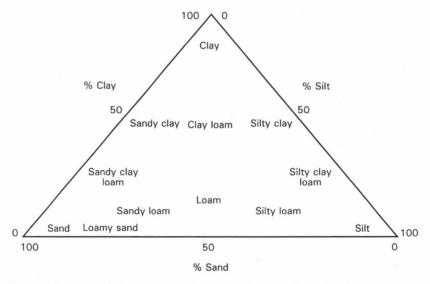

Figure 9.1 Textural class diagram. To use the diagram, determine the percentages of sand, silt and clay in the soils, and plot the data.

toxic at high concentrations (e.g. copper). It is often said that more can be learnt from soil pH than from any other single soil factor. Soil pH *per se* rarely affects plant growth, but it strongly influences the availability of plant nutrients and toxins. Aluminium and nearly all of the heavy metals are much more available for plant uptake in acid soils than in neutral or alkaline soils.

9.3.4 The soil profile and soil classification

In the context of an EIA, it is important to know what type of soil is present on and around the development site, as soil types differ considerably in properties such as liability to erosion and susceptibility to compaction. A pit dug into an undisturbed soil will usually show that the soil consists of distinct layers. Such a vertical section is called a soil profile, and each individual layer is called a horizon. Two different soil profiles are shown at Figures 9.2 and 9.3. Not all of the horizons are always present, and the horizons are frequently subdivided. The three layers L, F and H represent litter (leaves or needles), fermentation (where the breakdown of organic material contained in the litter largely occurs) and humus (where breakdown is largely complete). The A and E horizons are eluvial horizons, that is upper horizons which are depleted in nutrients because the nutrients have been washed down the profile in the process known as leaching. In contrast, the illuvial B horizons are often enriched with nutrients, iron or organic matter which has been leached down from above. Finally, the C horizon is the parent rock, which often considerably influences the chemistry of the soil above.

It is largely on the basis of the soil profile that soils are classified. This chapter concentrates on the soils likely to be found in Britain, using the classification system adopted by Avery (1990). Bridges (1978) is a useful starting point for considering soils outside Britain. Avery's terminology (or similar) is used in many British texts, and certainly seems to be the preferred terminology for British EISs. There are, however, many other systems, and the classifications of the US Soil Taxonomy and FAO–UNESCO are gaining ground, even in Britain. The American textbooks (Hausenbuiler 1985, Brady 1990, Miller & Donahue 1990), recommended here, all use the US classification, and in Table 9.1 this terminology is compared with the equivalent British terminology for major soils of the British Isles (Avery 1990).

Soils can be defined as zonal and intrazonal. The development of zonal soils is influenced by several factors (climate, rock type, vegetation, relief and time), but the undoubted overriding factor in their development is climate. Intrazonal soils can occur in all climatic zones, and are influenced by local factors such as geology and topography (not necessarily climate). The British Isles has only two zonal soils: podzols in the north and west, and brown soils in the south and east. Both these are typical soils of cool, wet temperate zones; they have many subdivisions, and are not found in all parts of the British Isles. The concept of a zonal soil is one usually applied to large land masses and it does not imply that the soil

Table 9.1 A comparison between the British soil classification (Avery 1990) and the US Soil Taxonomy.

Avery (1990)	US Soil Taxonomy	Notes
Podzols	Spodosols	Humid to per-humid temperate climates. Acidic soils characterized by grey coloured A and E horizons, and the deposition of humus and/or iron in the B horizon.
Brown soils	Mostly alfisols	Humid temperate climates. Leached and elluviated soils, but reasonably fertile. Argillic B horizon. Includes brown earths.
Lithomorphic soils	Mostly entisols	Thin (30 cm) soils with no diagnostic subsurface horizon. Includes rankers and rendzinas.
Gley soils	Aquic soils of a great variety of types	Soils characterized by saturation with water for at least part of the time. Reducing conditions are prevalent.
Peat soils	Histosols	Organic soils, bog and fen peats, forming in humid climates often in depressions.
Man-made soils	Plaggepts and arents	Ploughed and disturbed soils.

type cannot occur outside this area. For instance, there are many podzols found in the South East of England.

The *podzol* is typical of northern areas of Europe, where it is associated with the boreal coniferous forest and heaths, and the climate is characteristically cold and wet. These soils are highly leached and acidic (often the pH is in the region 3.0-4.5). They are little used for agriculture, but are very important for forestry. Because of their unsuitability for cultivation, they are often the substratum of remaining semi-natural vegetation such as heathlands.

There are many types of podzol, including humus-iron podzols (Fig. 9.2). The surface organic layers are of an extremely acid type known as mor. This is a fairly thick layer of pine needles and humus that develops as a superficial layer. It is broken down only slowly, and is incorporated into the lower horizons very slowly as a result of the absence of earthworms. Beneath the superficial organic layers there is then only a thin mixed mineral–organic horizon, the A horizon, followed by the characteristic grey eluvial E horizon, which is depleted in iron. In the illuvial B horizons, the colour of the soil changes – it is often orange-brown and rich in iron. Frequently there is a thin iron pan at the junction of the E and B horizons, and a layer rich in humus may also exist at this point. Podzols develop best on permeable sands and gravels, but can be found on a wide range of parent materials.

Brown soils are the zonal soils generally associated with those areas originally covered with deciduous forest. In Europe these are oak, beech and ash woodlands. Again there are many types of brown soil, and Figure 9.3 shows one example, an acid brown soil. Brown soils are the dominant zonal soils of northern France, Germany, central Italy and much of eastern Europe, as well as southeast Britain. Brown earths are the best known and most widespread category of this group, and are generally fairly fertile. They are mostly located in warmer and drier climates than are podzols, and the precipitation:evapotranspiration ratio (see Ch. 10) of the environments in which these soils develop is generally lower than that of the podzols. The amount of water percolating through the soil is sufficient to cause a moderate amount of leaching, but is not enough for podzol formation. Much of the original forest that grew on brown soils has been cleared for agriculture.

The brown soils are related to the podzols, as both are leached soils. In fact brown soils are often described as podzolic, implying that they have undergone leaching and could develop into a podzol, even though they look nothing like a true podzol. Another characteristic of brown earths is an argillic (clay-enriched) B horizon, where clay particles, which are washed down the profile, are deposited. The typical brown earth will be leached of carbonates, and it is generally neutral to moderately acid (pH 4.5–6.5). Annual leaf-fall from trees, shrubs, herbs and grasses makes for a much more varied litter than does that from conifers and heath. It is also more nutritive and more easily ingested by soil fauna, including earthworms. The presence of earthworms means that humus incor-

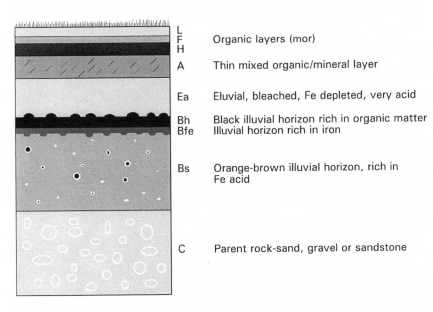

L F H	Organic layers (mor)
A	Thin mixed organic/mineral layer
Ea	Eluvial, bleached, Fe depleted, very acid
Bh Bfe	Black illuvial horizon rich in organic matter Illuvial horizon rich in iron
Bs	Orange-brown illuvial horizon, rich in Fe acid
C	Parent rock-sand, gravel or sandstone

Figure 9.2 Profile of a typical humus–iron podzol, showing superficial organic layers (mor), eluvial A and E horizons (which are bleached and often grey in colour), illuvial B horizons (rich in iron), and the parent material of the C horizon (redrawn from Bridges 1978).

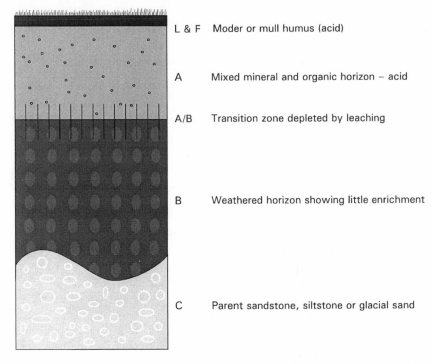

L & F   Moder or mull humus (acid)

A   Mixed mineral and organic horizon – acid

A/B   Transition zone depleted by leaching

B   Weathered horizon showing little enrichment

C   Parent sandstone, siltstone or glacial sand

Figure 9.3 Profile of a typical acid brown soil. Here the organic material is of the richer moder or mull type. The soil is leached, but not nearly to the same extent as the podzol. The A and B horizons are far less distinct (redrawn from Bridges 1978).

poration into the upper horizons is greater than in podzols, as a result of better mixing. The humus is of the moder or mull types (both of which are much richer in plant nutrients than is mor), which forms on freely drained base-rich soils with good aeration.

Several other zonal soil types may be found in Europe (Bridges 1978), namely: the tundra soils of Norway and Finland; the mountain soils of the Alps, Dolomites and Pyrenees; the chestnut and brown soils of the semi-arid grasslands in central Spain; and the red, brown and cinnamon soils of the Mediterranean woodlands in Spain, southern France, southern Italy and Greece.

Of the intrazonal soils, *lithomorphic soils* are thin soil types, where the parent rock is the dominant feature in soil development, representing an early stage in soil development. The best known lithomorphic soils are the calcimorphic soils, such as rendzinas, which develop on calcareous substrates (e.g. chalk or limestone). In a typical rendzina, the A horizon, which is generally fairly thin (e.g. 25 cm), rests directly on the parent C horizon. The soil is very dark brown or black in colour and is generally alkaline (pH 7.5–8.4). In contrast, rankers are young, acidic soils that develop over non-calcareous rocks. In southern Britain the climax vegetation on lithomorphic soils is deciduous forest (e.g. beech, oak), but the trees have been mostly cleared and these areas are now mostly used for grazing or cereal crops.

Gley soils are hydromorphic soils, in which water stands in the profile for at least part of the year. Poor drainage can occur in many different types of soil, but is most frequent in areas such as river flood plains that are subject to periodic flooding. Gleying occurs when water saturates a soil, filling all the pore spaces and driving out air. Any remaining air is soon used up by micro-organisms, causing the development of anaerobic conditions. In these reducing conditions iron compounds are reduced from the ferric (Fe^{3+}) to the ferrous (Fe^{2+}) state. In this ferrous form the iron is much more soluble and is removed from the soil, leaving colourless minerals behind. Thus, gley soils often have a characteristic grey coloration and are frequently mottled.

Peat soils include those that either have more than 20% (by weight) organic matter, or are continually saturated with water and contain 12–18% organic matter (Brady 1990). Accumulation of organic matter is encouraged by wet conditions resulting from heavy rainfall, surface flooding, and high levels of groundwater. Peat soils can be divided into two main types on the basis of hydrology: **ombrogenous** or **bog** peats, which are fed entirely by rainfall and are always acid; and **geogenous** or **fen** peats, which are fed by mineral groundwater and can be either acidic or alkaline, depending on the geology of the catchment area. Bog peats occur where rainfall is in excess of 1,000mm per year in upland and western areas of the British Isles. These soils commonly have a low pH (below 4.0), and are generally poor in nutrients. In contrast, the rich fen soils are generally nearly neutral or even alkaline in pH, and are strongly influenced by mineral-rich groundwater.

Almost all of the soils of the British Isles have been influenced by human activity to some extent. Avery (1990) restricts the term *man-made soils* to mineral soils where present or former management of the soil has resulted in distinctive features. The most common of these are cultosols (plaggepts in the US taxonomy), where an artificially produced A horizon is more than 40cm thick, and the soil has often been ploughed. Disturbed soils (arents in the US taxonomy) are produced in a range of conditions where soil has been dug out and restored, or where there has been gross disturbance by heavy machinery.

9.3.5 Land capability and agricultural classification

The classification of soils outlined above is based mainly on the soil profile and does not take into account the uses or potential uses of the soil. In an EIA context this may be more important than the soil type. Essentially, land-use classifications are based on the limitations that the land (including soils, climate and topography) imposes for agricultural and other usage. According to the classification of Bibby & Mackney (1969), soils can be divided into seven or eight classes. Class I has no limitations, any crops can be grown, and the soil is also suitable for a wide range of other uses. Classes I–IV are all used for cultivation, but Class IV is marginal. Classes V–VIII are generally unsuitable for arable crops, and become increasingly unsuitable for other uses as well. Subclasses are often used, which are based on the type of limitation affecting land use. For example, if wetness was the major limitation affecting an area of Class II land it would be classified IIw. In the UK the Ministry of Agriculture, Fisheries and Food (MAFF 1988) has adapted this classification for use on agricultural land classification maps (Tables 9.2, 9.3). According to this classification, approximately one-third of the agricultural land in England and Wales is in Grades 1, 2 or 3a, about a half is in grade 3b or 4, and the rest is in grade 5. Knowledge of the agricultural land classification of an area will be important in an EIS, particularly if the proposed development will involve the loss of more than 20 ha of high-grade (1, 2, 3a) agricultural land. If this is the case, there is a statutory obligation to consult MAFF (DoE 1989).

Table 9.2   MAFF agricultural land classifications (based on MAFF 1988).

Grade	Quality of agricultural land	Total UK agric. area (%)
1	Excellent quality, with only very minor or no physical limitations to agricultural use.	2.8
2	Very good quality, with minor limitations to agricultural use.	14.6
3a	Good quality land capable of growing a moderate to high yield from a limited range of crops.	48.9
3b	Moderate quality land capable of growing a restricted range of crops with lower yields.	
4	Poor quality land with severe limitations which restricts the range of crops and yields. Mainly used for grass.	19.8
5	Very poor quality land with severe physical limitations. Mostly used for permanent pasture or rough grazing.	13.9

Table 9.3   Other categories used on agricultural land classification maps.

Land category	Notes
Urban	Housing, industry, etc., with little potential for a return to agriculture
Non-agricultural	Uses where most of the land could be returned to agriculture (e.g. golf courses, parks, allotments)
Woodland	Includes both commercial and non-commercial woodland
Agricultural buildings	Includes glasshouses
Open water	Lakes, ponds and rivers
Land not surveyed	–

9.4 Legislative background and interest groups

Thompson (1990) discusses the moves towards legislative protection of soils. Although the European Soil Charter was adopted by the Committee of Ministers of the Council of Europe in 1972, only a few European countries have soil policies on their statute book. Soils are protected only when they form part of an otherwise valuable habitat. Several European countries are now either introducing legislation or considering ways of doing so.

Contaminated land has been much in the news in the UK recently, as section 143 of the Environmental Protection Act (1990) states that a public register of land that has been subjected to contaminative uses should be compiled by district councils. There was much concern that such a register would lead to widespread property blight. This led to a revised proposal, restricting the register to fewer contaminative uses, but this was also withdrawn in March 1993. The topic has been reviewed by Watson (1993). Although the contaminated land register is now unlikely to happen, there is no doubt that pressures to legislate in this area will continue, and that soil analysis for contamination will form an important part of many EIAs in the future.

With respect to agricultural impacts, Planning Policy Guidance note 7 (DOE 1992d: *The countryside and the rural economy*) seeks to protect the "best and most versatile" land from development, which is normally defined as Grades 1, 2 and sometimes 3a. Statutory consultees and source of information about impacts on soil include HMIP, the British Coal Corporation, NRA, and the Secretary of State for Energy.

9.5 Baseline studies

Soil baseline studies can be divided into three major headings: desk study, field work and laboratory work. A critical decision in the EIA scoping stage concerns whether a soil survey is required, or whether a desk study will suffice. A desk study will be relatively cheap, a field survey will be more expensive, and work involving the analysis of samples in the laboratory will be time-consuming and expensive. In some studies a brief description of the soil type may suffice, in others a soil map will be required, and in a few (e.g. where contamination is suspected) tables and maps documenting the levels of chemical components will be necessary. It is important to point out that some of the methods described below may have restricted application, and that all studies will not require all of the analyses.

9.5.1 Desk study

When carrying out an EIA for soils, the first question to be asked is what is known already? A literature survey should determine whether the soils are likely to be in their natural state. If the area under consideration is in the countryside, the soil may well have been influenced by agricultural activity (e.g. ploughing). In urban environments it will be important to determine if any parts of the site could be affected by industrial wastes. This kind of information will be important in deciding what, if any, analyses will be required. Much information will be gained from consulting old Ordnance Survey maps, books on local history, and local council officers.

Soil survey maps are less widespread in their coverage than geological survey maps, covering about 20% of England and Wales, but can be very useful if they are available for the site of interest. The maps are accompanied by explanatory booklets, giving details of the local geology, vegetation and land usage, but even if they are available, they may not give enough detail for EIA. The history of systematic soil surveying in Great Britain has been documented by Avery (1990).

9.5.2 Field work

The major problem with field work on soils is variability. The chemical and physical properties of a soil can change within 10cm. Sampling a whole site, and at all depths, can soon require thousands of samples. Ball (1986) provides a detailed account of the methodology involved in surveying a site, with particular reference to plant ecology.

For *soil sampling*, ideally large numbers of pits should be dug all over the site in question, and samples taken back from each horizon encountered in soil profiles for analysis in the laboratory. This is often not possible because of time constraints, and only where the soil forms a major part of an EIA will this occur in practice. It may also cause unwanted damage to soils, and could be very unpopular with a landowner. Permission should always be sought before digging a pit. Where possible, pits should be dug, but a nearby road cutting may suffice. The maximum depth to which a soil profile is dug is usually about 1m. The soil should be removed in order, and replaced as near as possible in the same order. Often, sampling is confined to taking samples of the surface soil (down to 20cm depth) for analysis using soil augers. This is, after all, the layer from which most plants will obtain their water and nutrients. An auger has a screw tip which is screwed into the ground to remove successively deeper layers of soil. Deciding on the layout of the sampling, and the numbers of samples required, is also very important, and both considerations will vary with the circumstances of the survey. Generally samples are taken systematically using a sampling grid (see §12.3.2), but random sampling, or sampling particular areas of interest, may also be carried out in some situations; the reader is referred to Webster (1977), who gives a very detailed mathematical account of this topic. Most soil profiles will be similar to those in Figures 9.2 and 9.3. Simple classification of the soil should be fairly easy in most cases. It should be remembered, however, that many soils have been disturbed, and that the original profile may have been lost. This applies particularly to agricultural soils that have been ploughed.

Field observations of colour and texture can be a clue to soil composition. A black or grey-brown soil is likely to have a high humus content. A yellow, orange or red-brown soil probably contains much iron. A white soil may contain abundant silica, aluminium hydroxide, gypsum or calcium carbonate. Soil colour is also used as an aid to classification: Munsell Soil Color Charts (Munsell Color Co., Baltimore, Maryland, USA) cover the whole soil colour range, and fresh, unrubbed soil can be compared with these standard charts. Some idea of texture can be gained in the field by observing it with a lens and by feeling it between the fingers. This requires much experience if an exact identification of the soil texture is needed, but even an inexperienced person should be able to classify the soil into the broad categories of clay, silt, sand or loam.

Field observations of soil strength (resistance to crushing), mineral content and pH are much faster and cheaper than physical and chemical analysis in the laboratory. Portable field apparatus can be used to obtain estimates of soil strength (very important where compaction may be a problem; see §9.6). Small hand-held penetrometers consist of a metal probe which is pushed into the soil until it reaches a certain mark. The probe is spring loaded, and the pressure required to push it into the soil is read off on a scale. These penetrometers give good results if the soil is bare or sparsely vegetated, but are difficult to use where there is lush vegetation. Portable pH meters and soil test kits produced for horticultural or agricultural purposes may be used to estimate pH and mineral status, although they require some practice before reliable results can be obtained. Field tests should never replace more precise laboratory work when it is necessary, but may well give indications of where problems are likely, and where more work will be required.

9.5.3 Laboratory work

If the desk study and the field work indicate that there is a need for laboratory work, careful thought must be given to the analyses required for the study in question. Not all of the analyses outlined below will be needed in every case. Ball (1986) and Rowell (1994) cover much of this material in detail.

Some measure of the *water content* of the soil will frequently be needed. There are several methods available, some of which require complex equipment. In practice, most workers use gravimetric analysis. This involves taking a sample of soil from the field and weighing it before and after heating in an oven at 105°C. The difference between the weights is then expressed as a percentage of the fresh weight. All samples must be taken on the same day (or at least in a dry spell over several days). The method is destructive and it is impossible to sample the same specimen of soil continually. Comparison with results from other sites may be very difficult, as the method depends on the prevailing weather conditions. For soil surveys on a single site at a single time, the gravimetric method yields good results, giving information on where the dampest parts of the site are, and where flooding is most likely. For other types of work where monitoring over a time period is required, more sophisticated machinery (e.g. neutron probes) can be used (Brady 1990).

The *organic component* of a soil is important for fertility and structure, and varies from soil to soil. Mineral soils will generally have 10–20% by weight organic matter in the surface layers, while organic soils such as peat can be 80% or 90% organic matter. The simplest method is to determine loss-on-ignition. The procedure involves heating the dried soil in a muffle furnace at high temperature (550°C) until all of the organic matter has burnt off and only the ash remains. The greatest problem is making sure that all the organic matter has been driven off without driving off volatile inorganic material (e.g. carbonates). This method gives good results for all soils other than those with a high calcium carbonate content: this should be removed by reacting it with dilute hydrochloric acid before ashing. A more precise estimate of organic carbon can be obtained by wet oxidation with potassium dichromate (Avery 1990).

Assessment of *soil texture* in the field has been outlined earlier. If greater accuracy is required, then longer and more expensive laboratory determinations can be carried out. Basically, the methods differentiate between the mineral fractions of soils on the basis of particle size. The usual method involves sedimentation of mineral particles in a water column. The disadvantages are that it takes a long time (several days), and at current (1994) prices each determination will cost in the region of £20 (+VAT).

Bulk density is the ratio of the dry weight of a soil sample to the total volume it occupies in the field. It can be used indirectly to assess differences in soil structure and porosity caused by natural processes or management. It is usually measured directly with the use of a volumetric corer. Essentially, a pipe is pushed into the ground to extract a core of soil on which measurements can be made. In EIA, this is a very useful measure if soil compaction is considered to be a problem during or after a development. Although a quick onsite measure of compaction can be taken using a penetrometer, more detailed additional laboratory analyses will often be needed.

In the context of an EIA, analysis of *soil chemistry* will be carried out only when soil fertility or toxicity problems are thought to be very important factors, as such analysis is very expensive. Allen et al. (1986) provide detailed methods of soil analytical chemistry techniques. The following measurements should be made in these circumstances:

(a) Soil pH: the measurement of soil pH is fairly easy with a pH meter, but care should be taken if reliable consistent results are to be obtained. These will be affected by the background solution used (usually distilled water or

$0.01\,M$ $CaCl_2$), stirring and settling time.

(b) Cation exchange capacity: this is the total number of cations adsorbed on soil colloids (clay and organic matter), and gives some indication of potential fertility.

(c) Soil nutrient status: among the elements that might be measured are potassium, calcium, magnesium, nitrogen and phosphorus.

(d) Heavy metals: these include lead, copper, zinc, cadmium and chromium.

(e) Organic pollutants: one of the most important groups are the polynuclear aromatic hydrocarbons which are found at high concentrations in the soils of many sites that have been contaminated by industry (Wilson & Jones 1993).

(b) and (c) should be attempted only if the nutritional status of the soil is anticipated to be a problem (e.g. when reclaiming some types of land after a development). (d) and (e) will be necessary only where contamination is suspected, or perhaps where a baseline study is required (e.g. when building a smelter, what is the metal concentration before operations begin?).

9.6 Impact prediction

After the baseline survey has been carried out, the likely impacts of the proposed development must be considered. A new motorway, a housing development, building a nuclear power station, or laying an underground pipeline will all have different degrees of impact on soils. In some developments, large amounts of topsoil will be transported away from the site, whereas in others this kind of disturbance will be minimal. The DoE (1989) suggest that the following effects of a development should be taken into account: physical (e.g. changes in topography, stability and soil erosion), chemical (emissions and deposits on the soil), and land-use/resource changes. Although there is much information available concerning the engineering properties of soils, and on the description of soils for engineering purposes (e.g. Hausenbuiler 1985, West 1991), EIAs are more concerned with the impacts of a development on the soil; these are likely to be erosion, disaggregation, compaction and pollution.

Almost all developments are likely to lead to some soil *erosion,* unless suitable mitigation procedures are adopted. The reader is referred to Hudson (1981) for more details on this topic. There are two major types of erosion, by water and by wind. The factors that most influence erosion by water are mean annual rainfall, and storm frequency and intensity (see Ch. 10). In areas with low rainfall there is little erosion by water. What rain does fall is mainly taken up by plants and there is little run-off (the exception being flash floods in desert environments). In areas of very high rainfall (more than $1000\,mm$ rain) the vegetation is usually dense forest, which protects and covers the soil. The greatest erosion will occur when the vegetation in middle rainfall areas is undisturbed or in areas of high rainfall when the forest is removed (Hudson 1981). Only dry soil is subject to blowing in the wind, and so rainfall must be fairly low for wind erosion to occur (less than 250–300 mm). Steady prevailing winds are generally found on large, fairly level land masses, and it is these that are most susceptible to wind erosion (e.g. East Anglia).

On a world basis, by far the greatest cause of erosion is agriculture. All other activities are much less important on this basis, but may be more important locally. In the UK erosion by water is most likely. In a new development the two factors most likely to cause erosion are the removal of vegetation, and steep slopes. When soil erosion by water occurs, damage will often not be restricted to the terrestrial environment. Soil removed will affect nearby water courses, causing an increase in turbidity and siltation. Moreover soil erosion will also lead to an increase in soil nutrient levels in water courses. As a result, it is not uncommon for the levels of certain nutrients (particularly nitrate) to exceed legal limits in streams and rivers (see §10.3).

Disaggregation is simply the mixing up of soils when soils are disturbed. It often occurs during development, especially where the development project involves the removal of soil from one location to another. There are two major effects: physical disaggregation, which is the physical disruption of soil structure; and chemical disaggregation, which may involve the release of toxic chemicals, or more likely the redistribution of elements within the soil profile. This is very likely in soils with distinct horizons (e.g. podzols). In these cases, toxic or infertile subsurface material may be brought to the surface, causing problems for the re-establishment of vegetation after the development is completed.

Compaction is an almost inevitable result of any development, especially during the construction phase. Vehicles driving over soil will compact it, as will the storage of soil heaps, bricks or other materials. The two problems compaction of soil causes are difficulty of root penetration and reduced infiltration of water. The latter will lead to increased runoff and the associated risk of erosion (see Ch. 10). Compaction may also cause waterlogging, which results from water being unable to move easily through the compacted soil (see Fig. 11.3). Most plants cannot tolerate waterlogged conditions, as these become anaerobic. A detailed account of the effects of wheel traffic on soils and the plants growing in them is provided by Voorhees (1992).

Most developments pose the threat of some *pollution* of the local soils during the construction phase (e.g. oil from vehicles, the building materials used). During operation, some developments will have the potential to cause further pollution (e.g. oil wells, refineries, chemical works). Roadside soils are often polluted with lead from exhausts and salt from winter de-icing operations. Far more serious impacts on soils are those caused by gaseous emissions, and these impacts may occur hundreds of miles away from the source of the pollution (see Ch. 8). In particular, it is now widely recognized that acidic precipitation ("acid rain") has major effects on soil, increasing available soil aluminium levels. Aluminium toxicity is almost certainly a major contributor to the dieback that has been observed in the forests of northern Europe and North America in the past 20 years (Godbold et al. 1988). Thus, in a strategic sense, any development that leads to the production of sulphur dioxide (e.g. power stations) or nitrogen oxides (the principal pollutants arising from vehicle exhausts) will have an effect on soils, but the effects will often be combined with those from other developments, and felt a considerable distance away.

All of the above impacts will have serious effects on soils, but the soil types outlined in §9.3.4 will be affected to different extents by each type of impact. These are well discussed in the North Western Ethylene Pipeline Environmental Statement (Shell Chemicals 1989), and only a brief outline of the major points is given here. Podzols are affected most by disaggregation, because they are the most "organized" soils, with distinctive layers. As they are already acidic, and have a low buffering capacity (the greater the buffering capacity, the more acid will be needed to change the pH of a soil), they are the soils that are most vulnerable to acidic precipitation. Gleys are also vulnerable to disaggregation, as they are also "organized". Brown earths are not particularly susceptible to any of the above impacts. Peats are extremely sensitive soils, especially to erosion (instability) and compaction. Susceptibility to erosion is also a feature of many sandy soils.

9.7 Mitigation

Mitigation methods can be considered in relation to the main impacts. It is not possible to cover here all of the mitigation measures necessary to prevent *erosion* problems during and after developments, but the following general guidelines are of use:

- Remove as little vegetation as possible during the development, and revegetate bare areas as soon as possible after the completion of the development.
- Where possible, create gentle gradients and avoid steep slopes.
- Install suitable drainage systems to direct water away from slopes.
- Avoid creating large open expanses of bare soil. These are most susceptible to wind erosion. If such large areas are created, then windbreaks may be a useful mitigation procedure.
- If the development is near to a water body, siltation traps may need to be installed to trap sediment and prevent damage to the freshwater ecosystem.

The main mitigation method against *disaggregation* is to take the soil out in order of horizons and keep each horizon in a separate pile. If the piles are to be stored for any length of time, they may need to be grassed over to prevent erosion. Some fertilizing and seeding of the soil, once it is put back in place, may also be necessary.

There are several possible mitigation measures against *compaction* both during and after the development:

159

- Use wide tyres to spread the weight of vehicles.
- Use a single or few tracks to bring vehicles to the working area.
- Till the area after compaction has taken place.

It is important to avoid runoff of *pollutants* carried in a liquid form and, if this is perceived to be a major problem, then procedures for the containment of the pollutants on site must be considered (see §13.7). Airborne pollutants can have serious effects on soil chemistry, and methods for the reduction of emissions known to affect soils should be recommended (see §8.5).

It is most unlikely that a development can take place without having some impacts on soil, and some of these will remain even after suitable mitigation measures have been carried out. Thus, for example, it is relatively easy to mitigate against the effects of chemical disaggregation by keeping the soil horizons in separate piles and placing them back in the correct order, but it is much more difficult to protect the soil's native physical structure. Some compaction of the soil is almost an inevitable consequence of a development. Tilling the soil afterwards will mitigate against this impact, but will also change physical structure. Local soil pollution can often be mitigated against, but, as indicated above, many developments have impacts on soils through acidic precipitation at a considerable distance from the site. Even if mitigation measures at those sources are adopted (e.g. flue-gas desulphurization at power stations burning fossil fuels), they are unlikely to be entirely successful. Monitoring of impacts to test the quality of site restoration after a development is recommended, but is rarely carried out in practice.

CHAPTER 10
Water

Peter Morris & Jeremy Biggs

10.1 Introduction

Water is an essential resource that sustains all life on Earth. It is also vital for many human activities, including agriculture, domestic and industrial use, transport, and recreation. Population growth, intensive agriculture, industry, urban development, and water-prodigal lifestyles all place increasing demands on water resources, and result in a wide variety of **hydrological** impacts – many of which affect people directly. For example:

- excessive demands can seriously deplete water supplies
- alteration of drainage systems can increase the risk of flooding in settlements
- deterioration of water quality causes health hazards, affects water-based leisure activities, and imposes financial costs for domestic users, industry, agriculture and fisheries
- water engineering projects can have socio-economic implications, and impacts on traffic, land use and landscape, amenity, heritage and archaeology.

In addition, hydrological systems interact with other components of **ecosystems**, and a hydrological assessment is bound to overlap with assessments of weather and climate (Ch. 8), soils and geology (Ch. 9), and the biota of terrestrial, coastal, and especially **wetland**, ecosystems (Chs 11–14).

Hydrology is the study of water on, above and below the Earth's surface, including: atmospheric water vapour, precipitation (rain, snow and dew), evaporation, and land-phase water, i.e. soil water, **groundwater** and surface waters (lakes, rivers, etc.). In principle, it includes the study of marine waters, and of water chemistry and hence water quality. In practice, marine science has developed as a separate major discipline (some aspects of which are referred to in Ch. 14), and hydrology is primarily concerned with *freshwater systems* – as in this chapter. Similarly, although the physical and chemical components of hydrological systems are inextricably linked, each is a major field of study involving different methodologies. For these reasons, some of the main topics in this chapter include separate sections on *physical hydrology* and *water quality*.

Hydrology is a complex science, and an adequate hydrological assessment will require the services of experts in physical hydrology, water quality, water engineering, and perhaps geomorphology. Clearly, this chapter cannot presume to give instruction in these specialized disciplines. Rather, it aims to provide an overview of the hydrological component of EIA, including relevant aspects of hydrological systems and associated study methods.

10.2 Definitions and concepts 1: physical hydrology

10.2.1 Introduction

Aspects of hydrological systems relevant to EIA are discussed briefly in the following sections. Further information can be found in introductory texts such as Leopold (1974) and Jones (1983), and more advanced texts such as Gregory & Walling (1973), Dunne & Leopold (1978), Shaw (1988), Bras (1990), Ward & Robinson (1990), and Wilson (1990).

10.2.2 Water cycles and budgets

The only input of water to land masses is *precipitation* (*Pn*) of atmospheric water vapour, much of which has evaporated from the oceans. The land loses water by **evapotranspiration** (*ET*) to the atmosphere, and **runoff** (mainly in rivers) to the oceans. The global circulation of water between ocean, atmosphere and land is a closed system with no significant gains or losses. Individual land masses and areas within these are open systems with inputs and outputs that control their **water budgets**. Most EIAs can be considered largely in the context of a *catchment*, which is a drainage basin bounded by a *watershed*, i.e. a ridge from which precipitation water is shed either side to different catchments (Fig. 10.1). It should be noted that the terms *catchment* and *watershed* are defined here according to UK usage; in the USA the term watershed is used in place of catchment. A good overview of the nature and measurement of catchment processes is provided by Gregory & Walling (1973).

Any catchment has a water budget in which the only input is *Pn*, and *ET* and runoff (in streams and rivers) are the main outputs. However, water is also lost by groundwater seepage, at least when the underlying bedrock is permeable. Thus, the catchment water budget is

$$Pn - (ET + Q_s + Q_g) = \Delta S \qquad (10.1)$$

where: Pn = precipitation
         ET = evapotranspiration

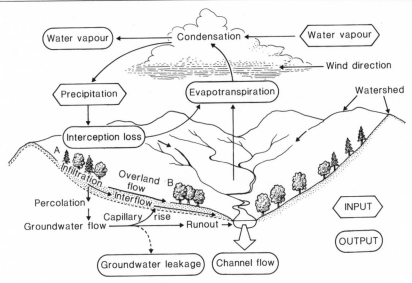

Figure 10.1 Hydrological fluxes in a catchment.

Q_s = surface discharge/runoff
Q_g = groundwater discharge
ΔS = change in storage

The main storage components are surface waters, groundwater and soil water.

Whereas *Pn* is the only input to a whole catchment, within-catchment sites may receive land-phase water from higher in the drainage system, e.g. site B in Figure 10.1 may receive water from higher in the catchment (including site A). In this case the term *site catchment* can be used to refer to that part of the system from which a site receives water, and the site's water budget equation must be modified to include two inputs (recharges) in addition to *Pn*.

$$(Pn + R_s + R_g) - (ET + Q_s + Q_g) = \Delta S \qquad (10.2)$$

where: R_s = surface recharge
R_g = groundwater recharge

How much inflow occurs depends on the site catchment, e.g. sites on or near a ridge are likely to have little surface recharge and no groundwater recharge, floodplains usually have periodic surface and groundwater recharge, and **geogenous mires** are dependent on groundwater. Ecosystems with appreciable land-phase inputs are particularly vulnerable to impacts from developments.

10.2.3 The meteorological water balance

Pn and *ET* bring about the interchange of water between the atmospheric and land-phase components of a hydrological system such as a catchment. Runoff increases when *Pn* exceeds *ET*, and decreases when *ET* exceeds *Pn*; so in simple terms a water regime can be characterized by the *Pn:ET ratio*, or the *meteorological water balance* (*Pn – ET*). When *Pn > ET*, there is a water surplus which is discharged as runoff; when *Pn < ET*, there is a water deficit which leads to a reduction in storage water. In addition to having greater runoff, areas with high *Pn:ET* ratios (such as northern and western parts of the UK) are more susceptible to soil leaching and erosion (Ch. 9) than those with lower ratios.

In Britain's seasonal climate, all areas normally have an appreciable winter surplus, and a summer deficit which is usually slight in northwestern areas and increases to the south and east. This meteorological deficit leads to the development of **soil moisture deficits** (SMDs) which inhibit plant growth and explain the frequent need to irrigate many crops in the drier areas.

Meteorological water balance also varies from year to year, and this affects storage components, including river and groundwater levels. For example, over much of the UK there was a severe drought in 1975–6, followed by an extended period of generally below-average rainfall (culminating in the drought years of 1989–1992) which resulted in a lowering of groundwater and river levels, especially in southern and eastern England. Such trends hinder the assessment of impacts such as water abstraction, especially since, once depleted, groundwater levels can take a long time to recover. Even short-term variations can be important, e.g. a few weeks without rain can result in severe soil drought, or a period of heavy rainstorms can cause flooding.

It follows that some potential impacts of a development will vary according to meteorological water balance patterns in the area, including seasonal and longer-term variations.

10.2.4 Interception, infiltration and overland flow

Much of the precipitation falling on a catchment is intercepted by vegetation canopies, and re-evaporates without ever reaching the ground. The resulting *interception loss* is often incorporated in *ET* (as in Eq. 10.1, Eq. 10.2 and Fig. 10.1).

Most rainfall reaching the ground normally infiltrates into the soil. However, if *Pn* exceeds the soil's ability to absorb water (i.e. its *infiltration capacity*), the excess collects in depressions or runs down inclines as *overland flow*, which is a major cause of soil erosion. Infiltration capacity tends to be:

- high on ground with good vegetation cover, especially on deep sandy soils and loams (see Ch. 9)
- low where vegetation is sparse, on clay soils (especially when these are already wet), and on shallow, frozen or compacted soil.

It is most likely to be exceeded under intense or sustained rainfall, or if heavy winter snowfall is followed by a rapid spring thaw. Infiltration also decreases with increasing slope.

Both interception loss and infiltration reduce the proportion of *Pn* that contributes to runoff. The importance of this in EIA is discussed in §10.2.7.

10.2.5 Water in the ground

Infiltration water may be stored in the soil, but more than 90 per cent normally returns to the atmosphere by *ET* or percolates downwards (Fig. 10.1). If percolating water reaches an impermeable layer, it accumulates or moves laterally down inclines as *interflow*. This is often rapid, and can contribute (together with overland flow) to *quickflow*, which refers to the *peak flows* in streams and rivers that are generated by rainstorms and can cause flooding. However, it is usually a minor and intermittent flux, compared with vertical percolation to the *water table* (surface of the groundwater).

Except shortly after rain, the ground above a water table is normally unsaturated, whereas that below is permanently saturated, so the subsurface system can be divided into an *unsaturated zone* and a *saturated zone* (Fig. 10.2). Immediately above the water table there is a thin *capillary fringe* that is semi-saturated by capillary rise, i.e. upward movement of water in narrow channels. The two main zones are roughly equivalent to soil water and groundwater respectively. However, (a) the unsaturated zone may include subsoil material such as bedrock and it is sometimes subdivided into a *soil zone* (or *rooted zone*) and an *intermediate zone*, and (b) in many wetlands the substratum is permanently waterlogged, so distinction between soil water and groundwater becomes meaningless. Floodplains characteristically have fluctuating water tables that may be near or above ground level in winter or after appreciable rainfall, but often retreat well below the surface in dry seasons.

Within the saturated zone, groundwater is held in **aquifers**, which may be

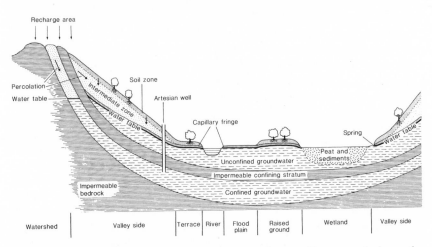

Figure 10.2 Groundwater relationships in a catchment. The ground above the water table is normally unsaturated, whereas permeable (porous) material below the water table is permanently saturated. The wetland is shown as a geogenous mire, but could also be taken to represent a lake (which would have originally existed at the site, and would have been gradually filled by the peat and inorganic sediments).

165

confined or unconfined. An *unconfined aquifer* has a free water table, and a well sunk into it will fill to the water table level. Water can move from a free water table to maintain the capillary fringe, and can be drawn farther upwards by **evaporative pull**. However, it cannot travel to the surface from deep water tables, and in these situations (e.g. on high ground in Fig. 10.2) the groundwater is not available to vegetation, which therefore relies on the meteorological water balance and soil water storage. A *confined aquifer* lies beneath an impermeable stratum, although it must fill from one or more unconfined areas. Confined groundwater is *artesian*, i.e. is under pressure, and water will flow from a well head if this is below the "effective" water table level, i.e. that of unconfined groundwater in the area. Similarly, *artesian springs* can occur where there are fractures in a confining stratum.

Groundwater flows down inclines. It may emerge as a spring or spring line, or seep directly into lakes or channels. It is largely responsible for *baseflow* (or *slowflow*), which is fair-weather flow (as opposed to quickflow) in streams and rivers. The potential water storage and flow rates in groundwater bodies depend largely on the nature of the aquifers, especially their *porosity*, *permeability* (or *hydraulic conductivity*), thickness, and *transmissivity* (permeability × thickness). Flow rates are normally slow, e.g. < 10m per day in sandstones and slower in less permeable rock types.

Groundwater is important in EIA for several reasons:

- It is a major water resource, both directly and because it maintains the supply in rivers from which water is drawn. It currently provides about a third of public water demands in the UK; in many areas it is the only practicable future source; and overabstraction can have serious consequences.
- It can become contaminated by pollutants, which may be very difficult to remove. The susceptibility of a groundwater body to pollution depends largely on the nature of the aquifer and overlying material, and the depth of the unsaturated zone. These factors are employed in a system developed by the National Rivers Authority (NRA) for classifying the vulnerability of groundwater supplies (NRA 1992a).
- It is essential for the maintenance of many wetland ecosystems.

10.2.6 Surface waters

Standing waters (lakes, reservoirs, gravel pits, ponds) constitute a major water storage/supply component, and are important wildlife habitats. They affect other hydrological components such as local water table levels and streamflows, and can influence the local climate, e.g. inducing fog. They are susceptible to various impacts including pollution.

Channel flow (in streams and rivers) is normally the main outflow component from a catchment, but the water in a channel system can also be regarded as a storage component. Indeed, during quickflow periods the normal *bankflow* (from adjacent land into the channel) may be reversed, or the water may spill over the channel banks and cause surface flooding. Flood risk is a major reason for considering channel flow in EIAs. However, flooding of a natural river floodplain is an intrinsic mechanism for accommodating flood waters, and flooding is only a problem when (a) developments or intensive agricultural lands sited in the floodplain are inundated, or (b) flood defences erected to protect these prevent the natural overspill onto the floodplain, so producing an abnormal surge that causes flooding farther down stream.

10.2.7 The importance of vegetation

Even regional water budgets are markedly affected by vegetation which increases the proportion of water returned to the atmosphere and hence reduces runoff. This is achieved by interception, **transpiration** and associated water uptake, and promotion of infiltration (§10.2.4).

Interception loss depends partly on the *interception capacity* of a vegetation canopy, which varies with vegetation type, e.g. conifer forests and tall grass intercept 25–35 per cent of annual *Pn*, broadleaved deciduous forests intercept 15–25 per cent, and short swards or sparse vegetation intercept much less. *Transpiration* can return over 50 per cent of rainfall to the atmosphere, and where there is appreciable plant cover, transpiration and interception loss can account for more than 75 per cent of rainfall, leaving less than 25 per cent to become runoff. Like interception loss, transpiration varies with vegetation type, e.g. it is higher from forest than from grassland. Vegetation enhances *infiltration*, mainly by maintaining a porous, uncompacted soil surface.

Although manmade structures intercept some *Pn*, both interception and transpiration are markedly reduced when vegetation is replaced by a built environment. Similarly, whereas some mature crops have high interception capacities and transpiration rates, cultivated land is usually bare or sparsely vegetated for much of the year. Infiltration is drastically reduced by factors such as soil compaction and replacement of vegetated ground by impervious surfaces. A consequence of these relationships is that increased runoff is frequently a major impact of changes in land use affecting semi-natural vegetation, and in particular of major developments.

10.3 Definitions and concepts 2: water quality

10.3.1 Introduction

Developments have considerable potential to pollute hydrological systems. Consequently, water quality is the subject of much legislation (see §10.4) and can be an important issue in EIAs.

Water in the environment is never pure, and natural waters vary considerably in the range and concentrations of dissolved substances present. They also differ in terms of variables such as suspended particulate material, **pH**, and temperature. Human influences on water quality include (a) changes in the concentrations of naturally occurring chemicals such as nitrates, phosphates, and metals, and (b) the input of totally new synthetic substances such as organophosphate pesticides. The aspects of water quality that are usually most relevant in EIAs are briefly discussed below. Further information can be found in Hutchinson (1975), Moss (1988), Meybeck et al. (1989), DoE (1992e) and Laws 1993.

10.3.2 Oxygen levels and organic pollution

The level of dissolved oxygen in water can have important implications for wildlife (see Ch. 13) and often for commercial fisheries. Oxygen levels in natural habitats vary considerably both within and between waterbodies. Fast-flowing streams and rivers normally have high levels, because turbulent flow enhances oxygen absorption from the atmosphere. Levels are lower in slower-moving water, especially at night, but should never be very low in most British rivers. Still waters such as ponds and stagnant ditches have highly variable oxygen levels that may range from supersaturated during daylight hours to zero at night. Large waterbodies such as lakes and reservoirs frequently stratify during the summer into an upper layer (the *epilimnion*), which is well oxygenated, and a lower layer (the *hypolimnion*), which is isolated from the atmosphere and tends to suffer oxygen depletion.

Oxygen depletion can occur through pollution, mainly by organic matter from sources such as sewage, soils, and agricultural or industrial effluents. High levels of organics may be discharged from sewage treatment works, cattle yards, silage clamps, most food-processing industries, and the wood and paper industry. Dissolved oxygen is consumed by the respiration of microbes that degrade the organic matter. Reduced oxygen levels can in turn lead to increased levels of potentially harmful chemicals (e.g. ammonia, methane, hydrogen sulphide, and heavy metals) by increasing their solubilities or production.

10.3.3 Thermal pollution

Freshwater systems have temperature regimes to which the aquatic life is adapted. Temperatures above the normal range can directly affect freshwater communities (see Box 13.1), and can lead to oxygen starvation because increasing temperature (a) promotes oxygen consumption by increasing rates of animal and microbial respiration, and (b) reduces the amount of dissolved oxygen held by water. The main source of thermal pollution is power stations.

10.3.4 pH and acidification

The pH of natural waters varies considerably between sites, and can change dramatically both seasonally and through the day. Many freshwater systems have naturally low pHs and should not be regarded as having poor water quality, even if, for example, they do not support a commercially desirable activity such as a salmon fishery. However, acidification of freshwater ecosystems by **acid deposition** is now a major problem in some areas, and many waterbodies that are currently highly acidic have become so only within the past century. The problem is promoted (a) globally by emission of gaseous oxides of sulphur and nitrogen from power stations, and (b) locally by afforestation. Low pHs affect many freshwater animals directly, but a major effect is that they increase the solubility of toxic pollutants such as aluminium (see Ch. 9). In consequence, freshwater systems can be seriously affected, as can commercial fisheries (see Box 13.1).

10.3.5 Nutrients and eutrophication

Excessive levels of nitrates and phosphates in freshwater systems can cause problems for human and environmental health. The main *health concern* is methaemoglobinaemia (blue baby syndrome), a condition associated with nitrate. In many areas, nitrate levels in waterbodies used for drinking water (particularly rivers and aquifers) are now sufficiently high to cause concern, and have led to protective measures such as the creation of aquifer protection zones (see §10.4.2).

The main cause of *environmental* damage is **eutrophication**. Aquatic systems become eutrophicated through leaching of chemical fertilisers from soils, animal excreta, and effluent from sewage treatment works (which contains nitrogen and phosphorus from human excreta and phosphorus from detergents). Eutrophication frequently has considerable nature conservation costs (see Ch. 13). It may also bring economic and social problems by causing fish kills, increasing drinking water treatment costs and, since it usually favours the growth of algae at the expense of higher plants, decreasing the amenity value of freshwater systems. It is generally perceived as a threat to standing waters, but it may go unnoticed in many rivers because of the overriding polluting effects of organic wastes. Certainly there is now increasing evidence of eutrophication effects on large slow-flowing and highly regulated rivers, such as the Volga, in central Europe.

10.3.6 Silt

Silt consists of fine organic and/or inorganic particles. These may remain in suspension in waters for some time as *suspended solids* or *suspended sediment* (the quantity of which is often called the *suspended sediment load*) or may sink to the

169

bottom of the waterbody to become **sediment**. Silt is a natural component of all freshwater systems, but when present in unnaturally high concentrations it can contribute to their pollution. Silts are derived from various sources including agricultural land (especially where soils are eroded) and urban surfaces. Those from eroded soils or sewage may have a high organic content, causing deoxygenation of rivers and streams. Where a catchment is used by industry or intensive agriculture, silt may often also contain high levels of phosphates, metals, pathogens and **biocides**. Deposition of polluted sediment is particularly significant in lakes and ponds, where it can accumulate and intensify in its long-term affects. In addition, high deposition rates may progressively seal waterbodies, isolating them from groundwater flows and changing the characteristics of the bottom substrate, with potentially damaging effects on wildlife (see Ch. 13).

10.3.7 Metals, micro-organics and other chemicals

Water pollution by these chemicals is largely attributable to deliberate export of industrial effluents, or runoff from roads and agricultural or urban areas.

The metals of greatest concern in freshwater systems include aluminium, cadmium, chromium, copper, lead, mercury and zinc. Metal toxicity frequently varies between different types of animal and plant; for example, zinc is relatively non-toxic to humans but very toxic to most fish, so levels of zinc acceptable in drinking water would be much higher than those acceptable for a fishery. Metals are most toxic when in solution, and metal solubility is influenced by the prevailing water conditions. Most metals are more toxic at low pH, and less toxic in hard water (which has high levels of calcium and normally a high pH). For these reasons, different water quality standards are often set for metals in hard and soft waters. Organic compounds often remove dissolved metals from water by binding with them. However, they may also release metals which would otherwise have remained insoluble, and this is the reason for concern over some water softeners such as EDTA (ethylene diamine tetra-acetic acid) which are added to many commercial detergents. Metals may also act synergistically (the effect of one is exacerbated by the other).

In addition to the many naturally occurring toxins such as heavy metals and oils, about 100,000 synthetic chemicals are currently in use (Meybeck et al. 1989). About 3,000 chemicals account for 90 per cent of bulk production (IRPTC 1983). An important group of compounds is the *micro-organics*, which includes most biocides. Unfortunately, adequate toxicity data exist for only a tiny proportion of these compounds; long-term ecotoxicology and environmental fate is known for only 20–30 chemicals. Further, recent research has shown that many chemicals have detrimental effects on organisms at levels far below those that cause immediate death, and often far below legal limits. *Oils* are commonly washed into freshwater systems from roads and industrial and development sites; and motorized pleasure boats also cause oil pollution. The toxicity of oils is complex. In addition to blanketing objects and organisms, they can cover the water surface, reducing oxygen absorption from the air. Oils also contain many harmful chemicals (see Box 13.1).

10.3.8 Pathogens

There are four broad categories of human pathogens in temperate fresh waters: *viruses, bacteria, protozoans* (microscopic animals) and *helminths* (flatworms), although the latter are not normally a problem in Britain. Viral pathogens tend to have a limited host range, so sources are usually limited to waters containing human wastes such as sewage. The potential sources of bacterial and protozoal pathogens are greater because they tend to have less specific requirements. Few pathogens are specifically monitored in the UK, but human health limits are set for bacteria indicative of faecal contamination (see §10.7.3).

10.4 Legislative background and interest groups

10.4.1 Consultees and interest groups

The statutory consultees for the water component of an EIA are normally *Her Majesty's Inspectorate of Pollution* (HMIP) and the relevant water authority which in England and Wales is the NRA. The rôles and importance of the NRA and HMIP are outlined in Box 10.1. In Scotland, the River Purification Boards (RPBs) are responsible for water resources, flood warning and water pollution control, and contact should also be made with the Scottish Office Environment Department. In Northern Ireland, responsibilities for the various aspects are split between the DOE's Environmental Service and the Department of Agriculture for Northern Ireland.

A hydrological assessment may involve *other interested parties*. Where there are potential impacts on wildlife and landscape, the relevant *nature conservation authority* (NCA) and the *Countryside Commission* will have an interest and must be informed. The *water undertakers*, including the water service companies in England and Wales (which supply water and treat sewage), and the statutory water companies (which supply water), have a clear interest in potential impacts on water supply and quality. *Landowners* with land adjoining a watercourse can have a direct, and sometimes legally enforceable, interest in any development that may affect that watercourse. Under civil law, such landowners have "riparian rights"; they normally own the river bed and have a right to receive water in its natural state, subject only to reasonable use by an upstream owner for ordinary purposes. *Owners of other property rights,* such as fisheries, have the same right, and have frequently been known to exercise this in the courts. For exam-

171

Box 10.1 The rôles and importance of the NRA and HMIP

The NRA has responsibilities covering management of water resources, pollution control in freshwaters, groundwaters and coastal waters, inland water navigation, flood defences, land drainage, fisheries, water recreation activities, and nature conservation in inland waters and associated lands.

The NRA will be contacted if the proposed development involves any one of the following: retention, treatment or disposal of slurry, trade or sewage effluent; carrying out works on the bed or banks of a stream or river; mining operations; refining or storage of oils; or waste deposition.

The NRA is highly influential in EIAs with a water component, and may significantly affect the outcome of development proposals in three main ways:

1. A proposed development process involving the discharge of polluting effluent (e.g. trade or sewage effluent) into **controlled waters** will require a consent licence from the NRA, which has the power to grant, refuse, or attach conditions to licences. It is an offence knowingly to allow pollutants, other than those already under HMIP control, to enter controlled waters without NRA consent. Whether the NRA grants a licence can depend on criteria such as: (a) the water quality standards set for the receiving waters, including potable water supplies, fisheries or irrigation water; and (b) the NRA's other duties under the Water Resources Act 1991 e.g. furthering nature conservation.

2. The NRA controls licences for (a) land drainage schemes, and (b) water abstraction, determining whether water may be withdrawn from groundwater or surface waterbodies for industrial, agricultural or domestic use.

3. As a consultee to planning applications for some development types, the NRA's views influence the LA's decision to authorize a development. It frequently seeks to use the planning system to implement its policies and impose controls that cannot be readily achieved under its own powers, e.g. flood prevention, land drainage and nature conservation.

The NRA can have a rôle throughout the development process – from the construction phase (e.g. in relation to a proposal to pump silt-laden waters from building foundations) to the post-operation phase (e.g. monitoring streams adjacent to a waste tip to ensure minimal pollution by leachate). In addition to its rôle as statutory consultee, the NRA may (a) provide hydrological information (see §10.5.2), and (b) give advice on aspects such as the scope required for the hydrological assessment, suitable personnel to undertake assessment work, and potential mitigation measures.

The NRA has the power to set and maintain Statutory Water Quality Objectives (SWQOs) for controlled waters (enforceable under the Water Resources Act). These are currently being established, and are likely to become one of the most influential water quality standards in Britain. They will be set for individual stretches of water, the quality objectives for which will have to comply with any standards set by EU Directives, including those for fisheries and potable water supplies. Standards may also be set for other water uses, including (a) the general quality of the ecosystem based on an ecological index derived from RIVPACS predictions (see §10.7.3), and (b) measures of amenity, gauged by aesthetic acceptability. Initially most SWQOs will be set for running waters, but they can apply to lakes if these form part of a river system. In addition the NRA suggests that quality objectives should in future be set for all standing waters over 0.4 ha.

The NRA has published proposals for policies to protect groundwater throughout England and Wales (NRA 1992a). They include: measures to protect aquifers from pollution and interference of water flows; and source protection zones around abstraction sources (with detailed proposals indicating the degree of acceptability of a wide range of activities in each zone). At present these policies are non-statutory, but the NRA has expressed its intention to seek designation of water protection zones. In response to its remit for the supervision of flood defences, the NRA has set target standards of service for rivers and coastal systems.

HMIP will be notified if a proposed development involves any of the list of manufacturing, mining or waste disposal operations, or release of any of the list of pollutants, which specifically come under its jurisdiction (see Table 10.1). It has responsibility in England and Wales for operating a system of Integrated Pollution Control, which seeks to regulate pollution from hazardous activities. Integrated Pollution Control Regulations list approximately 105 prescribed processes (e.g. petrochemical processes and timber processes) covering some 5,000 different installations. In addition there are controls over the release of prescribed substances, including 23 chemicals which it is an offence to release to water. Where the responsibilities of HMIP and the NRA overlap, control principally rests with HMIP.

ple, it is estimated that the Anglers' Cooperative Association have been involved in over 1,000 cases involving water pollution since the Second World War (Ball & Bell 1991). *Other non-governmental organizations,* such as Greenpeace and Friends of the Earth, campaign on water issues and often carry out water quality monitoring exercises that supplement those of the statutory enforcement bodies.

10.4.2 Legislation, policies and standards

The main EU and UK legislation relevant to the hydrological component of EIAs is outlined in Table 10.1. Most EU Directives relate to water pollution, and the standards set usually aim to protect health or the economic uses of water (including fisheries). Other relevant legislation on nature conservation and environmental protection is outlined in Table 11.1. In some cases, legislation affecting coastal developments may be relevant also (see Table 14.2).

EU Directives are implemented in the UK, either directly under the Acts and associated Statutory Instruments (SIs), or through the policy and powers of the NRA (see Box 10.1). Where a project requires planning permission, an EIA may be required under the general regulations (SI 1988/1199). A project not subject to planning regulations (e.g. work carried out under the terms of a General Development Order) may still require an EIA under SI 1988/1217. In relation to potential hydrological impacts, developments can be divided into those such as river engineering projects, which involve direct manipulation or utilization of water systems, and those with less direct influences (see Table 10.4).

All major projects are likely to have hydrological impacts, and the DOE guidelines (DoE 1989) indicate that screening for the water component (involving collection of relevant information and assessment of potential impacts) is mandatory in any EIA (see Appendix A). Further guidance on the regulations can be found in DoE (1988) and CIRIA (1992), or can be sought from the NRA.

Most river engineering projects are covered by Annex 2 of EEC/85/337, for which EIA is discretionary rather than mandatory. However, such projects are normally supervised by the NRA, which may decide to conduct its own EIA. Alternatively, one or more of the other bodies with statutory environmental responsibilities (e.g. the NCA) may request that an EIA be undertaken. In any case, a preliminary (screening) hydrological assessment should be carried out. This may demonstrate that no significant hydrological impacts are likely. Even then, however, a written justification must be produced (e.g. by the NRA) giving the reasons why an EIS is not needed.

Table 10.1 Current UK and EU legislation relevant to hydrological assessments.

Legislation	Aims and aspects
Control of Pollution Act 1974	Licensing of waste disposal operations (provisions will be replaced by those of the Environmental Protection Act 1990).
Surface Waters Directive 75/440/EEC	Control of the quality of surface waters intended for abstraction of drinking water.
Bathing Waters Directive 76/160/EEC	Control of the quality of inland bathing waters. Implemented by SI 1991/1597 (England and Wales) and SI 1991/1609 (Scotland).
Dangerous Substances in Water Directive 76/464/EEC	Control of very toxic and persistent pollutants. Requires member countries to establish a consent system or set emission standards for a prescribed list. Other Directives for specific pollutants include 73/404/EEC, 78/176/EEC, 82/176/EEC, 82/883/EEC, 83/513/EEC, 84/156/EEC, 86/280/EEC, 89/428/EEC, 90/415/EEC.
Fisheries Directive 78/639/EEC	Control of the quality of **designated waters** for supporting fish. Sets limits for substances such as suspended solids.
Groundwater Directive 80/68/EEC	Protection of groundwater against pollution caused by certain dangerous substances.
Drinking Water Directive 80/778/EEC	Control of the quality of water intended for human consumption. Sets limits for total coliforms and substances such as nitrate.
Environmental Assessment Directive 85/337/EEC	Directive on the assessment of certain public and private projects on the environment.
Agricultural Sewage Sludge Directive 86/278/EEC	Sets limits on heavy metal levels in sewage sludge applied on agricultural land.
SI 1988/1199, SI 1988/1221, and SR 1989/20	The Town and Country Planning (Assessment of Environmental Effects) Regulations (for England & Wales, Scotland, and Northern Ireland respectively). Implement 85/337/EEC.
SI 1988/1217	The Land drainage improvement works (Assessment of environmental effects) Regulations.
Environmental Protection Act 1990	Regulations for a prescribed list of particularly polluting industrial activities and substances that are subject to authorisation by HMIP (see Box 10.1). Waste regulation, collection, and disposal authorities established.
Urban Wastewater Directive 91/271/EEC	Requirements for collection, treatment and disposal of urban, and some industrial, wastewaters.
Nitrates Directive 91/676/EEC	Reduction of nitrate pollution from agricultural sources.
Water Resources Act 1991	Protection of water resources and aquatic habitats. Definition of duties, and consolidation of powers, of the NRA.
Land Drainage Act 1991	Consolidation of land drainage legislation. Powers of LAS and Drainage Boards
Water Industry Act 1991	Duties of water companies; standards set for quality and adequacy of water supplies and wastewater treatment
SI 1992/337 and SI 1992/575 (Scotland)	Surface Waters (Dangerous Substances) Regulations. Implement 86/280/EEC, and subsequent amendments, e.g. 90/415/EEC.
Water Consolidation Act 1991	Amendments to the 1991 Water Acts

10.5 Scoping and study options

10.5.1 Scoping

The first stage of a hydrological assessment is to establish the scope of the investigations needed. This is usually achieved by a combination of a desk study, brief site visits and discussions between the interested parties. The scoping process should aim to provide a preliminary assessment of the:

- impact area, and potential impacts
- parameters that should be included in the study, and the level of accuracy and precision needed
- availability and usefulness of existing information
- most appropriate field survey methods for collecting new information
- time and resources available for field surveys
- possible mitigation measures
- need and potential for monitoring.

A key issue at the scoping stage is the selection of relevant parameters and appropriate study methods. As indicated in Tables 10.2 and 10.3, there is a wide range of potentially relevant parameters. Clearly, it is usually inappropriate to consider many of these parameters at more than a superficial level, and vital to concentrate resources on a carefully judged selection relevant to the expected impacts of a particular development. Moreover, as indicated in Table 10.2, collection of field survey data may be limited in the context of EIA; although it can become more feasible, and is often essential, if monitoring is to be undertaken.

As indicated in §10.1, a hydrological assessment is almost certain to overlap with other EIA components, and there should be liaison between consultants to prevent duplication of effort, and to ensure that all have the information they require. For example, to assess impacts on freshwater ecosystems, ecological consultants may need information on the existing state of ecosystems, including the degree to which they are already degraded or polluted, and their sensitivity to further change (see Ch. 13). Similarly, mitigation proposals may include creating or enhancing waterbodies for water sports, fisheries or conservation purposes. Hydrological consultants need to be aware of such requirements at an early stage, so that they can make any necessary adjustments to the assessment plan.

10.5.2 Study options

It is usually possible to compile some of the information needed for a hydrological assessment from existing data sources. A thorough search is usually less expensive and time-consuming than obtaining new data, and it is pointless to undertake new work that merely duplicates existing information. In addition, only existing data can provide information on past conditions and trends.

Table 10.2 Checklist of physical parameters for the hydrological component of an EIA. The ticks are a rough guide to the likely ease of obtaining information in an desk study or field survey, absence indicating very low feasibility or non-applicability, and three ticks indicating high feasibility. However, feasibility will vary considerably in relation to: the project; the study area, e.g. catchment or on-site only; and the time, expertise and resources available.

Categories	Subcategories	Feasibility Desk	Field
Topography	Relief; site catchment size and character; drainage areas	✓✓✓	✓✓
	Floodplain extent, nature and current use	✓✓✓	✓✓
Geology & soils	Types, characteristics and depths	✓✓✓	✓✓
Rainfall	Seasonal, annual, storm frequency, drought frequency	✓✓✓	✓
	Short term to check correlation between on-site and weather-station local patterns		✓✓
	Interception	✓	✓
Snow	Seasonal, annual, extremes	✓✓	
Dew and fog	Seasonal	✓✓	✓
Evaporation	Actual evapotranspiration – seasonal and annual	✓	✓
	Potential evapotranspiration – seasonal and annual	✓✓✓	✓
Infiltration	Average and during storms	✓	✓
Soil water	Soil moistures levels – short term	✓	✓✓
	Soil moistures levels – seasonal, annual, long-term	✓✓	✓
Groundwater	Aquifers – location, porosity, permeability, yields	✓✓	✓
	Recharge/discharge areas and rates; flow rates	✓✓	✓
	Water table levels/storage: short-term	✓	✓✓
	Water table levels/storage: seasonal, annual, long-term	✓	
	Patterns of flow	✓✓	✓✓
Standing waters (ponds, lakes, gravel pits, reservoirs, etc.)	Location and elevation	✓✓✓	✓✓
	Dimensions and capacity	✓✓	✓
	Recharge/discharge – mean, seasonal, annual, low, peak	✓	
	Reservoir purposes and operating schedules	✓✓✓	
Runoff	Interflow and overland flow – frequency and magnitude	✓	✓
	Channel flow – seasonal, annual, long-term, storm peaks	✓✓	✓
	General (using rainfall–runoff models)	✓✓✓	
Flood risk	Flood frequency and magnitude	✓✓	✓
	Flood control facilities	✓✓✓	✓✓
Erosion and sedimentation	Soil erodabilities and location of erosion problems	✓✓	✓
	Channel erosion and location of problems	✓✓	✓✓
	Stream loads and location of sedimentation problems	✓✓	✓✓
Drainage	Effect on wetlands and location of problems	✓✓	✓✓
Water use	Total, agricultural, urban, industrial (seasonal, annual)	✓✓	
	Abstraction sources (rivers, reservoirs, groundwater)	✓✓	✓
	Water importation/diversion	✓✓	✓
Wastewater systems	Capacity and condition of sewers and stormwater systems	✓✓	✓
	Current flows (daily, seasonal); infiltration/inflow	✓✓	
Water budgets	Components of budget equation	✓	✓

Table 10.3 Common parameters of water quality surveyed in EIAs.
C: conservation; H: human health; F: fisheries.

Parameter	System	C	H	F	Notes
Nutrients					
Phosphorus	Rivers	⊕	–	–	Several different forms. Much of load transported in sediment.
	Lakes/ponds	•	–	•	Varies between hypolimnion and epilimnion (§10.2.3). Detection often difficult.
Nitrate	Rivers	⊕	•	⊕	Naturally higher in summer, but usually higher in late autumn/winter.
	Lakes/ponds	•	•	⊕	Levels generally increase with amount of flow through system.
Chlorophyll a	All systems	⊕	–	⊕	Used as a general index of standing crop of algae.
Organic matter					
Biochemical Oxygen Demand (BOD)	Rivers	•	⊕	•	Measures amount of oxygen used by microbes decomposing organic matter. One of the principal **determinands** in monitoring sewage works, etc.
Chemical Oxygen Demand (COD)	Rivers	•	–	•	Measures total amount of organic matter which *could* use up oxygen. An alternative to BOD, useful where **non-labile organics** are suspected.
Metals					
Aluminium, cadmium, copper, lead, mercury, zinc	All systems	⊕	⊕	⊕	Often serious pollutants of freshwaters. Toxicities usually increase with decreasing pH and water hardness.
Calcium, magnesium, sodium, potassium	All systems	•	•	•	Used to assess type but not quality of water. Useful in conjunction with other determinands in assessing likely toxicity of other metals.
Others	All systems	⊕	⊕	⊕	Industry-specific surveys may be needed (e.g. silver for electroplating, tin from old mines) but most not routinely covered.
Micro-organics	All systems	⊕	⊕	⊕	Difficult to identify unless potential source suspected. For these reasons, though potentially important, rarely included in standard surveys.
Oils					
General effects	All systems	•	⊕	•	Most oils are easily detected by sight and smell. Not normally a human health problem as polluted waters unlikely to be imbibed. Taste can ruin fisheries.
Carcinogenic effects	All systems	–	–	–	Rarely routinely measured, as particular carcinogen will vary with type of oil, geographic source and batch.
Ammonia	Rivers	⊕	•	•	Toxic to fish, and toxicity increases as pH increases.
	Lakes/ponds	⊕	⊕	•	In large water bodies, only likely to be high when used for intensively stocked fisheries. Small stagnant waterbodies may naturally have high levels of ammonia.
Hydrogen sulphide	Rivers	⊕	⊕	⊕	Generally as for ammonia.
	Lakes/ponds	⊕	⊕	•	
Cyanide	All systems	⊕	⊕	⊕	Very toxic but occurrence limited to particular industries.
Sediment	Rivers	•	–	•	Part of routine monitoring of all rivers particularly in association with sewage outfalls.
	Lakes/ponds	–	–	⊕	May be of concern in fisheries, and in reservoirs where filters may be blocked.
Pathogens	All systems	–	•	–	Especially for water-based recreation areas. Most surveys simply estimate the amount of faecal contamination.
Dissolved oxygen	Rivers	•	–	•	A routine determinand because many river animals need high levels (see Chapter 13)
	Lakes	⊕	–	⊕	Levels vary with depth, time of day and season.
	Ponds	–	–	⊕	Levels often highly variable.
pH	All systems	•	•	•	Used in all systems, but interpretation in terms of water quality is very use-related. Used to qualify other data.
Alkalinity	All systems	•	•	•	**Alkalinity** is used to qualify pH data.
Conductivity	All systems	•	•	•	A general index of overall amount of dissolved chemicals in water. Useful to get an impression of the likely levels of other major determinands.
Temperature	All systems	•	•	•	Important in assessing thermal pollution; mainly used to qualify other data.

– infrequently measured   ⊕ fairly frequently measured   • frequently measured
Note: an infrequently measured determinand may be important in specific circumstances.

In the case of development types for which EIA is mandatory, it is obligatory for NRA and/or HMIP to provide the developer (on request) with any information in their possession that is likely to be relevant to the preparation of the EIS. The NRA and RPBs hold extensive information, particularly for watercourses. They regularly monitor river water quality in terms of both ecological attributes and water chemistry, and hold a wide range of physical and flood-control related data. The NRA is also beginning to produce *catchment management plans*. Where available, these provide information about (a) aspects such as water abstraction, effluent disposal, SWQOs, flood defence, amenity and recreation, and (b) information about the NRA's approach to catchment issues such as low flows, pollution, and wildlife conservation.

Other useful sources of baseline information include: the water companies, the Meteorological Office (MO), the Institute of Hydrology, the Institute of Freshwater Ecology (IFE), the NCA, LAs, the Royal Society for the Protection of Birds (RSPB), county wildlife trusts, angling and sailing clubs, local universities, consultants with experience in the area, previous EISs, and scientific papers. Ordnance Survey maps (topographic, geological, and soil survey), aerial photographs and satellite images can provide essential information, including indications of changes that have taken place.

Although time and resource restrictions in many EIAs may necessitate heavy reliance on existing information, this is unlikely to be fully adequate for accurate impact prediction, the design of mitigation measures, and monitoring (which is often severely limited by lack of good baseline data). For example, the NRA will often decide that the information it holds is inadequate or out of date, and will require the developer to collect additional or new data. Thus, it is normally necessary to update and/or supplement existing information, e.g. by appropriate field surveys.

Ideally, baseline data for any biotic or abiotic component of a system should be gathered over the longest possible period of time, and with the greatest possible frequency of sampling, since limited data are likely to be misleading. Moreover, as indicated in Table 10.2, a hydrological assessment will normally require seasonal, annual, and in some cases long-term, data on many parameters.

10.6 Baseline studies on physical parameters

10.6.1 Introduction

This section considers appropriate methods for collecting data on the physical hydrological parameters discussed in §10.2. It cannot provide more than an overview, and frequent reference is made to relevant texts and other sources of information.

10.6.2 Precipitation and interception

Information on the collection and analysis of precipitation data is widely available, e.g. see Gregory & Walling (1973), Dunne & Leopold (1978), Newson (1979), Bannister (1986), Brassington (1988), Shaw (1988), Ward & Robinson (1990), and Schwab et al. (1993).

Rainfall data from the nearest weather station should be adequate for most EIAs, and can be obtained from the station or from the MO. However, rainfall can vary appreciably even within small areas, and it may be desirable to obtain onsite rainfall measurements to check the correlation between the site rainfall pattern and that at the weather station. Rainfall is measured in *rain gauges* (for periodic readings) or *rainfall recorders* (which provide continuous records). These are sited on open ground to avoid interception by buildings and tall vegetation (§10.2.4), and measurement of *gross precipitation* (above the vegetation canopy) is then normally acceptable. In some cases it may be desirable to measure *net precipitation*, i.e. that reaching the ground. The difference between gross and net precipitation is a measure of interception loss.

Because of flooding and erosion risks, the frequency, intensity and duration of storms is an important aspect of rainfall. These are studied by rainfall frequency analysis and the use of **design storms**.

Snowfall is difficult to measure because wind causes erratic deposition and drifting, and resort is usually made to measuring depths and melting sample cores to determine rainfall equivalents. *Dewfall* can be measured, but is not likely to be important for EIA in temperate climates. *Fog* can be estimated by comparing the water collected in an ordinary rain gauge with that in a gauge fitted with a mesh screen that intercepts fog.

10.6.3 Evapotranspiration and meteorological water balance

Evaporation from a free water surface is controlled by the water temperature and the prevailing meteorological conditions – principally humidity, air temperature and wind. A complication in the estimation of evapotranspiration (*ET*) is that its rate may be limited by additional factors, including a shortage of soil water and the nature of the vegetation. To allow for the effect of soil moisture conditions, distinction is drawn between (a) *potential evapotranspiration* (*PE*), which is only achievable when there is no shortage of soil water, and (b) *actual evapotranspiration* (*AE*), which is affected by meteorological conditions *and* soil moisture status. *AE* is equal to *PE* when the soil is saturated, but falls below *PE* when a soil moisture deficit (SMD) develops, because plants then tend to wilt and reduce transpiration by **stomatal closure**.

Evaporation from a free water surface, and *AE* or *PE* from a vegetated surface, can be measured at point sites by using *evaporation pans, lysimeters* and *irrigated lysimeters* respectively (described in Gregory & Walling 1973, Newson

179

1979, Bannister 1986, Brassington 1988, Ward & Robinson 1990). However, installation of a network of these is rarely practicable, and measurements for an area are not generally available. Models have been developed to estimate *ET* from meteorological data (see Dunne & Leopold 1978, Lee 1983, Schwab et al. 1993, Ward & Robinson 1990). One of these, the *Penman method*, is used by the MO to calculate *PE* values from data collected at local weather stations, and these should be adequate for EIA.

Pn and *ET* data can be used to calculate (a) meteorological water balance (*Pn* – *ET*), and (b) SMDs, which are assumed to occur when *ET* exceeds *Pn* over a period of a week, and are calculated by a simple accounting method (see Newson & Hanwell 1982). Although not specific to particular soil types or sites, SMDs can provide a general estimate of soil moisture at given times, and hence of the frequency of soil drought in a catchment. During the summer months, MAFF produces bulletins with SMD data to alert farmers to drought conditions.

10.6.4 Infiltration, overland flow, and interflow

Point measurements of these parameters can be made with fairly simple equipment (see Gregory & Walling 1973, Dunne & Leopold 1978, Newson 1979, Shaw 1988), and may be justified for small areas of particular concern in a project, e.g. on a steep slope. However, it is not practicable to obtain direct field measurements over large areas, and resort is usually made to approximate indices (based on factors such as soil properties, vegetation cover, and slope) that can indicate runoff potential. These are incorporated in runoff models (see §10.8.4).

10.6.5 Standing waters and channel flow

The desk study can make an inventory of waterbodies, which can usually include details of elevation, size, and reservoir capacities and operating schedules.

Measurement of channel flow is often called *stream gauging*. Its main purposes are to estimate surface water discharge and/or recharge for an area, *peak runoff rates* (quickflows) and hence flood risk, and flow relationships in a stream network. The three main methods are explained in various texts (e.g. Newson 1979, Brassington 1988, Shaw 1988, and Gordon et al. 1992), and only a brief outline is given here.

The v*elocity–area method* is the only feasible method for large rivers. It involves measurement of (a) the cross-sectional area of the channel, and (b) flow rates at different locations and depths across the channel, obtained with a *current meter,* which consists of a flow-driven propeller with an electrical connection to a calibrated revolution counter. *Dilution gauging* involves injecting a tracer such as salt into a stream and measuring its dilution at suitable distances down stream.

It is not very accurate, but it is useful when other methods cannot be easily employed, e.g. in small turbulent streams. Both the above methods require repeat measurements on different occasions to obtain average flow values for given time periods. The s*tream gauging structure method* is the most accurate procedure. It employs a physical structure, such as a weir, which is installed in the stream and permits flow rates to be calculated from water depth measurements. Individual measurements can be made, but a water-level recording instrument that can produce continuous flow records is usually installed next to the gauging structure. This can provide data on seasonal and annual flow rates, and peak flows associated with storms. The results are plotted as *hydrographs*, i.e. plots of streamflow against time.

Two potential applications of channel flow measurement for EIA are (a) rainfall and streamflow data can be used to predict runoff and flooding at a *gauged site*, and (b) predicted changes in runoff, and the effectiveness of mitigation measures can be monitored. A problem with the first of these applications is that accurate flood prediction requires a long record of rainfall and channel-flow data, or a strong correlation between a short record and a longer record from another catchment in the same area (Newson 1979). These requirements are unlikely to be met for the majority of EIAs, in which runoff and flood-risk predictions will have to rely on the use of models (see §10.8.4). A worthwhile monitoring programme could include periodic channel-flow measurements, and may justify the installation of one or more stream gauging stations in the project's impact area.

10.6.6 Water in the ground

Measurement of soil water is dealt with in Chapter 9. Important aspects of groundwater include *storage, flow*, and *water table level* (WTL). Field evaluation of aquifers is possible (see Brassington 1988), but storage and flow estimates are usually based on models (explained in most hydrology texts) and/or published data on the characteristics of aquifers, in conjunction with geological survey data including maps (see Dunne & Leopold 1978, Bowen 1986, Brassington 1988).

Changes in storage can be estimated from changes in WTL, e.g. at monthly intervals. In unconfined aquifers, WTL is measured as the depth from ground level to the water table. In wetlands such as marshes and peatlands (see Appendix E) where changes in WTL can be critical, the water table is normally near ground level, and can be measured easily in perforated tubes (e.g. made from plastic wastepipe) inserted vertically in the ground. Where the water table is not near the surface, tube installation involves drilling with a mechanical auger, and is likely be too expensive and time-consuming for EIA. The problem can be solved by finding existing bore holes and wells. The NRA and relevant water company will know the locations of many of these, and wells can be found on 1:25,000 or 1:10,000 Ordnance Survey maps. In all cases, the depth to the WTL can be measured using a *dipper*. This consists of a probe attached to a conductor cable that is marked with a scale and can be wound onto a portable drum that has a light or buzzer to indicate when the probe is in contact with water (see Brassington 1988). For more precise or long-term monitoring, a water-level recorder (similar to those used in stream gauging) can be installed.

Water table readings taken at a network of stations can also provide information on *groundwater contours*, and hence on likely flow patterns. For this, the recorded WTL depths are subtracted from the relevant ground-level altitudes (measured, with a surveyor's level, from the nearest bench marks) to calculate the absolute heights (altitudes) of the water table. A WTL contour map can then be produced which will show the groundwater slope(s) and hence the likely direction(s) of flow. As illustrated by Figure 10.3, such information may be useful for assessing the vulnerability of a geogenous wetland to potential impacts such as pollution or water abstraction associated with a proposed project in its catchment.

10.6.7 Water budgets

The water budget of a catchment or site can be estimated using Equations 10.1 or 10.2 respectively. This is possible only if all but one of the variables can be measured or neglected, and it normally involves measurements taken over at least one year. It may be particularly important for estimating the degree to which a wetland depends on surface and/or groundwater recharge. For example, calculation of the annual budget for Cothill Fen SSSI indicated that this is maintained by a massive groundwater input (*c.* four times the annual rainfall), and is therefore highly dependent on the quantity and quality of groundwater in its catchment (Morris 1988).

10.7 Baseline studies on water quality

10.7.1 Introduction

Water quality can be assessed by chemical or biological/biotic methods. *Chemical methods* involve analyzing water samples for relevant **determinands** (nitrate, oxygen, pH, etc.). They have the advantage of giving (usually) accurate estimations of levels, which can be readily compared with relevant statutory standards. There are, however, three major disadvantages:
- there are many possible pollutants in any given situation and each has to be assayed separately
- many pollutants (e.g. the hundreds of micro-organic compounds) are both difficult and expensive to monitor

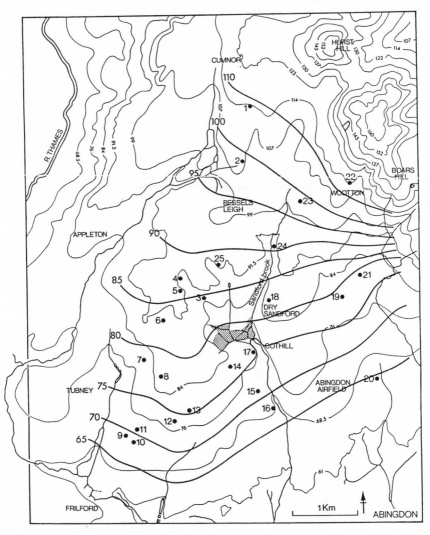

Figure 10.3 Groundwater contours (m) in the catchment of Cothill Fen SSSI (stippled area) in Oxfordshire. The contours were drawn from mean absolute WTLs at 25 wells (numbered), each mean being derived from measurements taken at monthly intervals over a period of two years. The SSSI is a geogenous wetland and was thought to be threatened by a proposed extension of sand extraction workings (and subsequent landfill) near to its western boundary. The results suggest that the groundwater flow in the area of the proposed project largely bypasses the site, but that appreciable water abstraction or pollution in the area directly to the north could have serious consequences. (Morris 1988; data of Morris & Finlayson).

- the sample will only reflect the chemical conditions at the time of sampling.

Biotic methods use plants or animals as an indirect way of measuring water quality. The two main advantages are:

- a single type of survey will often detect the net effects of one or more possible (often unknown) pollutants
- they can be used to give an impression of long-term environmental health, e.g. pollution inputs that affect a river only occasionally may be detected by sampling the biotic community, even if the pollutant is not present at the time of survey.

Two disadvantages are:

- it is not possible to determine the exact pollutant
- they are not very well developed for still-water systems.

10.7.2 Chemical methods of assessment

Many *chemical determinands* could be measured for water quality studies in EIAs. Table 10.3 lists the most common, highlighting those most relevant to studies of human health, conservation, and fisheries.

Chemical determinand levels often vary considerably, not only seasonally and throughout the day, but within the same waterbody at the same time, sometimes over quite short distances. In addition, many elements occur in different forms, only one of which may be of interest. For example, phosphorus may be measured as soluble reactive phosphorus, soluble unreactive phosphorus, particulate phosphorus, or a combination of these. Metals are often present in many forms, including organometallic forms, measurement of which is often technically difficult. Understanding the inherent variability of chemical parameters is critical not only for selecting a relevant analysis and sampling programme, but also for interpreting the results. This topic is wide and cannot be discussed in detail here; an overview is given in Hunt & Wilson (1986).

The level at which individual parameters are monitored can also markedly influence the cost and/or extent of the survey, and care is needed to avoid selecting levels that are either too precise or too crude. For example, it would be pointless to stipulate a detection limit of $5 \mu g/l$ for monitoring nitrate in lowland rivers, where levels are never likely to fall below $1 mg/l$; conversely, there is little point in conducting a survey only to find that the assays have failed to detect the determinand under study. Results of water chemistry monitoring around the world are given in Meybeck et al. (1989), and may help in formulating a broad strategy. However, water analysis will usually be carried out by an independent analyst (a public analyst if the results are to be legally accepted), who should be consulted about suitable procedures. Details of assay methods can be found in Golterman (1978), Mackereth et al. (1978) and relevant HMSO standards (Standing Committee of Analysts).

Methods of *collection, preservation and transport of samples* are given in Golterman (1978) and Hunt & Wilson (1986), and the latter includes an extensive discussion of sampling strategies. Various samplers exist for taking samples at depth (Hellawell 1986); most other samples can be taken using a suitable bottle. Field test kits and meters are available for many determinands. In general, kits using titrations are usually reliable, whereas those that involve assessing colour differences are more variable (although some are good). Kits and meters should not be employed without a standard for use in the field.

10.7.3 Biological indicators of water quality

Hellawell (1986) and Rosenberg & Resh (1993) provide extensive reviews of the use of *biological indicators* in assessing water pollution, and summary papers describing the various types of biological monitoring of river water quality throughout Europe are given in Newman et al. (1992). Most groups of freshwater organisms have been used as indicators of given pollution problems; but in Britain and Europe, **macroinvertebrate** families (not species), are by far the most widely used group. For rivers, large databases are available from the NRA and other regional water authorities.

In Britain the main biological assessment method is the *Biological Monitoring Working Party* (BMWP) index (NWC 1981a), which is widely used to assess water quality in streams and rivers. The BMWP method awards points to different invertebrate families according to their perceived tolerances to low oxygen levels (low points for tolerance, high for intolerance). This and the associated index, **average score per taxon** (ASPT), are used as a broad indication of the level of water pollution. Different rivers (or parts of rivers) have different indices derived by this system, so a recent modification has been the development of a computer program RIVPACS (**R**iver **I**nvertebrate **P**rediction and **C**lassification System; Moss et al. 1987), which allows actual BMWP and ASPT values in a river to be compared with those predicted for an unpolluted site of similar physical characteristics. BMWP scores are sometimes used incorrectly in EIAs, e.g. it is often wrongly assumed that macroinvertebrate data at the family level can be used directly to assess the conservation value of freshwater invertebrate communities. River water quality indices are also frequently misused by applying them to still water habitats. DoE (1992e) gives a useful overview of the present, and probable future use, of BMWP, ASPT and RIVPACS in Britain.

Several *other bioindicators* are used. Higher plants have been advocated as indicators of water and sediment quality (Holmes & Newbold 1984), but they are generally held to be less obviously sensitive to pollution than invertebrates are, and this approach is not routinely used. Fish are sometimes monitored to assess incoming water quality at inlets to reservoirs (Hellawell 1986), and changes in fish populations with time can give information about long-term pollution trends such as acidification and eutrophication. Various plant and animal species *bio-accumulate* toxins, and many have been advocated as means of assessing continuous and intermittent pollution (Hellawell 1986). Some micro-organisms, such as the bioluminescent bacterium *Photobacterium phosophoreum,* have been used to assess *water quality* (Calow 1993).

Pathogens in waters can be detected by two broad methods; detection of species/strains or detection of an indicator groups/species. Detection of individual species would be ideal, but there are several problems:

- there are many different pathogens in fresh water, all of which would need to be assessed
- many species and strains of bacteria and virus require sophisticated culture and detection methods, often taking long periods of time
- for some, detection techniques have not been developed
- protozoan parasites are difficult or impossible to grow in culture, so large samples are often needed (e.g. a tonne of water for *Cryptosporidium*).

For these reasons, most routine monitoring involves the use of indicator groups, and relies on two broad assumptions: (a) that the principal concern is with human faecal contamination of water, and (b) that the indicators used will be present in proportion to all pathogenic species of interest. In practice these two conditions are never fulfilled, and there has been much debate over which indicator organisms should be used and how much faith should be placed in such assessments. Nevertheless, in the absence of any other practicable method, human health limits for fresh water are set in terms of the number of indicator organisms per unit volume. The most common organisms used are coliform bacteria, some species of which are a natural (largely non-pathogenic) component of the biota of the human gut. In Britain, assessment is made for total coliforms (which will include many species that are not necessarily of faecal origin), and faecal coliforms (which should correspond more closely to the extent of faecal contamination of the water). Bathing waters, and surface waters used for potable water extraction, are also monitored for faecal streptococci and *Salmonella.*

10.7.4 Baseline changes without the development

In order to predict accurately the impacts of a development, it is necessary to consider changes in the baseline conditions that may occur in its absence. These are likely to arise from (a) climatic change, (b) natural changes such as sedimentation and ecological **succession**, and (c) impacts from other current or proposed developments in the area.

Climate change cannot be predicted with any certainty, and the best procedure is to allow for changes similar to past extremes, e.g. of wet and dry periods. Many wetlands gradually become drier as they undergo succession. The process can be delayed by management, and the likelihood of significant change occurring during the lifetime of a development can be assessed from the management regime and the current state of the system.

To assess the impacts of other developments, it is necessary to identify these and consider each in terms of the potential impacts discussed below. This can be an onerous task, and the assessment may have to be limited to general predictions and/or major developments. An additional problem is that many impacts are *cumulative*. For example, the additional water abstraction requirement of a development can be estimated, but this must be considered in the context of other current and potential needs in the area (see Ch. 15).

10.8 Impact prediction

10.8.1 The impact area

An early task in impact prediction is estimation of the impact area. This will depend on the nature of the project, the nature of the catchment, and the position of the project within this. In some cases, impacts may be confined to the project site and its immediate surroundings, but factors affecting hydrological systems can have much larger impact areas. For example, wetlands and domestic water resources located some distance from the project may be affected by changed water flows, groundwater levels, or water quality. Estimating the impact area is bound to involve some guesswork, and the original estimate may have to be revised in the light of information that emerges during the assessment process.

10.8.2 Types of impact

Hydrological impacts can be divided into those arising from the direct manipulation or utilization of hydrological systems (outlined in Table 10.4a), and those from projects not directly associated with such manipulation or utilization (outlined in Table 10.4b). General reviews of these impacts can be found in texts such as Gregory & Walling (1987) and Goudie (1990); impacts of river management, and dams are discussed in Petts (1984) and Brookes (1988); impacts of roads are reviewed in Box & Forbes (1992), DOT (1993) and Brookes & Hills (in press); hydrological consequences of urbanization are discussed in Miller (1977), Hall (1984), Shaw (1988) and Walesh (1989); and EIA in relation to waste treatment and disposal facilities is discussed in Francis (1992) and Petts & Eduljee (1994).

Some impacts are particularly intensive during, or are largely confined to, the *construction phase* of projects. A general list of construction-phase impact factors is given in Table 11.7. Impacts of particular relevance to hydrological systems are indicated in Table 10.5.

Table 10.4 (a) Impacts from direct manipulation or utilization of hydrological systems.

Activities	Potential impacts
River engineering/manipulation Resectioning/channelization (widening, deepening, realigning/straightening); usually to increase channel capacity for flood defence and/or land drainage, or to facilitate project layout.	Loss of channel and bank habitats in channelized section (especially slow-flow habitats). Enhanced erosion and hence silt production (especially during construction phase, when pollution risks also increase). Increased flood risk and siltation downstream. Lowering of WTL, and associated loss of wetland habitats in floodplain caused by deepening.
Diking/embanking and bank protection, e.g. with concrete; usually for reasons as above.	Natural floodplain inundation prevented, with consequent silt deprivation, loss of wetlands and risk of soil drought. Drainage from floodplain prevented (unless sluices installed) with consequent risk of waterlogging.
Clearing bank vegetation, usually associated with the above.	Destruction of vegetation and wildlife habitats. Loss of visual/amenity value.
Dredging and filling (deposition of dredgings) to (a) maintain flood capacity or (b) improve navigation.	Increased suspended sediment load and hence turbidity. Physical damage to channel habitats and biota. Smothering of estuarine ecosystems, e.g. marshes and swamps, with consequent increased erosion risk.

Diversion:
  to increase water supply to receptor area;
  as a relief channel (in flood defence).
	Decreases supply in donor area. Usually involves channelization and evaporative loss from open channels.
	May destroy habitats in main river corridor.

Development on the river floodplain	Usually requires (or follows from) flood protection, e.g. embanking (often with channelization). Marked increase in flood risk down stream (see §10.2.6).
Reservoirs and dams: General	Loss of terrestrial habitats, agricultural land and buildings. local rise in WTL. Local climatic changes, e.g. increased fog. Visual impacts of retaining walls. Water-borne pathogens. Earthquake or landslip risk. Failure risks. Ancillary works, e.g. pipelines to supplement lowflows down stream or to water treatment works.
On-stream dams: Above dam/within reservoir	Loss of original river section; changes in flow regime, detention time, depth and stratification potential; siltation.
Below dam	Changes in flow regime (reduced frequency, velocities and volumes); changes in water chemistry (including reduced nutrient availability, sediment and oxygen levels); WTL, flooding and siltation reduced in floodplain.
Barrier effects	Migration of fish and invertebrates blocked.
Off-stream dams (not constructed across a major channel)	Changes in groundwater recharge areas and flow directions, often with reduced WTL below the dam.
Barrages	Estuarine ecosystems lost. Coastal erosion risk increased (see Ch. 14).
Irrigation	Water abstraction (often from rivers to which only 20–60% of diverted water is returned). Increased *ET* and local runoff. Risk of waterlogging and salination.
Drainage (including drainage of agricultural land and wetlands, e.g. for afforestation)	Often involves channelization Soil moisture reduced, and soil drought risk increased. Organic soils oxidised. WTL lowered and wetlands lost. Increased flood and erosion risk downstream, especially when uplands drained.
Water abstraction	Water storages/supplies depleted. WTL lowered. Wetlands lost or damaged. Risks of soil drought and subsidence.

Table 10.4 (b) Impacts from projects not directly associated with the manipulation or utilization of hydrological systems.

Projects	Potential impacts
Roads	Changes in drainage systems due to landscaping (e.g. gradient changes, bridges, embankments, channel diversion or resectioning). Changes in groundwater flow, e.g. when deep cutting through an aquifer. Increased runoff velocities and volumes due to replacement of natural vegetated surfaces by impermeable surfaces, with consequent risks of flash floods and erosion. Increased sediment loads, e.g. from road and tyre wear, vehicle corrosion, and erosion of new embankments or cuttings. Deoxygenation through organic content of silt. Pollution of water courses by de-icing salt (and associated impurities), metals (mainly from corrosion of vehicle parts), organics (e.g. oils, bitumen, rubber), plant nutrients, biocides (from verge maintenance), and accidental spillages of toxic materials.
Urban and commercial development	Changes in drainage systems due to landscaping. Changes in groundwater flow, e.g. when deep foundations cut into aquifers. Increased abstraction and hence lowering of WTL. Reduced groundwater recharge. Increased runoff velocities and volumes (as for roads) with consequent rapid changes in stormflow, and flood and erosion risks. Pollution of water courses as for roads but with additional metals and other pollutants from corrosion of construction materials – and often biocides, pathogens and plant nutrients from amenity areas. Groundwater pollution.
Industrial development	As above but with: greater runoff effects associated with high proportion of hard surfaces; higher pollution levels and a wider variety of pollutants (from impurities in raw materials, harmful by-products and unrecovered products) including metals and microorganics from heavy industry and refineries, biocides from wood treatment works, and nutrient-rich or organic effluents from breweries, creameries etc.. Thermal pollution from power plants with consequent changes in microbial activity and dissolved oxygen levels.
Wastewater treatment works	Usually sited close to watercourses. Increases in silts, nutrients (especially where phosphate stripping is poor), heavy metals from urban and industrial sewers, organics, and pathogens, e.g. faecal coliforms (especially where storm water reservoirs have storm overflow problems or sludge is stored on site).
Landfill	Increased runoff if site is clay-capped. Contamination of groundwater and near-surface runoff by toxic leachates (depending on the types of infill materials).
Quarrying and mining	Changes in drainage systems and groundwater flows. WTL and streamflows often lowered locally. Increased waterlogging and flood risk downstream (especially from opencast methods involving water use, e.g. hydraulic, solution mining, and dredging). Increased siltation downstream, e.g. through erosion of spoil heaps, rock dumps and tailings piles which lack vegetation cover. Chemical pollution from spoil heaps etc., including acid pollution through oxidation of sulphides and consequent production of sulphuric acid. Oil pollution from vehicles, machinery and stores.
Forestry and deforestation	Reduced interception, ET and infiltration after felling, during logging and before planting – with consequent (a) decreased local groundwater recharge (possibly affecting local wetlands), (b) increased runoff causing rapid changes in river flow rates and consequent channel erosion and flood risk, (c) enhanced soil erosion and nutrient losses, and (d) high stream-sediment loads and consequent siltation. Pollution by biocides, especially herbicides used to prevent regrowth after clear felling. Possible reduction of local and regional Pn from large scale deforestation.
Intensive agriculture	Enhanced runoff and erosion from bare soils. Drainage or irrigation impacts. Pollution of soils, groundwater and surface waters by a variety of pollutants including: nutrients (especially N and P) from fertilisers; biocides from crop protection and sheep dips; organics from soil erosion, silage clamps and muck spreading; heavy metals from slurry runoff, and pathogens in animal wastes.

Some water engineering projects can have *positive impacts*. For example, reservoirs can provide water-based amenities and new wetland habitats (although variable water levels can be a problem), and river projects can lead to enhancement of the existing river corridor for wildlife and amenity (including access by footpaths, etc.), especially where the existing conditions are poor (as in many urban locations).

Table 10.5 Hydrological impacts particularly associated with the construction phase of projects.

Category	Potential impacts
Runoff	Appreciable increases often associated with disturbance, vegetation removal, soil compaction, and pumped water (when excavations, e.g. for foundations, extend below the water table); can markedly affect streamflows and stream sediment loads.
Sediment	Export of large quantities of sediment in runoff, with suspended-sediment loads up to $50\,\mathrm{g}\,\mathrm{l}^{-1}$ recorded in streams draining from construction sites (Milliman & Meade 1983). Oxidation of organics in the sediment can threaten fish and invertebrate populations through deoxygenation of the water.
Nutrient release	Considerable quantities, especially of nitrogen, can be released through disturbance of established grassland soils, and the effect can persist for several years. Amounts are difficult to predict, but nitrogen release can be in the range 50–$200\,\mathrm{kg}\,\mathrm{ha}^{-1}\,\mathrm{yr}^{-1}$ (DoE 1986). Nitrogen entering streams and groundwater may lead to enrichment of water bodies away from the site.
Oils and other pollutants	Waters frequently polluted from spillages on construction sites, especially when on-site pumps and storage tanks of petroleum products are located near water, and vehicles are filled from these. Can also lead to groundwater pollution.

10.8.3 Methods of impact prediction

Impact prediction can be assisted by the use of checklists and matrices, flowcharts and networks, and quantitative predictive models. These methods are discussed in §11.5.3 in relation to ecological assessments. Their advantages and limitations also apply in hydrological assessments. Some impacts such as landtake can be readily measured, but many are much less amenable to precise quantitative prediction. Two major impacts associated with the majority of projects are increased runoff (and associated flood risk), and deterioration in water quality. Prediction of these is discussed in the following sections.

10.8.4 Prediction of runoff and flood risk

Numerous models, many of which employ computer programs, have been developed to predict runoff and associated flood risk. They include linear rainfall–runoff models, more complex *catchment models*, and specific *urban runoff mod-*

els. These are reviewed in sources such as York & Speakman (1980), Anderson & Burt (1985), Shaw (1988), Walesh (1989), Bras (1990), Ward & Robinson (1990) and Schwab et al. (1993). A thorough review of flood prediction methods was undertaken by the Institute of Hydrology (NERC 1975), with later modifications for the urban context (Institute of Hydrology 1979). Floods are also discussed briefly by Newson (1975), and in more depth in Beven & Carling (1989).

Prediction is facilitated by flow records for gauged sites and catchments (§10.6.5), but in spite of major advances in recent years, accurate predictions for ungauged sites are still difficult. The main limitations are:
- runoff depends on a complex system of processes and conditions; "methods of runoff estimation necessarily neglect some factors and make simplifying assumptions regarding the influence of others" (Schwab et al. 1993)
- no matter how sophisticated a model is, accurate prediction will depend on the availability of good input data
- "few [models] can be applied indiscriminately beyond the conditions for which they have been developed" (Ward & Robinson 1990).

The use of models is also expensive and, given the time and resource restrictions common in EIA, and the fact that many proposed developments will be in areas for which appropriate data are not available, their application is likely to be limited in the majority EIAs. A competent hydrologist and/or the NRA should be consulted as to the need and feasibility of applying sophisticated methods. Whether or not these are employed, there is a case for obtaining an approximate estimation of the increased runoff that a development will generate, using a relatively simple rainfall–runoff model such as the rational method.

The *rational method* is a long-established procedure for estimating peak runoff, and predicting how this will respond to changes in land use. It requires relatively little information on the characteristics of a catchment, and can employ existing rainfall data for the area. The procedure is explained in Appendix C. Because of its relative simplicity and applicability to urban situations, the rational method is still one of the most popular models in the USA (Walesh 1989), is widely accepted for use in the design of storm drains (Dunne & Leopold 1978), and is an attractive option for use in EIA. However, it has several limitations that must be recognized.
- The predicted runoff values are imprecise because various simplifications and assumptions are involved.
- It should be normally limited to catchments under 800ha (Schwab et al. 1993).
- It is often misused, and should only be employed under supervision by a competent hydrologist (Walesh 1989).
- The predicted runoff values give no information on the form of a flood, other than the peak flow, and therefore cannot predict *total runoff volume* during a design storm. This is a serious limitation if mitigation requires provision for impounding flood water in reservoirs. It is rectified in the *modified rational method*, which incorporates the facility to produce runoff

hydrographs (Shaw 1988, Walesh 1989). For the design of floodwater reservoirs, it is even more important to predict *annual runoff volume* (often called the *water yield*). Appropriate methods for estimating this are discussed in Dunne & Leopold (1978) and Schwab et al. (1993).

10.8.5 Prediction of changes in water quality

The likely effect of a development on water quality will depend not only on the development type but also on the *type and quality of the receiving waters*. For example, rivers export most of their pollutants down stream, so the effect at any one point may be transitory. On the other hand, polluted water and silts may be carried considerable distances before they are sufficiently degraded or diluted to have no effect. Standing waterbodies such as lakes and ponds are sediment sinks, and their water turnover rate is usually slow. As a result, sediment and pollutants tend to accumulate, and the associated impacts may intensify with time. Groundwater pollution is increasingly recognized as a threat from developments. Urbanization affects the quality of runoff in two ways: (a) the presence of many different urban activities and development types creates a multitude of potential pollutant sources (see Table 10.4b), and (b) pollutants are rapidly transported to receiving waters by the increased volume and speed of runoff usually generated.

In general, pollution sources can be divided into types, point source and non-point source. *Point source pollution* arises from discrete points, usually the outfalls of manufacturing or sewage works. All point source pollutants discharged to water require a consent licence (see Box 10.1). *Non-point source* (or *diffuse*) *pollution* typically results either from surface runoff, or from many minor point sources e.g. land drains. The most usual origins are urban or intensively farmed agricultural areas.

Estimating the amount and effect of diffuse pollution loads is generally more difficult than for point source pollutants; there are relatively few methods, and they tend to be limited in capability. Hand calculation methods most commonly used include: the unit load method, the universal soil loss equation, and concentration times flow method. Walesh (1989) gives a useful overview of the applications and drawbacks of these and other methods. Wherever possible, more than one technique should be employed to undertake any non-point source calculation, both to guard against the uncertainty inherent in many of these methods and to check for gross errors.

Computer-assisted models for predicting changes in water quality are discussed by Steele (1985), and models specific to groundwater are considered by Konikow & Patten (1985). Their current capabilities are often limited by lack of adequate data and incomplete understanding of the hydrological systems, and their application in EIA is limited also by lack of time and resources. Quantitative assessment of *critical nutrient loading* can be used to estimate the potential for eutrophication in waterbodies.

10.8.6 Uncertainty in hydrological predictions

As indicated in the previous sections, various aspects of a hydrological assessment are bound to involve uncertainties, and these must be admitted in the EIS. Improvements in the precision of prediction data (including those from modelling) will depend on the acquisition of good baseline and monitoring data that can contribute to a cumulative and reliable database.

10.9 Mitigation

10.9.1 Introduction

Mitigation can aim to avoid, minimize, reverse, or compensate for, an impact. In addition there may be opportunities to enhance the existing conditions. Two principles central to EU policy on the environment are (a) preventive action is preferable to remedial measures, and (b) where possible, environmental damage should be rectified at source. Britain's environmental strategy *This common inheritance* (DOE 1990c) states that the Government encourages a precautionary approach to the control of pollution, including:

- prevent pollution at source
- minimize the risk of harm to human health and the environment
- encourage and apply the most advanced technical solutions
- apply a critical loads approach to pollution, in order to protect the most vulnerable environments
- ensure that the polluter pays for the necessary controls.

Scheduled processes or pollutants that are highly damaging require authorization from HMIP, who will be interested in ensuring that operator will use best available techniques not entailing excessive cost (BATNEEC) to minimize release of harmful substances. Where a process is likely to involve the release of more than one medium, the best practicable environmental option (BPEO) must be adopted to minimize damage to the environment as a whole.

10.9.2 Mitigation measures

General mitigation procedures related to ecological impacts during the construction and operational stages of developments are discussed in Chapter 11. Measures specifically related to the hydrological impacts outlined in Table 10.4 are discussed below.

For *river projects*, mitigation measures undertaken by NRA apply to the river corridor, although this is normally limited to a narrow strip on each side of the channel rather than the natural floodplain. They include measures to minimize impacts in the managed section, e.g. avoiding straight and concreted culverts, landscaping and planting within the corridor, and providing bank habitats for wildlife (Brookes 1988). In addition, increasing use is made of measures to simulate the rôle of a natural floodplain, and hence to reduce the risk of downstream flooding. The principal methods are:

- creation of an *artificial floodplain* by erection of banks along the boundaries of the river corridor
- use of *flood relief channels* and *overspill areas,* such as gravel pits that simulate a floodplain, and may be surrounded by land otherwise used for amenity purposes.

Although these measures can be effective in specific cases, much flooding in the UK is associated with general overdevelopment in river floodplains, and NRA is likely to oppose further development in these areas.

Mitigation of the long-term impacts of *dams* and *reservoirs* can involve (a) adjusting the size or location to avoid sensitive areas, (b) minimizing the height and slope of retaining embankments, and (c) planting embankments, e.g. with native trees. The scale of *irrigation* and *drainage* schemes can be controlled, and particular attention should be paid to these when wetlands are likely to be affected.

The *siting of urban and industrial developments*, including the general *routing of roads*, is usually dictated by socio-economic factors rather than environmental constraints. However, every effort should be made to ensure that serious hydrological impacts are not generated by the siting of a project. Account should be taken of NRA's groundwater source protection zones, the locations of major watercourses, the need to avoid excessive development on river floodplains, and the need to avoid damage to valuable wetlands. Similarly, mitigation may involve *modifications to the project layout*, including the location of components and boundaries. The *hydrological impacts of roads* can be reduced by measures such as careful routing, design features to minimize the impacts of viaducts on river corridors (not just the channels), and measures to control storm runoff (see DoT 1992b, Brookes & Hills (in press)).

As outlined in Table 10.6, a variety of techniques have been developed to reduce or delay *urban runoff*, and to protect against local flooding by storm runoff. Some of these measures can serve also to alleviate the problem of *water abstraction*, e.g. by promoting infiltration and hence groundwater recharge, and this can be further addressed by design features that minimize water use, e.g. metering and the installation of water-efficient equipment (including domestic appliances). Computer-assisted modelling systems, such as that developed by the National Water Council (NWC 1981b), can be used for designing and analyzing *storm water drainage systems* (see also Hall 1984, Shaw 1988, Walesh 1989, Schwab et al. 1993).

Water detention basins/reservoirs, which are also used to control runoff from major roads, have several alternative names, including detention ponds, balancing ponds and storm reservoirs. They can have retaining embankments, or can

Table 10.6   Mitigation measures relating to urban runoff and flood protection.

Areas	Measures and structures	Purposes and limitations
Large flat roof	Cistern storage; rooftop gardens; sod roof cover; pool or fountain storage.	Reduce runoff by interception loss and storage.
	Constricted drainpipes; rough surfaces (rippled or gravelled).	Delay runoff by ponding; some reduction by interception loss.
Car park	Porous surfaces (gravel, brick, porous or punctured asphalt); gravel trenches	Reduce runoff and increase recharge to groundwater.
	Underground cisterns and vaults.	Reduce runoff by storage and use.
	Vegetated marginal ponding areas.	Delay and/or reduce runoff.
Residential and commercial	Porous drives and pavements (as car parks); porous or perforated pipes; sand/gravel filled trenches; dry wells.	Reduce runoff and increase recharge to groundwater.
	Rippled pavements.	Delay runoff.
	Cisterns for single/grouped homes.	Reduce runoff by storage and use.
General	Contoured landscape and appropriate siting of roads and buildings	Increased length of runoff route by diversion channels.
	Vegetated areas	Reduce runoff, and increase groundwater recharge.
	Channels:	
	General	Local flood defence; runoff routing.
	Vegetated	Delay runoff; good visual and habitat value; susceptible to erosion.
	Hard lined (e.g. concrete)	Robust; can have large capacity; flow relatively rapid; poor habitat and visual value (can be improved, e.g. by use of gravel on base).
	Weirs, sluices, drop spillways (steps in channel with stilling ponds below)	Reduce flow rate by ponding above step and absorbing energy in stilling pond.
	Terraces and diversion gullies	Aligned across slopes; divert runoff to protect land below; may spill over.
	Embankments:	
	General	Local flood defence.
	Hard (e.g. concrete)	Permanent; can be large; negative habitat and visual impacts.
	Earth	Can be vegetated above water level; susceptible to erosion and seepage.
	Reservoirs/detention basins; can be on-stream or off-stream (filled via pipes or open channels)	Delay and/or reduce runoff by storage and evaporation; can increase groundwater recharge.
	Natural or created vegetated ponds or wetlands	As above + habitat and visual value.

Sources include: Goudie (1990), Hudson (1981), USDA–SCS (1972), Viessman et al. (1977).

be largely below ground level. Reservoirs need some means of minimizing siltation, but storm water retention can be achieved also by *siltation ponds* and *lagoons* which are primarily intended to trap silt.

The export of suspended solids, especially from construction sites, can be a major threat to freshwater ecosystems. Measures to mitigate against this and other impacts on water quality are discussed in Chapter 13, together with possible measures for wetland restoration.

10.10 Monitoring

Lack of monitoring is a major deficiency in current EIAs in the UK. Monitoring should be stipulated in the EIS whenever there is uncertainty about (a) the level, extent or duration of impacts, and/or (b) the effectiveness of proposed mitigation measures or the proficiency with which they will be carried out. It is important also for assessing *residual impacts*, and for contributing to a *cumulative database* for use in future EIAs. Monitoring should be conducted throughout the construction, operation and decommissioning phases of the project, and should be funded by the developer. Parameters used in establishing the baseline conditions can be monitored, and their inclusion in a monitoring programme can help to justify the collection of field survey data in the baseline studies. Account should be taken of changes that may have occurred without the development. This may be difficult and it requires the comparison of impacted sites with *reference sites*.

CHAPTER 11
Ecology – overview
Peter Morris

11.1. Introduction

The EC Directive on environmental assessment (CEC 1985) uses the terms *fauna* and *flora* to refer to the biological component of EIAs. Standard dictionaries usually define these terms simply as "all the animals and plants in a given place or time, and a description of them". This is not sufficient for EIA which requires an understanding of how plant and animal species will be affected by impacts, and hence how they are affected by **environmental factors**. The scientific study of the relationships between living organisms and their environments is *ecology*, so "fauna and flora" really refer to the *ecological component* of EIA. Ecology includes the study of *species populations*, (biotic) **communities**, and **ecosystems** – all of which should be considered in an ecological assessment. It is important to assess the threats to individual species populations, and the loss of a single **dominant** or **key species** can drastically affect a whole community. Conversely, the impacts of developments usually affect whole communities and ecosystems, and loss of these is the commonest threat to species.

In a survey of 37 recent EISs relating to proposed road developments, Treweek et al. (1993) found that the ecological assessment was often woefully inadequate. Examples of deficiencies in the statements were:

- only 35 per cent indicated that new surveys were conducted (suggesting that impact predictions in the majority were based on out-dated or inadequate data), and 31 per cent of these were carried out at inappropriate times, e.g. in winter for herbaceous plants;
- the majority made no mention of the survey methods used;
- over half made no specific reference to species, and in 73 per cent the flora and fauna survey was apparently very superficial;
- 54 per cent gave lists of species in terms of presence/absence, but only three included any measures of **species abundances**;
- five made no mention of potential ecological impacts, and these were only quantified in three; most made only vague references to potential impacts

on the flora and fauna without any attempt to relate these to associated impacts such as changes in the local hydrology;

- 73 per cent acknowledged the need for mitigation, but 51 per cent lacked any description of proposed measures and only 8 per cent included detailed prescriptions;
- in most cases the emphasis in mitigation proposals was on cosmetic measures such as landscaping (63%) or tree planting (85%).

These findings suggest that there is an urgent need to improve the general standard of ecological assessment for EIA. Two common constraints are inadequate funding and lack of sufficient time, and these are linked to a need for wider understanding of the requirements for good ecological assessments and the opportunities and problems involved.

11.2 Definitions and concepts

Ecology is a broad and complex discipline, and this has led ecologists to specialize on different taxonomic groups (e.g. plants, animals, or subgroups of these) and/or types of ecosystem (e.g. terrestrial, freshwater or marine). Partly for this reason, distinction is drawn between *plant communities* (which can be recognized as *vegetation* types) and *animal communities*. However, species and groups (plants, animals and microbes) in a community affect one another. For instance, **food chains** depend on plants, and vegetation is a major component of animals' **habitats**, so removal of the vegetation inevitably destroys associated animal communities. Conversely, animals affect plants in various ways, and some vegetation types such as grasslands are largely maintained by grazing. In addition, communities and ecosystems intergrade, and the gradients transcend categories such as terrestrial, freshwater and marine. For example, freshwater **wetlands** include ecosystem gradients from open waters to semi-terrestrial systems such as peatlands and marshes, and these intergrade with terrestrial systems such as grasslands, heathlands and woodlands. It follows that:

- no single ecologist can be expected to deal with all aspects of an ecological assessment at more than a superficial level, and it may be necessary to employ specialists for different taxonomic groups and/or different ecosystems;
- particular taxonomic groups or ecosystem types cannot be considered in isolation, so the work and findings of the team members must be co-ordinated;
- the ecological assessment should be co-ordinated with other components of an EIA, particularly those dealing with environmental systems such as climate, soils and water (Chs 8–10), which are major ecosystem components. Indeed, if an environmental system is not being surveyed independently, it should be included in the ecological component.

Similarly, although it is convenient in this book to subdivide the ecological material into several chapters (11–14), these are all based on common ground. This chapter provides an overview for the whole ecological component.

The main aims of the ecological component are to assess the *conservation value* of species and communities within the *impact area* of a development, the likely *impacts* of the development, and the *mitigation measures* that can be taken to avoid or minimize these impacts. They are therefore related to the aim of *nature conservation* which, in broad terms, is to maintain **biodiversity**. In Britain as elsewhere, biodiversity is under increasing threat. Recent surveys indicate that: (a) only 3 per cent of the meadowlands that existed forty years ago still retain any wildlife interest and 50 per cent of native broadleaved woodlands and lowland heaths have disappeared during the same period; and (b) between 1978 and 1990 wildlife diversity fell by 14 per cent in woodland, 13 per cent in meadows and 11 per cent in uplands (Harding 1992, WWF 1993).

Justification for the conservation of species, communities and their habitats can be made on various grounds, including:

- the moral imperative
- their value for amenity, recreation, education and scientific investigation
- economic and medical benefits
- conservation of genetic resources
- the rôle of natural communities, especially vegetation, in maintaining viable ecosystems on which people depend; for example, vegetation destruction can disrupt the hydrology of whole catchments, with consequences such as soil erosion and/or flooding (Ch. 10).

11.3 Legislative background and interest groups

11.3.1 Legislation and protected sites

The main legal requirements governing the ecological component of EIAs are the EC Directive (CEC 1985) and the implementation of this in Britain by the environmental assessment regulations. The guidelines given in DOE (1989) (see Ch. 1 and Appendix A) draw attention to: (a) sensitive and protected areas such as National Parks, National Nature Reserves (NNRs), and Sites of Special Scientific Interest (SSSIs) or the equivalent Areas of Special Scientific Interest (ASSIs) in Northern Ireland; and (b) the requirement to consult the relevant *nature conservation authority* (NCA). This was the Nature Conservancy Council (NCC), which is now replaced by English Nature (EN), the Countryside Council for Wales (CCW), Scottish Natural Heritage (SNH) and the Northern Ireland Environment Service: Countryside and Wildlife (ES: CW). Work done by these bodies involving UK and international conservation issues is co-ordinated by the Joint Nature Conservation Committee (JNCC).

Ecological EIA is also affected by UK and international legislation on conservation, as outlined in Table 11.1. Other relevant legislation relating to air pollution, fresh waters and coastal systems is outlined in Chapters 8, 10, 13, and 14.

Much of the legislation affords protection to nature reserves, but there are also nature conservation sites that lack statutory protection. Wildlife sites can be classified according to three levels of importance: *sites of international importance*, *sites of national importance*, and *sites of regional or local importance*. In addition, some types of area and site are considered to have indirect importance for nature conservation. Examples of these categories and their UK designations are given in Table 11.2.

In theory, statutory sites have a high degree of protection, but WWF claim that up to 300 protected areas are destroyed or damaged in the UK each year (WWF 1993). International reserves and NNRs are afforded the greatest protection. SSSIs have some legal protection under the *Wildlife and Countryside Act* (1981); the landowner/occupier must inform the relevant NCA about all potentially damaging operations (some of which are permitted under certain conditions) and may be prosecuted if the conditions are not met (details in EN 1992). They should receive additional protection if designated as SACs. LNRs are usually designated for planning purposes only, and although the LA's policy may be to avoid development in them, they have no statutory protection. Non-designated sites normally have little legal protection unless they contain protected species. Further information on the legislation and types of protected site listed in Tables 11.1 and 11.2 can be found in Lyster (1985), Ball & Bell (1991), EN (1991), Hughes (1992), Spellerberg (1992), Collis & Tyldesley (1993), DoT (1993), Haigh (1993), or can be obtained from DoE, JNCC, or the relevant NCA.

11.3.2 Consultees and interest groups

The statutory consultee for all ecological aspects of EIA is the relevant NCA, which has several important rôles:
- It must be notified by the LA about a development application and will assist in the screening and scoping procedures.
- It must be provided with a copy of the EIS for comments which may include: an appraisal of the EIS in terms of its scope, technical competence, validity, and proposed mitigation measures; and an opinion on whether planning consent should be granted.
- It has a duty to provide non-confidential information if requested, and is normally willing to give advice on all aspects of the EIA.

Initial contact with NCA should be made through the relevant regional office which will:
- hold information on rare or protected species and ecologically important ecosystems within the impact area;
- employ experienced ecologists and have contacts with other experts (EN is compiling a Directory of Expertise);

Table 11.1   UK and EU legislation, and international conventions, on nature conservation.

Legislation	Aims and aspects
Town and Country Planning Act 1947	County council control over development.
National Parks and Access to the Countryside Act 1949	Formation of NCC, National Parks Commission (NPC), National Parks, Areas of Outstanding Natural Beauty (AONBS), NNRS and Local Nature Reserves (LNRS), rights of way and access to open country.
Countryside Act 1968	NPC replaced by Countryside Commission; provisions on access to the countryside and regard for conservation.
Ramsar Convention on Wetlands of International Importance 1971	Conservation of Wetlands of International Importance (WIIS), especially as waterfowl habitats.
UNESCO Convention for the Protection of World Cultural and Natural Heritage 1972	Identification and protection of natural and cultural areas of outstanding international value.
CITES (Convention on International Trade in Endangered Species of wild fauna and flora) 1973	Control of trade in endangered species of plants and animals.
Wild Birds Directive 79/409/EEC	Protection of wild bird species and their habitats in Special Protection Areas (SPAS).
Bonn Convention 1979	Conservation of migratory species of wild animals.
Wildlife and Countryside Act 1981	Designation of SSSIS, NNRS, Marine Nature Reserves (MNRS), and Areas of Special Protection for Birds (AOSPS); Nature Conservation Orders, Limestone Pavement Order, protected species.
Berne Convention on the Conservation of European Wildlife and Habitats 1982 (protected plant species list revised in 1991)	To conserve wild flora and fauna, including migratory species, and their natural habitats; contains lists of protected species; used as the basis for UK wildlife legislation and EC Habitats Directive 92/43/EEC.
Wildlife and Countryside (Amendment) Acts 1985, 1990	Revision of protected species designations; SSSI designation operative immediately on notification by NCA.
Nature Conservation and Amenity Lands (NI) Order (Northern Ireland) 1985	Duties of public bodies; declaration of NNRS, MNRS, ASSIS, and District Council Nature Reserves.
Regulation 797/85/EEC and Agriculture Act 1986	Environmentally Sensitive Areas (ESAS) for protection of wildlife by adoption of suitable agricultural methods.
Environmental Protection Act 1990	Replacement of NCC by regional NCAS and JNCC; various provisions for environmental protection, including additional protection for SSSIS, a prescribed list of polluting activities and substances and Integrated Pollution Control.
Town and Country Planning Act 1991	Requirements for planning permission.
Planning and Compensation Act 1991	Provision for additions to classes of project requiring EIA; LA powers to safeguard conservation areas strengthened.
National Heritage (Scotland) Act 1991	National Heritage Areas (NHAS), afforded special protection for both wildlife and landscape.
Habitats Directive 92/43/EEC	Special Areas of Conservation (SACS) for protection of habitats of wild fauna and flora; lists priority habitat types and species for SAC designations.

Table 11.2   Designation of sites for nature conservation in Britain.

Importance	Site designation and features	UK statutory designation
International sites	**World Heritage Sites** – designated under the UNESCO Convention for the Protection of World Cultural and Natural Heritage.	No single designation
	Biosphere Reserves – designated under the UNESCO "Man and the Biosphere Programme" to conserve the diversity and integrity of communities and the genetic diversity of species.	NNR
	WIIs (Ramsar Sites) – designated under the Ramsar Convention; recognize the special importance of wetland ecosystems.	NNR or SSSI
	SPAs – designated under EC Wild Birds Directive 79/409/EEC to protect important European habitats for birds.	NNR or SSSI
	Biogenetic Reserves – designated by the Council of Europe for conservation of a representative selection of habitat/community types.	NNR or SSSI
	SACs – to be designated under EC Habitats Directive 92/43/EEC for protection of important European habitats in addition to SPAs.	SAC/European Site
National sites	**NNRs** – primarily for research and conservation of communities; can include sites with special geological or physiographic features.	NNR
	MNRs – for conservation of intertidal and shallow-sea ecosystems, and physiographic coastal features.	MNR
	SSSIs – to protect a representative cross section of British habitat types, and ecosystems of international importance, on sites less important than NNRs.	SSSI (ASSI in N. Ireland)
	AOSPs – mainly to project vulnerable groups of birds.	AOSP
	Ancient woodlands (evidently continuously woodland since about AD 1600, and considered particularly valuable if largely composed of native, non-planted trees and shrubs).	None
	NHAs – areas afforded special protection (for wildlife and landscape) in Scotland	NHA
Regional or local sites	**LNRs** – designated by LA (which must have some legal control over the site) in consultation with relevant NCA; for education and amenity in addition to conservation.	LNR
	Non-statutory sites of importance for nature conservation – usually designated by LAs for land use planning purposes; recognized by LAs and NCAs as having local conservation importance.	None
	Non-statutory nature reserves – established and usually owned by non-government organizations (NGOs) such as County Wildlife Trusts, Royal Society for the Protection of Birds (RSPB), Woodland Trust, and National Trust.	None
	Forest Nature Reserves – designated by Forest Enterprise on its land.	FNR
Areas and sites related to conservation	**Statutory areas of conservation or scenic interest**, e.g. National Parks, AONBs, National Scenic Areas (Scotland), Heritage Coasts, Conservation Areas, Country Parks.	As listed
	ESAs – grants available to owners/ occupiers to promote ecologically beneficial management.	ESA
	Historic parks and gardens; local landscape designations.	None

Note: A site may have more than one designation, and international sites such as WIIs and SPAs must first be designated as NNRs or SSSIs. The term "European Site" includes SACs and SPAs.

- provide advice on aspects such as the scope of the EIA and appropriate mitigation measures.

It is vital that the NCA is consulted early. Failure to do so may lead to problems later in the EIA caused by unforeseen ecological constraints, objection to the planning application as a whole, or failure to identify and mitigate impacts through lack of time.

The *National Rivers Authority* (NRA) performs similar rôles to the NCA when ecological impacts are likely to be associated with hydrological impacts, e.g. in river floodplains (Ch. 10). Other statutory consultees include the Countryside Commission and Her Majesty's Inspectorate of Pollution (HMIP).

Close consultation with the relevant local authorities is highly desirable, since these may have specific policies on nature conservation and on the implementation of relevant national legislation, including that relating to EIA. Other interest groups who should be informed and consulted include the NGOs such as county wildlife trusts and the RSPB.

11.4 Scoping and baseline studies

11.4.1 Scoping

The scope of the ecological assessment will depend on: the nature of the proposed development and of its likely impacts; the size and nature of the impact area; the time and resources available; and the degree to which environmental systems such as soils and waters are being dealt with by other members of the EIA team. Scoping should involve a desk study, together with brief site visits, with the aim of providing a preliminary assessment of:
- the impact area and the ecological constraints within this
- the potential impact factors and associated impacts (§11.5)
- the species, groups, habitats, communities, and environmental variables that should be included in the study
- where, when, and by what methods field surveys should be conducted
- the level of accuracy and precision needed in the information
- possible mitigation measures (§11.7).

All these aspects should be re-assessed as the EIA progresses (see Fig. 1.1), but it is important that they are considered at an early stage when the project design is relatively amenable to modification.

The *impact area* will depend on the location and nature of the project (including secondary developments such as the construction of access roads), and the nature of the surrounding area. In some cases, impacts may be confined to the project site and its immediate surroundings, but factors such as air pollution, water pollution, and changes in the hydrological regime can have much larger impact areas (see Chs 8 & 10). Estimating the impact area is bound to involve some guesswork, and the original estimate may have to be revised in the light of information that emerges during the assessment process.

The most obvious *ecological constraints* are the presence of species and eco-systems of regional, national or international importance, and a primary task will be to identify these. However, (a) the general ecological importance and sensitivity of the impact area should be considered, and (b) the initial inventory should seek to itemize all nature reserves and other habitats that may warrant further investigation, including small sites and linear features such as hedgerows and roadside verges that may be valuable in their own right or as **wildlife corridors**. The technique of *constraint mapping* is outlined in §11.7.2.

11.4.2 Study options

An ecological assessment can usually make good use of existing information. However, much of this will be sketchy and/or out of date, and hence inadequate for the accurate assessment of impacts and the formulation of mitigation measures. Most nature conservation sites will have been surveyed to some extent, but even here the information is unlikely to be comprehensive. Consequently, it is normally essential to undertake new fieldwork.

An important consideration is ease of comparison between survey data, and this is a strong argument in favour of adopting the strategy recommended by NCC (1990) for ecological surveys. This involves three phases in terms of intensity of study and hence detail of information sought.

- *Phase 1 habitat survey* seeks to provide a general description of habitat/ vegetation types within a study area, and to fit these to a standard classification so that they can be readily compared. Unless a good, recent phase 1 habitat survey already exists for the impact area, it will always be necessary to conduct a new survey at this level.
- *Phase 2 survey* seeks to provide further information, usually on selected sites. In phase 1 habitat survey, information on the **species composition** of communities is normally restricted to species lists. Although these are useful, they give no indication of **species importance** in a community. For example, a species recorded on a site may be abundant or merely represented by a few individuals. Phase 2 survey involves the collection of (a) quantitative vegetation data, again with the aim of applying a standard classification such as the *National Vegetation Classification* (NVC) (Rodwell 1991a et seq.) to facilitate comparative evaluation, and (b) abundance data on selected animal species and/or groups. It will be appropriate for some sites in the majority of EIAs.
- *Phase 3 survey* involves more intensive sampling to provide detailed quantitative information on species populations and/or communities. In many EIAs, it will be omitted, or restricted to small areas.

The phases can be carried out in order, e.g. as the need becomes apparent during an EIA, or concurrently, and the latter may be necessitated by time constraints. Procedures for phase 1 are discussed in this chapter, but those appropriate to phases 2 and 3 are left to subsequent chapters.

11.4.3 The phase 1 habitat survey method

The main features of the method are outlined in Table 11.3.

Table 11.3   Features of the NCC Phase 1 Habitat Survey Method.

Aspects		Features
Aim		Rapid survey to provide a record of semi-natural vegetation and wildlife habitat.
Scope	(a)	Primarily designed for survey over large areas of countryside but applicable to specific sites and/or vegetation/habitat types.
	(b)	Largely restricted to vegetation and associated environmental features, e.g. of topography and substratum. Because animals are mobile, fugitive and generally small, large-scale faunal surveys are not considered practicable.
Form		Habitat classification; hierarchical (major types subdivided); types are identified by alphanumeric codes (see Appendix E) and standard colour codes for maps.
Survey methods	(a)	Collection of existing information
	(b)	Field surveys based on Ordnance Survey maps (1:10,000 or 1:25,000). Plots are classified and areas measured, or estimated by the line-intercept method. Habitat types and areas are recorded directly on maps and/or on map record sheets. Target note record sheets are used for additional information, e.g. species lists (plants and animals), notable species (dominant, rare, indicator etc.), vegetation features, topography and substratum (soils, geology, wetness), site history/antiquity, site ownership, management and protection.
	(c)	Aerial photography and satellite imagery. These are not considered adequate substitutes for field survey, but useful for checking and/or supplementing data.
Records retained	(a)	Completed maps (usually 1:10,000 scale); digitised maps if facility and time available, and possible use of GIS (see Appendix D)
	(b)	Map record sheets and target-note record sheets
	(c)	A written report
Evaluation		Not primarily intended for evaluation, but considered adequate for categorizing sites on a 3 point scale: 1. Site of high conservation value; 2. Site of lower priority for conservation; 3. Site of limited wildlife interest.
Limitations	(a)	The maps are not 100% accurate (error estimates should be provided).
	(b)	Small sites may be omitted (< c.0.5 ha with 1:25,000 scale maps, and < c.0.1 ha with 1:10,000 scale maps).
	(c)	Sites are only visited once, so seasonal variations may be missed.
	(d)	Species lists may not be complete, and rarities may have been overlooked.
	(e)	Changes may have occurred since the survey.

The rationale and recommended procedures are set out in the *Handbook for phase 1 habitat survey* (NCC 1990), in which the stated aim is "to provide, relatively rapidly, a record of the semi-natural vegetation and wildlife habitat over large areas of countryside". However, the method can be applied also to specific vegetation types or sites.

The *NCC habitat classification* is summarized in Appendix E. Like other habitat classifications such as that used in the **CORINE biotopes** project, it is really a habitat/vegetation classification because the types are defined using traditional terms that refer to a combination of environmental and vegetation characters, e.g. acid grassland, heathland, and **swamp**. It owes much to Tansley (1939) and, although some categories have been re-classified (e.g. in the case of peatlands such as **fens** and **bogs**), Tansley's two-volume work is still an excellent source of information on British vegetation. Similar information can be found in Polunin & Walters (1985) (which includes continental Europe) and Rieley & Page (1990).

The *line-intercept method* was originally designed to measure the cover (%) of species in a community (see Box 12.2), but can be used to estimate the proportions of habitat/vegetation types as the percentage of land occupied by each in an area. For large-scale surveys, **transects** are superimposed on maps or aerial photographs, e.g. using overlays. The length of each transect occupied by each vegetation type is recorded, and the cumulative length occupied by a type, i.e. its percentage of the total length of the transects, is taken to represent its cover. An equivalent field-survey method for use on sites is discussed in §11.4.6.

11.4.4 Types and sources of existing information

As indicated above, phase 1 habitat survey aims to make maximum use of existing information, and this objective should apply to the ecological component of an EIA. The desk study, initiated at the scoping stage, should be developed into a thorough search of existing information on baseline conditions in the impact area. Suitable types and potential sources of such information are outlined below.

Historical information and maps

A good historical record can provide valuable information about a site, and can enhance its conservation value (§11.6.3). Historical information on sites may be available from the relevant NCA, records and maps in museums, county records, and libraries. Very old maps are sometimes available, but early Ordnance Survey maps are probably the first reliable cartographic records. Current Ordnance Survey maps provide information on local topography and the location of woodland and some other vegetation types. Geological and soil survey maps can indicate the influence of substratum conditions on land use and vegetation in the area (see Ch. 9).

Aerial photographs and satellite images

Aerial photographs and satellite images can help to identify land-use patterns, vegetation types, and features such as hedgerows. Aerial photographs can be purchased from commercial firms. EN holds photographs covering much of England, and those of relevant areas may be available from the regional offices. Other possible sources include MAFF and LAs. Parts of Scotland are covered and the photographs have been interpreted to produce a digitized land-use map (1:25,000) by the Macauley Land Use Research Institute. Landsat satellite images are available for the whole of Britain.

Previous surveys

Large areas of the UK have now been covered by phase 1 habitat surveys. They are often carried out by County Wildlife Trusts, and copies are held by LAs and regional offices of the NCAs. These bodies should hold inventories of local nature-conservation sites and information on previous surveys, including any conducted for EIA, and some LAs now have extensive information on local land use. Information is also held by organizations such as the RSPB, the Woodlands Trust, and local natural history societies. More detailed information may be available for some NNRs, SSSIs and sites used for scientific research. The regional office of the NCA will usually hold or be aware of such information, but a literature search may be desirable.

Information on species

Records are normally included in survey reports, and will be held by the relevant organizations which often have local (e.g. county) recording groups for specific taxonomic groups such as invertebrates. Information on local plant and animal species distributions is held by the Local Biological Records Centre (LBRC). This is usually organized on a county basis (as the County Records Office), often works closely with and advises the LA planning department, and will probably have produced a county atlas of the flora (Appleby 1991). LBRCs also supply species and habitat data to the National Federation for Biological Recording (NFBR) at the national Biological Records Centre (BRC). This holds an extensive database, and produces atlases of species distributions throughout Britain (Harding 1993). Further sources of information on species distributions and habitat requirements are given in Appendix F; information on the conservation status of species (rare, protected, etc.) is reviewed in §11.6.2.

11.4.5 Field surveys

It is essential that ecological fieldwork is conducted, or at least closely supervised, by experienced ecologists who understand the principles and problems of sampling, and are competent in the identification of species. Mis-identification can be a major source of error, and different specialists may be needed for different plant groups, and particularly for different animal groups.

The first task in the field survey of a site must be a visual inspection to gain an overall impression of the habitats and communities present. This can then be compared with existing information and used to assess whether, where and in what detail new survey work is needed. In many cases an experienced ecologist can quickly decide that all or most of a site is not worthy of field survey. For instance, the only features of wildlife interest on an arable or improved-grassland site may be hedges and ditches, and a rapid inspection (at a suitable time of year) should reveal whether these warrant further study.

The timing of fieldwork can be a critical factor. Some groups (such as resident birds, aquatic **macroinvertebrates**, trees, **bryophytes**, and lichens) can usually be sampled throughout the year. However, as indicated in Figure 11.1, accurate data on other groups can be obtained only during short sampling seasons, and the optimal sampling periods for whole communities are even more restricted. Moreover, most communities contain species that are short-lived, or are inconspicuous or absent during part of the normal sampling season, and some have components with distinctly different seasonalities. For example:

- most woodland herb-layer plants grow and flower in March–April before the tree canopy comes into leaf
- some sand dune animals and annual plants should be sampled in April to early May, whereas most of the vegetation is best surveyed later in the summer
- bird surveys in summer will document the presence of residents and summer migrants, but miss winter visitors.

Implications of these seasonal complexities are that:

- little meaningful data can be obtained on the majority of species or communities unless sampling can be undertaken during the late spring and summer
- in many cases, failure to carry out repeat sampling on at least two occasions can lead to serious error
- ideally, the survey should start a full year before the submission date of the EIS
- the accuracy of the ecological assessment can be compromised in EIAs with short completion periods.

11.4.6 Use of the phase 1 habitat survey method

Use of the line-intercept method for habitat/vegetation types discernible on maps and aerial photographs was outlined in §11.4.3. A similar procedure can be applied in field surveys (e.g. of a site or part of a site) using transects across the study area, usually with a common baseline such as a path. For accurate measurements, lines or tapes are laid out, the length of each transect occupied by each habitat/vegetation type is recorded, and the cumulative length occupied by a type, i.e. its percentage of the total length of the transects, is taken to represent its cover. A more approximate and rapid survey can be achieved by simply walking the transects and estimating distances from the numbers of paces.

The precision of the results will be affected by the number of transects sampled. There can be no fixed rule for this except that the larger the sample, and hence the more transects, the better. Statistical considerations prescribe that the spacing between transects should be randomized. However, unless a large number of transects are sampled, their random locations may not yield a repre-

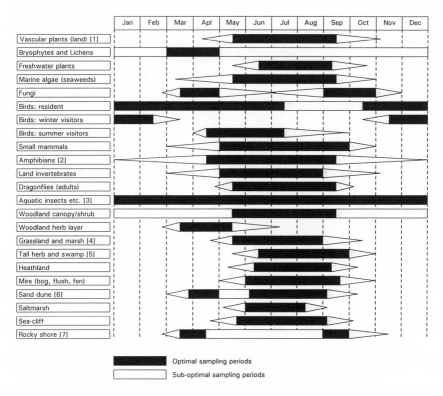

Notes:

[1] There are some exceptions, e.g. see woodland herb layer.

[2] Amphibian surveys are usually conducted during the breeding season. However: (a) this varies between species and in different areas, e.g. the common frog typically spawns in late January in Cornwall but not until early April in parts of the Pennines (Swan & Oldham 1993); (b) juveniles and some adults remain in water during the summer, so late surveys can be valuable.

[3] The optimal sampling period for freshwater macroinvertebrates is often said to be February-March. Pond Action have found that they can be sampled successfully throughout the year, but that because activity periods vary between species, at least two samples should be taken, preferably with one of these in early Spring (see Ch. 13).

[4] Hay meadows should not be sampled after cutting (usually in June).

[5] Short vegetation in tall herb and swamp should be sampled in March-April (as woodland herb layer).

[6] Some sand dune animals and Spring annual plants should be sampled in March-April.

[7] Rocky shores are best sampled during 'spring' tides when the lower shore becomes accessible (see Ch. 14).

Figure 11.1 Possible sampling periods for taxonomic groups and habitats. The periods shown are approximate only and vary somewhat in different locations. For example, growing seasons generally start later in northern Britain than in the south.

209

sentative sample of the study area; and this is more likely to be achieved with regularly spaced transects, provided that care is taken to avoid coincidence between these and periodic features such as the ridge and furrow systems of some old meadows.

Some results obtained on part of Snelsmore Common SSSI, Berkshire, are shown in Table 11.4. If the locations of the transects are known, such results can be used to produce a habitat/vegetation map based on the local Ordnance Survey 1:10,000 sheet.

Table 11.4 Proportions of NCC Phase 1 habitat types on part of Snelsmore Common. (a)Line-intercept values (lengths of habitat types in m) along 10 transects located at 30 m intervals.

Transect 1		2		3		4		5		6		7		8		9		10	
A1.1	15	A1.1	15	A1.1	14	A1.1	20	A1.1	20	A1.3	16	A1.1	17	A1.1	16	A1.1	22	A1.1	9
WW	20	WW	28	WW	26	WW	10	A2	18	A2	15	A1.3	10	A1.3	17	C1	15	A2	21
B5	10	B5	12	B5	12	C1	10	E3.1	37	E3.1	37	E3.1	32	E3.1	28	E3.1	22	B5	8
E3.1	35	E3.1	31	E3.1	23	E3.1	40	D2	11	D2	14	D2	12	A1.3	25	A1.1	30	E3.1	26
D2	27	D2	26	D2	30	D2	27	A1.1	10	A1.1	35	A1.1	13	A3	31	A3	26	A1.3	35
C1	9	WW	22	A1.3	20	A1.1	20	A3	28	D1	150	A3	21	A1.3	38	D1	140	C1	25
WW	18	C1	20	C1	24	C1	26	D1	51			D1	67	D1	45			D1	45
C1	22	D1	100	A3	40	A3	35	A3	44			A1.2	50	A1.1	53			A1.2	30
D1	97			D1	60	D1	66	D1	50			D1	54	D1	30			A1.3	80
Total 253		254		249		254		269		267		276		283		255		279	

(b) Key to habitat codes, total lengths of habitat types, and proportions of these as % of the grand total.

Habitat type		m	%	Habitat type		m	%
A1.1	Broadleaved woodland	309	11.7	B5	Marshy grassland (*Molinia* dominated)	42	1.6
A1.2	Coniferous woodland	80	3.0				
A1.3	Mixed woodland	253	9.6	C1	Tall herb and fern – bracken	151	5.7
A2	Scrub	54	2.0	D1	Dry dwarf shrub heath	945	36
A3	Scattered trees (mainly on heath)	225	8.5	D2	Wet dwarf shrub heath	157	5.9
				E3.1	Valley mire	299	11
WW	Wet woodland (*Betula* & *Molinia*)	124	4.7		Grand total	2639	

The NCC habitat classification (Appendix E) should cover the main habitat/ vegetation types likely to be encountered in an EIA in Britain, but it does not follow that all those found in a survey will match given categories precisely. This is partly because the classification has a fairly limited number of defined types, but the problem applies to all formal classifications because communities intergrade. The natural pattern in an area consists of **community gradients** rather than discrete communities; sharp boundaries are usually man made. Moreover, in terms of species composition and other features, communities intergrade regardless of their relative locations, and are infinitely variable. Consequently, no two examples of a designated type will be precisely the same, and a given sample is likely to include either more than one fairly well defined type or an intermediate community that is part of a gradient.

This problem arose in collecting the data of Table 11.4, with the result that some of the matches with NCC habitat types are approximate only. Moreover, (a) wet woodland dominated by *Betula* species (birches) and *Molinia caerulea* (purple moor grass) was designated as an additional category (WW) because it did not match any NCC habitat type, and (b) the valley mire included several subcommunities of a gradient from poor fen at the lower wet end of the valley to *Betula/Molinia* woodland at the drier end. Such information should be included in the target notes and written report.

11.4.7 Description of the baseline conditions

It is important to present the survey information in a form that is useful to decision-makers, but without undermining the ecological rationale. The structure and content of the description will vary according to the information collected, but the following guidelines will always apply:

- maps are normally essential, and photographs can be informative and attractive
- charts, graphs and tables can be very informative and can allow greater economy of text (which can be structured around them)
- details of survey methods, times of sampling, etc., should be given.
- primary data and species lists are normally best placed as appendices
- it is important to indicate the level of uncertainty attached to the information, e.g. in relation to the accuracy and completeness of data
- as much as possible, and certainly the general conclusions, should be intelligible to the non-expert.

11.4.8 Baseline changes without the development

Changes in ecosystems may occur without the development, and can take place prior to its initiation or during its lifetime. They can result from natural processes such as ecological **succession**, or from impacts generated by other current or proposed developments.

The natural **climax vegetation** in most of Britain is broadleaved forest, and the maintenance of other vegetation types such as grassland and heathland depends on management to prevent succession. In these cases, predictions can be made on the basis of management regimes and evidence such as the present extent of scrub development. Current or proposed management aimed at enhancing the ecological value of sites can also be taken into account. Even woodlands are not static, and changes in the proportions of dominant tree species can be predicted from the current age structure (see Ch. 12).

	Landtake/habitat loss	Habitat fragmentation	Trampling	Soil compaction	Change in water table	Change in stream flows	Change in site water budget	Change in flooding regime	Change in silt deposition	Soil erosion	Soil leaching/acidification	Substrate eutrophication	Water-borne pollution	Air-borne pollution	Dust deposition	Change in microclimate	Species/vegetation removal	Competition by 'alien' species
LINEAR DEVELOPMENTS Road, rail	●	●	○	○	●	●	●	○	○	○	●	○	●	●	●	●	●	○
Underground pipelines	○	○	○	○	○						○							○
Overhead lines	○	●	○	○														○
HEAVY INDUSTRY Chemical, combustion	●	●	○	○	●	●	●	●	●	○	○	●	○	●	●	○	●	○
Other	●	●	○	○	●	●	●	●	●	○	○		○	●	○	○	●	●
LIGHT INDUSTRY AND URBAN Light industry	●	●	○	○	●	●	●	●	●	○	○		○	●	○	○	●	●
Residential	●	●	○	○	●	●	●	●	●	○	○		○	●	○	○	●	●
AMENITY AND TOURISM Land management	●	●	○	○	○	○	○	○	○	○	○	○				○	○	●
Visitor pressure	○	○	●	●							●	○						○
EXTRACTION Open-cast	●	●	○	●	●	●	●	●	●	●	●	●	○	○	●		●	○
Deep mines	●	●	○	○	●	●	●	●	●	●	○	○		●		○		●
LANDFILL Special	●	●	○	○								○	●	○	○			
Domestic	●	●	○	○									●	●		○		
FORESTRY Afforestation	●	●	○	○	●	●	●	●	●	○	○	○	●	○	○		●	●
Deforestation	●	●	○	○	●	●	●	●	●	●	●	●				○	●	●
AGRICULTURE Drainage, irrigation	○	○			●	●	●	●	○	○	●	○					●	
Fertilisers, farm effluents											●	●	●	●	●			
Pesticides & herbicides													●	●			●	
Reclamation	●	●	○	○		○	○		○	○	●	○	●			●	●	
Cultivation/tillage			○	●	○	○	○			○	●	○	●		○		○	○
WATER MANAGEMENT Reservoirs, barriers	●	●	○	○	●	●	●	●	●				○				●	●
River manipulation	○	○	○	○	●	●	●	●	●	●	○			○			○	●

● = often major and/or extensive and/or permanent
○ = usually minor and/or localised and/or temporary

Figure 11.2 Potential ecological impacts associated with various types of development. For clarity, some impacts are described simply as changes in a parameter or system; in reality, subdivision of these into increases and decreases would be necessary. The entries are intended as a rough guide for illustrative purposes; in practice, impacts will vary considerably in relation to specific developments. Information is available on impacts of some development types including cross-country pipelines (DTI 1992), roads (Box & Forbes 1992, Brookes & Hills 1994, DoT 1993, Forbes & Heath 1990), wastewater treatment and disposal facilities (Petts & Eduljee 1994), and opencast mines (Walsh et al. 1991).

Changes in the baseline that may occur through impacts from other developments can be itemized using a matrix like Figure 11.2. In principle, these impacts can be predicted as for a specific project (§11.5), but the magnitude of the task may limit the assessment to general predictions or consideration of a few major developments.

The predictions may have differing implications in relation to the project, e.g. that (a) it would generate cumulative impacts, or (b) its impacts on a site would not be significant because the site's ecological value will decline anyway. Impacts from other developments also have implications in relation to monitoring (§11.8).

212

11.5 Impact prediction

11.5.1 Introduction

The aims of impact prediction are to (a) identify all likely impacts associated with a development, and (b) provide authoritative and, if possible, quantitative predictions about their effects. It is relatively easy to itemize potential impacts, but much more difficult to arrive at precise quantitative predictions of how these will affect species or communities. The two main problems are that precise data on impact factors and/or baseline conditions are often unavailable, and the interplay of biotic and abiotic factors in ecosystems is complex and far from fully understood. The EIA legislation (see Ch. 1 and Appendix A) requires that attention be paid to (a) direct and indirect (secondary, tertiary, etc.) impacts, (b) short-term and long-term impacts, (c) intermittent (including accidental), periodic and permanent impacts, and (d) cumulative impacts (including cumulative effects of different impacts and/or of the development in association with others). However, such distinctions are often not clear cut.

11.5.2 Types of ecological impact associated with developments

As indicated in Figure 11.2, there are several common ecological impacts generated by various types of development. Industrial, urban and road developments are regular sources of many *pollutants* that are harmful to plants and animals. These include gaseous emissions, air-borne particulates, a wide variety of effluents containing toxins such as heavy metals and harmful organic compounds, oil, and de-icing salt (see Chs 8–10 for further details). Pollution from landfill sites can include a variety of leachates depending on the nature of the material deposited. *Accidental pollution* can be a major threat, especially from heavy industry and transport, and developers should be asked to provide a worst-case scenario for this type of impact.

Serious impacts can be generated by activities not covered by the Town and Country Planning Regulations. For example, *commercial forestry* can involve landtake, changes in drainage patterns, and soil erosion and/or deterioration (Goldsmith & Wood 1983, Good 1987), and *agriculture* is responsible for a wide range of cumulative impacts, including habitat and vegetation destruction, hydrological changes associated with drainage or irrigation, soil erosion associated with cultivation, pollution by **biocides**, and **eutrophication** by fertilisers.

Some of the impacts listed in Figure 11.2 are clearly direct, e.g.:

- landtake/habitat destruction, which is a feature of most developments
- habitat fragmentation, which is particularly associated with linear projects such as roads, and progressive landtake by different developments
- vegetation damage through trampling by people, livestock or vehicles.

In other cases the direct impact is on an environmental system such as the soil or local hydrology (Chs 9 & 10), and consequent impacts on the biota are therefore indirect.

Figure 11.2 is a general matrix relating potential impacts with a range of development types. The same impacts can be itemized in a similar matrix for a specific project, but this must itemize impact factors specifically associated with the construction and operational phases of the project. These could include the factors listed in Table 11.5.

Table 11.5 Construction phase and operational phase factors that are likely to have ecological impacts.

Construction phase	Operational phase
Access roads	Buildings, roads etc.
Offsite vehicles and plant	Offsite traffic
Onsite vehicles and plant	Onsite traffic
Drainage	Drainage systems
Earth moving	Sewage systems
Soil storage	Water extraction
Temporary buildings	Use of adjacent area
Storage of materials	Maintenance
Presence of workforce	Workforce/residents
Gaseous and particulate emissions	Gaseous and particulate emissions
Effluents	Effluents
Accidental spillages, etc.	Accidental spillages, etc.

The *temporary and/or localized impacts* indicated in Figure 11.2 are most commonly associated with the construction phase of a development, and the relevant impact factors will cease at the end of this phase. However, restoration of the original communities may be protracted or impossible (see §11.7.4). *Cumulative impacts* (which include many forms of pollution) often involve other developments, and potential effects of a given project must be estimated in this context.

11.5.3 Methods of impact prediction

Some direct impacts such as habitat loss and fragmentation by landtake can be readily measured in terms of the areas of sites affected; and if the distributions of species populations, habitats and communities within the sites are known, the proportionate losses from these can be calculated. However, most impacts are much less amenable to direct measurement, and predictions have to rely on methods such as those outlined below (see also Ch. 15).

Checklists and matrices are discussed in sources such as Wathern (1984, 1988) and Glasson et al. (1994). Matrices (similar to Fig. 11.2) are particularly useful at the scoping stage to identify and cross-reference project features, impact factors, and impacts, including key impacts on which resources may be concentrated. However, they do not assess the nature, magnitude or significance of the impacts.

Flowcharts and networks can be hand-drawn (e.g. Sorensen 1971) or computerized (e.g. Thor et al. 1978). Their function is mainly to identify chains and webs of impacts, and they are therefore a useful approach for identifying knock-on effects from primary impacts. However, this may serve largely to demonstrate the complexity of ecosystems and associated difficulties, including the classification of indirect impacts into secondary, tertiary, etc. For instance, soil compaction can be considered to involve a simple chain: compaction (primary impact), reduced vegetation cover (secondary impact), and reduced animal abundance (tertiary impact). However, soil compaction is often associated with trampling and involves a complex of indirect impacts, including damage to the soil biota and potential soil loss by erosion. The main interactions (some of which are explained in Chs 9 and 10) are illustrated in Figure 11.3.

The changes that will occur in a particular community depend also on a complex of biotic and abiotic factors. For instance, species replacement and consequent changes in the species composition of a community often occur because an impact alters the competitive balance in favour of species that are more tolerant of the new conditions. Two examples of this are:

- eutrophication can markedly influence the vegetation of a bog or poor fen because it favours species that thrive in nutrient-rich conditions
- salt spray from de-icing salt is not normally lethal to road-side herbs, but can lead to the replacement of salt-intolerant species by salt-tolerant species (DoT 1993).

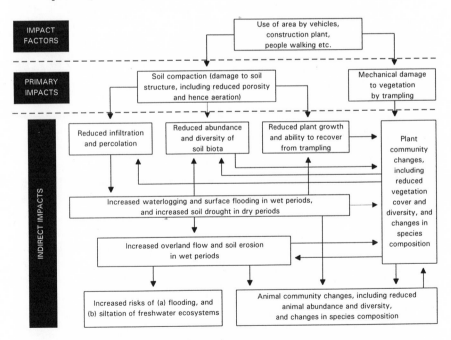

Figure 11.3 Interrelationships between ecological impacts associated with trampling and soil compaction.

Flowcharts and networks provide more information than checklists and matrices, but they cannot quantify the magnitudes of impacts, and in most cases their main rôle will be to justify or explain an assessment made by an experienced ecologist.

Quantitative predictive models have been used as research tools in ecology for many years, with applications in a variety of topics including conservation and wildlife management (Starfield & Bleloch 1986). Principles of computer-aided modelling and simulation are discussed by Spriet & Vansteenkiste (1982), and a recent review of environmental modelling can be found in Melli & Zannetti (1992). The use of simulation modelling in EIA was advocated by Holling (1978) and has been supported on the grounds that it should be able to provide impact predictions that can be rigorously tested, instead of the vague generalizations presented in many EISs (Beanlands & Duinker 1984). However, the application of complex models has limitations, especially in the context of the resource and time constraints in the majority of project-specific EIAs:

- Modelling is time-consuming and expensive.
- Most models cannot readily be used "off the peg" (North & Jeffers 1991); they should be developed and modified in relation to given situations, ideally in conjunction with experimental studies such as field trials, and in consultation with the EIA and project design teams who would need to have an understanding of the modelling processes.
- Accurate predictions require more detailed and reliable data than are often available, especially in relation to impacts on whole ecosystems, and there is little point in using a sophisticated model with inadequate data. For instance, in the case of soil compaction (Fig. 11.3) data would be needed on (a) the magnitude of the impact factors, (b) environmental factors such as slope, soil texture, rainfall patterns, and the local land-phase water regime (see Chs 9 and 10), and (c) the nature (including the sensitivity) of the existing communities.
- Ecosystems are so complex and little understood that even sophisticated models frequently fail to predict ecological surprises such as the effects of acid rain on forests and the persistence of radioactive caesium (^{137}Cs) in grassland following the Chernobyl disaster (North & Jeffers 1991).
- The potential for rigorous testing cannot be realized unless predictions are validated by an effective monitoring programme (see §11.8).

In spite of these problems, the use of computerized models is likely to increase, at least for major projects with a large EIA budget and a long lead time, and may involve developing subjects such as (a) *risk assessment*, including estimation of minimum critical areas and minimum viable populations (Burgman & Akcakaya 1993, Soulé 1987); (b) the study of *metapopulations*, i.e. groups of local species populations that are connected by the movement of individuals (Gilpin & Hanski 1991); and (c) *ecotoxicology* (Forbes & Forbes 1993, Moriarty 1988) including the estimation of *critical loads*, i.e. thresholds at which pollutants harm species or ecosystems (Hornung & Skeffington 1993). Increasing use is also likely to be made of GIS to provide models of spatial relationships, e.g.

for constraint mapping (see §11.7.2).

Information from previous projects can be useful in relation to the types and magnitude of both impact factors and impacts, especially if these were quantified and monitored. However, it must be remembered that (a) technology such as that for pollution control may have advanced since the case in question, and (b) the complexity of ecosystems may invalidate the extrapolation of information from one case to another.

Expert opinion is always needed for the interpretation of data and, in the absence of adequate quantitative data, impact prediction has to rely on judgements based on a knowledge of impact factors and ecological systems. Such predictions may be rather general, but can still be authoritative. Clearly, no one specialist is likely to have sufficient expertise, and the prediction process should involve consultation, e.g. in workshops attended by the EIA and design teams, and perhaps by other experts.

11.6 Impact significance

11.6.1 Introduction

It is important to assess the significance of predicted impacts, and this is usually considered to be a function of the *magnitude* of the impacts and the *sensitivity* and/or *value* of the ecological systems affected (EN 1994). For instance, the impact of a factor such as physical destruction clearly increases with the size of the impacted area, but its significance will vary according to the nature of the ecosystems affected. For example, the significance of impacts on many wetlands will be high because these have high conservation value and are particularly sensitive to various impacts; by contrast, improved grassland is generally more resistant to impacts and has relatively low conservation value. It follows that the EIA should include an assessment of the sensitivity and value of the species, communities, habitats and sites in the impact area; and it is logical to employ the criteria employed in evaluation for nature conservation.

Methods of evaluating species and communities for conservation are discussed in NCC (1989b), Spellerberg (1992), and Usher (1986). An overriding principle normally employed is evaluation of their *local, national and international importance*, but this incorporates a wide range of criteria.

11.6.2 Evaluation of species

The value of species is usually made in terms of:
- *rôles* such as those of dominants or key species
- *amenity value*, e.g. attractiveness

- *conservation status,* which includes protected species and criteria used to assess the need for protection, e.g. IUCN categories such as endangered species (threatened with extinction) and vulnerable species (likely to become threatened), which are in turn largely based on rarity.

Protected species in Europe are listed in Appendices I–III of the Berne Convention, and in the UK in Schedules 1 (birds), 5 (animals other than birds) and 8 (plants) of the Wildlife and Countryside Act 1981. The UK designations are revised periodically, and it is important to ensure that the current lists are consulted. These can be found in DoT (1993) and IEA (1994). Badgers are protected under the Protection of Badgers Act 1991, and seals and deer are afforded some protection under separate legislation. Endangered and vulnerable species are listed in IUCN, British and Irish Red Data Books. These are listed in Appendix F, together with further sources of information on the conservation status of species. A site register for invertebrates with endangered, notable or habitat indicator status is provided by Ball (1986).

Rarity is a multi-facet term (Rabinowitz et al. 1986). In the context of local, national or international areas, a species may show any of the permutations indicated in Table 11.6. The conservation value of rarity clearly differs between the different cases. For instance, there would normally be little point in highlighting the local rarity of a species when it is common elsewhere and the local area is outside its normal range. Species may also have different types of distribution, e.g. locally abundant but restricted to a few specialized habitats, sparse in scattered habitats, or widespread but locally infrequent.

Although the protection of rare species must be given a high priority in EIAs, there is a tendency to highlight their presence at the expense of other considerations. For example, the impression may be given that the main value of a fen lies in the presence of one or two fairly rare species of orchid. This may have the desired effect of impressing the non-expert, but the message that should be emphasized is that the majority of fen species are endangered in Britain because fens are endangered ecosystems.

Table 11.6 Possible permutations of commonness and rarity of a species, community or habitat when considered locally, nationally or internationally. In general, conservation value increases from local through national to international contexts, and if respective scores of 1, 2 or 4 are applied (with *common* scoring 0), ratings representing an order of importance can be attached to the categories as indicated.

Criteria	Categories							
Internationally	common	common	common	common	rare	rare	rare	rare
Nationally	common	common	rare	rare	common	common	rare	rare
Locally	common	rare	common	rare	common	rare	common	rare
Rating	0	1	2	3	4	5	6	7

Notes
1. Records of local rarity usually refer to the relevant county or **vice-county**.
2. National rarity is often assessed (e.g. in Red Data Books) on the basis of the number of 10km grid squares in which a species is found.

218

11.6.3 Evaluation of ecosystems

Ecosystems are usually evaluated in terms of the habitats and communities present in sites. Current sites of international or national importance (Table 11.2) have been evaluated, and the Habitats Directive (92/43/EEC) lists habitat types that should be given priority for SAC designation. Clearly, such sites and habitat types are the most valuable, but this should not detract from evaluation of others in the impact area.

In Britain, the most widely used criteria for evaluating sites are those proposed by Ratcliffe (1977) and adopted by NCC for selection of SSSIs (NCC 1989b). These consist of *six primary criteria* and four *secondary criteria*, as outlined below.

Ratcliffe's primary criteria

Size of a habitat or site is generally considered to be important, especially when the habitat type is fragmented within an area. The ecological implications of habitat size and fragmentation are discussed in several texts, including Newman (1993) and Spellerberg (1992). The three main factors usually cited are: (a) a large site is likely to contain greater *habitat diversity* and **species diversity** than a small site; (b) the viability of some species populations may be threatened on a small isolated site because its area is inadequate for their *range requirements*; and (c) small isolated sites are more vulnerable to *impacts from adjacent areas*. However, the concept is relative; for instance, a 5 ha species-rich meadow would be considered more valuable and sustainable than an equivalent area of woodland or moorland.

Diversity is considered in the general sense of *biodiversity*. Thus, a site containing a variety of habitats and associated species and communities is considered to have relatively high value. Similarly, in a comparative evaluation of similar communities, e.g. two grasslands or woodlands, greater value would generally be assigned to the one with greater **species richness** or *species diversity*. However, caution is needed in the use of these criteria for the reasons outlined below:

- simple species richness, which is commonly quoted, can be misleading because it takes no account of the abundance of species
- species diversity is area dependent, so recorded values normally increase with increasing area sampled, and data from surveys of similar communities are strictly compatible only when obtained from identical sampling areas
- because species are not apparent or, as with migratory animals, not present at some times of the year, diversity values will vary in relation to the time of sampling
- diversity studies are usually restricted to plants or selected groups of animals (e.g. butterflies or birds), and the diversity of the various groups may differ in response to given conditions
- there are valuable ecosystems such as heathlands, reedbeds and bogs, that

219

are intrinsically **species poor**; so the importance of species diversity as a criterion differs widely in different communities, and it should be used only to compare like with like.

Naturalness refers to the degree to which a community or habitat has been modified by human activity. There are now few if any totally natural ecosystems in Europe, and communities can be broadly classified as *semi-natural, improved*, or *artificial* (intensively managed). In the strict sense, estimation of naturalness therefore relies on an assessment of the degree to which the species composition of communities represents the probable native vegetation of the area, which over most of Britain was broadleaved forest. However, the concept can be applied using relatively natural communities, such as those defined in the NVC, as the standard against which to compare the vegetation in question. It may be particularly useful in the evaluation of semi-improved vegetation and that on derelict land in urban areas.

Rarity of a community or habitat type is taken to confer high conservation value. As for species, the assessment of rarity must be made in relation to local, regional and international contexts (Table 11.6); but local rarity may be given more weighting than would be the case for a single species.

Fragility/sensitivity is strictly an indication of need for protection rather than a criterion of value. It refers to ecosystems that are particularly sensitive to environmental change. For example, many wetlands are easily damaged by trampling, decreasing water availability (by direct drainage or disruption of the local hydrology), and changes in water quality (Ch. 10). Because of their sensitivity to previous impacts, fragile ecosystems also tend to be rare.

Typicalness assesses the degree to which a habitat or community is a good example of those that are or have been characteristic of an area, and may be considered particularly important when such types are subject to increasing fragmentation and loss. Like naturalness, it can be assessed by comparison of samples with NVC communities found in the area in question.

Ratcliffe's secondary criteria

Recorded history is considered valuable because a well documented historical record can enhance a site's potential for research, education and use as a model for management. In many cases the communities on these sites have evidently changed little in historical times (often being maintained by traditional management practices such as coppicing or mowing), and some evaluation systems include *antiquity* as an additional criterion. Such sites are usually **species rich**, and they often contain species rarely found elsewhere because of two attributes that contribute to their increasing rarity: poor dispersal ability and dependence on stable conditions. Information on ancient woodlands and the historical importance of habitats can be found in Rackham (1980, 1986).

Position in an ecological/geographical unit refers to the spatial relationship of a site with other habitats in the locality. The value of a site may be enhanced if it is near or adjacent to similar habitats, or is connected to these by linear features such as hedges or river banks that may act as wildlife corridors. It may then be considered to belong to a larger ecological unit.

Potential value normally refers to the potential for enhancement of the existing ecological value of a site by management or natural change, including colonization by species not currently present. Examples in this category are derelict urban sites, disused quarries and gravel pits, and semi-improved grasslands. In principle, however, it could also represent a negative criterion when a site seems likely to deteriorate, e.g. as a result of impacts other than those associated with the project.

Intrinsic (or aesthetic) appeal refers to public perception rather than ecological features, and is thus related to aspects such as visual and landscape criteria that are dealt with in Chapter 6.

Other criteria

Although Ratcliffe's criteria are still widely accepted, alternative or additional approaches have received increasing attention in recent years.

A criterion considered by NCC (1989b) to be "probably a better integrating measure of nature conservation value than any other single factor" is *non-recreatability*, which is usually related to naturalness because "the more natural an ecosystem, the greater the difficulty of re-creating it in original richness and complexity once it has been destroyed" (see §11.7.4).

Ecological evaluation methods have been developed, most of which aim to increase the objectivity of the process and/or increase comparability by the use of numerical values. They include the use of *priority ranking*, *habitat evaluation* (usually for specific animal species), and *composite indices* – which are discussed by Spellerberg (1992). However, some of these methods are subjective or have limited application, and none is universally accepted. **Indicator species** can be used, for example, to assess the flora of ancient woodlands (Peterken 1974, Rackham 1980), but Kirby (1988) argues that it is counterproductive to concentrate on these at the expense of a thorough survey of the whole woodland flora.

Cultural and social criteria (in addition to intrinsic appeal) are often highly valued, especially in the context of urban environments where criteria such as size and naturalness may be considered less important than features of value to the local residents (Goldsmith 1991a, Collis & Tyldesley 1993). These include *amenity value, educational value, closeness and accessibility to residents, presence in an area of deficiency*, and *existing and potential protection*. Assessment of these aspects can be facilitated by public participation, e.g. by means of meetings and/or questionnaires, and can be quantified by variables such as number of visitors, including school parties.

Another approach that has gained ground is the evaluation of ecological resources in terms of *economic value* (e.g. see Costanza 1991, Kula 1992, McNeely 1988, Pearce & Turner 1989). Criteria include notional monetary value, economic benefits, and replacement value (see Spellerberg 1992). These seem likely to find increasing use in strategic environmental assessment, but are

221

arguably impracticable and inappropriate for project-based EIA. A related single criterion is *utility* (Reid & Miller 1989). This can be applied to species or communities in terms of criteria such as genetic, pharmaceutical or economic value, or to the rôle of vegetation in stabilizing ecosystems and the potential socio-economic impacts associated with its destruction. For example, a strong argument in favour of retaining forest cover might be that its removal could seriously affect the livelihoods of people living lower in the catchment through impacts such as flooding and channel siltation. Similarly, coastal ecosystems such as mangrove swamps, salt marshes and sand dunes may be highly valued because they are known to afford protection to the local human populations.

11.6.4 Uncertainty in ecological predictions

The ecological components of EISs have been criticized for lacking scientific rigour and relying on generalizations (e.g. Beanlands & Duinker 1984), and Duinker (1987) argues that in predicting impacts it is better to be "quantitative and wrong than qualitative and untestable". However, given the present understanding of ecosystems, the predictive methods available, and the time and resource constraints in most EIAs, predictions about both the magnitude and significance of ecological impacts must frequently rely on qualitative judgements. Moreover, exact impact predictions (a) cannot be verified unless they also predict the effects of mitigation measures and/or can be tested against concurrent control experiments, and (b) are likely to detract from confidence in the EIA process if frequently proved wrong. Consequently, although predictions should be as exact as possible, a degree of uncertainty should be acceptable, provided that this is clearly stated in the EIS.

11.7 Mitigation and enhancement

11.7.1 Introduction

Mitigation can aim to avoid, minimize, reverse or compensate for an impact. Two primary considerations are (a) the susceptibility of impacts to these solutions, and (b) the likely significance of post-mitigation *residual impacts* (on which the ecological acceptability of the project may depend). Care is needed to ensure that a mitigation measure does not generate another significant impact, and measures proposed for other components such as visual impacts, hydrological impacts or noise should be examined to check that they will not create significant ecological impacts.

Two principles central to EU policy on the environment are (a) preventive action is preferable to remedial measures and (b), where possible, environmental damage should be rectified at source. On these bases the best mitigation measures should involve modifications to the project rather than containment or repair at the receptor sites or compensatory measures such as habitat re-creation. Modifications to the project can include adjustments to siting, site design (including boundaries and layout), and procedures employed during the construction phase, operational phase and (if relevant) decommissioning phase.

11.7.2 Site selection

The location of a project can be a key factor. It is usually determined largely by social, economic and technical criteria, rather than environmental considerations, and the choice of sites is often restricted. However, early consideration of ecological constraints is highly desirable. A technique that can assist in identifying acceptable areas is *constraint mapping* in which all known ecological constraints within a selected area of search are mapped, perhaps with the use of overlays or GIS (see Appendix D). In making such assessments, it is important to remember that ecosystems can be affected by impacts such as pollution or hydrological changes generated from a development sited some distance away.

Linear developments such as pipelines and roads can have a relatively large area of search for site selection and associated avoidance of ecological impacts. However, all the alternative routes of long-distance linear developments are likely to have various impacts, and the best that can be hoped for is selection of the ecologically least damaging option.

11.7.3 Site design and operations

The site design may be modified in various ways, including those listed below:
- incorporation of features to minimize (a) energy use and emission of air pollutants (Ch. 8), and (b) hydrological impacts (including water use, water pollution, and runoff) and associated impacts such as soil erosion (Chs 9 & 10)
- reduction of overall and onsite landtake and habitat/vegetation destruction, e.g. by altering site boundaries and/or layout
- modification of the layout to avoid impacts to sensitive and valuable ecosystems, and to retain existing semi-natural vegetation (including hedgerows) rather than destroying and then attempting to re-create it
- creation of **buffer zones** around the project, project components or protected areas
- location of project components where they will have least impact.

Disturbance during the construction phase can be minimized by practices such as (a) restricting the extent of access roads, temporary buildings and materials stores, and exercising care in the routing/siting of these, and (b) the use

223

of wide tyres on vehicles, and restrictions on size of vehicles and plant. Although mitigation during the operational phase will depend largely on the project design, it may also involve aspects such as maintenance procedures and the management of amenity features.

11.7.4 Restoration and compensation

When destruction or serious damage to habitats and vegetation cannot be avoided, it may be possible to (a) restore general habitat conditions (or create these in a different location), and (b) re-instate or create a community that bears some resemblance to that lost. Various methods of *restoration* and *habitat creation* are discussed in Beeby (1993), Bradshaw & Chadwick (1980), Buckley (1989), and Quickley (1989). Wells (1983) describes methods specific to grassland, including the use of appropriate seed mixes of wild flower and grass species.

Restoration is an option only in relation to temporary impacts, e.g. associated with the construction phase of the project, and can be facilitated by measures such as the storage of soil or turf during this phase. Reinstatement of the original communities is often difficult, especially if the site has been severely impacted; and in practice restoration frequently means simply the reinstatement of vegetation cover (with tree planting as a popular option) for visual or amenity purposes. Moreover, the more natural and complex a destroyed or damaged ecosystem was, the less chance there is of restoring its original richness, i.e. the greater is its non-recreatability (§11.6.3).

In some cases, an ecologically valuable community may become established by natural regeneration, and simply allowing this to occur should be considered as an option. Natural succession to woodland can take place quite rapidly, at least when a soil is present and there are suitable species in the locality. The process can be inhibited by management such as grazing or mowing, or can be accelerated by tree planting, although care is needed to ensure that this does not itself create impacts. All tree planting should be restricted to native species that are tolerant of local climatic and soil conditions. Indeed, where possible, it is considered good practice to use stock grown from seed obtained from local (preferably ancient) woodland.

If permanent destruction of a valuable ecosystem is unavoidable, several *compensatory options* are available including:

- provision of funding for the protection or management of local nature conservation or amenity sites, or the purchase of new sites
- enhancement of onsite areas that will not be built on (although as with restoration, visual or amenity criteria may take precedence over ecological value)
- transplanting (e.g. of seeds or turf) to onsite or offsite locations on similar soil

- habitat creation on sites of low conservation value, e.g. improved grass-land.

All new habitats should be as large as possible, and suitable conservation management should be applied. As with restoration, habitat/community creation has limited potential. Some communities, such as secondary woodland and tall swamp vegetation, can be created quite easily, but these are relatively common anyway, and creation of species rich communities equivalent to those that are now rare is probably impossible.

11.8 Monitoring

The developer is normally responsible for implementing agreed mitigation measures. However, a major failing in the current EIA process in Britain is that there is no statutory requirement to monitor the proficiency with which mitigation is carried out, or to assess its effectiveness. In consequence, monitoring is usually sadly neglected, with the result that there is no guarantee that mitigation was taken seriously, and little or no information on residual impacts. Monitoring information should also provide a valuable cumulative database for use in future EIAs. Therefore, good practice dictates that ecological monitoring, funded by the developer, should be prescribed in the EIS, and conducted throughout the construction, operational and decommissioning phases of the project. Monitoring of ecological impacts can be achieved by the methods used for baseline survey. Photo sites and permanent quadrats can be useful to record changes (Goldsmith 1991b). Account should be taken of trends that may occur without the development (§11.4.8). This is the most difficult aspect of monitoring, and it can be achieved with reasonable certainty only if changes in impacted sites are compared with those in *control/reference sites* (Bisset & Tomlinson 1988).

11.9 Conclusions

Several conclusions can be drawn about the ecological components of current EIAs, and ways in which these could be improved:

- Too many current ecological assessments are based on scanty or outdated information, make few and/or vague impact predictions, fail to make adequate mitigation proposals, and lack any proposals for monitoring.
- Experienced ecologists should be employed to carry out the assessment, and this requires adequate funding.
- Careful scoping should be undertaken to ensure that all relevant aspects are covered, and that resources can be concentrated where most needed.

- The baseline studies should make use of existing information, but this should be supplemented by new field surveys wherever necessary.
- The ecologists and statutory authorities should be consulted at an early stage, and the lead-time for the development should be long enough to allow an adequate survey (including the collection of field data at suitable times of year).
- Species lists should be compiled, but it must be recognized that these are of limited value for the prediction or assessment of ecological impacts; abundance data are needed to indicate the status of species' populations and communities (this aspect is discussed further in Ch. 12).
- To facilitate assessment of the significance of potential impacts, species, habitats and communities should be evaluated.
- Impact predictions and proposed mitigation measures should be as precise and quantitative as possible, although the difficulty of achieving this in the context of complex ecosystems must be recognized, and a degree of uncertainty must be accepted.
- Monitoring to assess residual impacts should be prescribed wherever necessary, and an aim should be to provide good baseline and monitoring data that can contribute to national and international databases for use in future project-specific EIAs and for strategic EIA.
- The information should be presented in a form that can be readily understood by decision-makers without compromising its ecological integrity.

CHAPTER 12

Terrestrial ecology

Peter Morris, David Thurling, Tim Shreeve

12.1 Introduction

The structure of this chapter differs somewhat from the others, because much of the relevant material on ecological assessments for terrestrial **ecosystems** is covered in Chapter 11. In particular, there is no need for a separate section on legislative background. The emphasis is on British terrestrial ecosystems that are susceptible to impacts, methods of collecting fairly detailed baseline information on the flora and fauna by field surveys, and impact prediction and mitigation in relation to the fauna.

12.2 Definitions and concepts

12.2.1 Introduction

The majority of surviving semi-natural terrestrial ecosystems in Europe have high conservation value, both intrinsically and because they are becoming increasingly fragmented and rare (see §11.6). In Britain, many are now in SSSIs, statutory nature reserves, or reserves managed by NGOs, but this does not currently afford adequate protection from development pressures and other factors (see §11.3.1).

Terrestrial ecosystems are usually characterized in terms of habitat/vegetation types, such as those identified in the NCC habitat classification (NCC 1990 and Appendix E). In the following sections, alphanumeric values refer to primary categories of this classification.

12.2.2 Woodland and scrub

The NCC habitat classification makes a simple division of woodland (A1) into *broadleaved* and *coniferous* types. However, the species composition of broadleaved woodlands varies considerably, both regionally and locally, and the *National Vegetation Classification* (NVC) includes many woodland communities and subcommunities (Rodwell 1991a). Similarly, the NCC's "semi-natural" category can be subdivided in relation to woodland histories. For example, Rackham (1986) divides British woodlands into *primary woodlands,* which are thought to be remnants of the primeval forest (often called the *wildwood*) that covered most of the country after the last ice age, *secondary woodlands*, which have regrown on land that was cleared at some time, and *plantations*, which date largely from the 19th century, and many of which are recent.

It is almost impossible to distinguish primary woodland from very old secondary woodland, so for practical purposes distinction is drawn between *ancient woodland* (usually defined as that thought to have been in existence before AD 1600) and *recent woodland* (Peterken 1993). Almost all ancient woodlands have been managed for centuries, e.g. by coppicing. They are relatively **species rich** in both plants and animals, and are dominated by native trees. In most parts of Britain and Ireland, these are broadleaved deciduous species (hardwoods), i.e. oaks, ash, elms, beech, hornbeam and small-leaved lime. In central Scotland the dominant species is Scots pine, and these woodlands are important habitats for some rare animals, e.g. red squirrel, pine marten, capercaille, crested tit, and Scottish crossbill. In much of northern Scotland the natural climax is birch woodland. Classifications of ancient woodlands are given in Rackham (1980) and Peterken (1993), and an inventory of ancient woodland in England and Wales is provided by Spencer & Kirby (1992).

Plantations usually consist of closely set, even-aged stands of trees, many of which are non-native species. They tend to be relatively **species poor**, but should not be discounted, because many have quite high ecological value or the potential to improve – and even commercial conifer plantations provide wildlife habitats, e.g. for birds such as crossbill, chaffinch and siskin. However, recent secondary or planted woodlands are relatively common compared with many semi-natural terrestrial ecosystems, and are unlikely ever to develop the full characteristics of ancient woodland.

Woodland clearance dates from the early Neolithic period (*c.* 4500 BC). Half of England had probably ceased to be wildwood by the early Iron Age (*c.* 500 BC), and by AD 1250, only a small proportion was wooded (Rackham 1986). Further clearance phases included a large loss to agriculture between 1840 and 1870, but the greatest threat to ancient woodland for a thousand years has come from the agricultural and forestry policies and practices since 1945 (Rackham 1986). The Woodland Trust estimate that, since 1947 in England and Wales, 40 per cent of the then remaining broadleaved forest, and almost half of the ancient woodland, has been destroyed.

Scrub is a natural **seral community**. If allowed to spread, it will destroy valuable communities such as species-rich grassland. However, it can contain a variety of shrub species, it provides a habitat for many passerine (perching) birds and some characteristic butterflies such as the green hairstreak, and it will eventually change to woodland. Thus, scrub should not be undervalued, and should certainly not be discounted on the excuse that it may look untidy, e.g. on derelict sites.

12.2.3 Grassland and marsh

Western European grasslands are **anthropogenic climax communities** maintained by grazing and mowing, and they will revert to scrub and then woodland if these controlling factors are removed. Grassland ecology and management are discussed in Duffey et al. (1974). The NVC (Rodwell 1993) contains a detailed classification of grassland plant communities, but they can be broadly divided into the three main categories of the NCC habitat classification, namely acid, neutral and calcareous grasslands.

The majority of *acid grasslands* (B1) are species-poor hill **pasture** on thin, acid soils overlying hard, **oligotrophic** rock, e.g. in the uplands of northern and western Britain, where they frequently intergrade with **moorlands**. They are still quite extensive, although large areas have become "improved" by intensive grazing and/or application of fertilizers. Changing practices, including overgrazing, have also led to the spread of bracken (C1) and gorse scrub (A2).

Neutral grasslands (B2) are mainly found on lowland loams and clays (see Ch. 9). In general, the most valuable are **meadows**, which are species-rich in both plants and invertebrates. A large proportion of Britain's meadowlands has been lost in recent years through development pressures and "improvement", including drainage schemes (WWF estimate that only 3 per cent of those that existed forty years ago still retain any wildlife interest). Wet meadow intergrades with **marsh** (B5), which has also suffered serious losses for similar reasons, including development in river floodplains and associated river engineering projects such as channelization (see Ch. 10).

Calcareous grasslands (B3) occur on chalk or limestone substrates, and are typified by the English downlands. Traditionally, they were maintained by low-intensity sheep-grazing which, together with rabbit-grazing, encouraged a species-rich sward and associated animal communities (see Smith 1980). Again, a large proportion has been lost since the Second World War, largely through conversion to arable farming, or scrub development following the removal of sheep and the effects of myxomatosis on the rabbit population.

12.2.4 Heathland

Like grassland, heathland is an anthropogenic climax community that was created by forest clearance and maintained by grazing, fire, and the use of materials for thatching and fuel – and many UK heathlands probably date from the Bronze Age.

The ecology of heathlands is discussed in Gimmingham (1972, 1975), Packham (1989) and Webb (1986), and British heathlands are classified in Rodwell (1991b). The substratum is always acid and oligotrophic. *Dry dwarf shrub heath* (D1) is dominated by **ericoids**, especially heather (*Calluna vulgaris*), and dwarf gorses. *Wet dwarf shrub heath* (D2) is similar but contains high proportions of **hydrophilous** plant species, and intergrades with **mire**. Lichen/bryophyte heath (D3) is a largely mountain community type, but variants are also found on sandy soils in some lowland areas such as the Brecklands of Cambridgeshire and Suffolk.

Heathlands used to be extensive in both upland and lowland areas of Britain. *Upland heaths* are part of the moorland complex and generally occur on fairly deep peat. Some are maintained as grouse moor, but many have been destroyed by commercial forestry, or have become degraded through practices such as overgrazing and drainage, which encourage the spread of bracken, gorse and grasses. *Lowland heaths* are found on thin peaty soils overlying sands (as in the Dorset heaths), or gravels (as in the Berkshire heaths). Lowland heaths provide habitats for some rare animals, including the Dartford warbler, the woodlark, the nightjar, the sand lizard, the smooth snake, the natterjack toad, and the silver-studded blue butterfly.

The RSPB estimate that half of England's heaths have been lost since 1950 (and 90 per cent of Surrey's heaths have been lost since 1820). Moreover, the remaining heaths are becoming increasingly fragmented. The causes are (a) urban, transport, agricultural, and forestry development, and (b) lack of regular management to prevent invasion by bracken, shrubs and trees (especially birch and pine).

12.2.5 Mire

The ecology of mires is discussed in Moore & Bellamy (1973) and Moore (1984), and British mires are classified in Rodwell (1991b). Mires support specialized plant and animal communities, and are refuges for increasingly rare species, including some birds of prey. **Bogs** (E1) are confined to high-rainfall areas. In particular, *blanket bog* (that forms a continuous cover, except on steep slopes or exposed ridges) develops only where there is a meteorological water surplus for most of the year (see §10.2). The growth of bogs evidently reached a peak around 500 BC, and much of the present moorland is degenerate blanket bog that is no longer growing because of climatic change and upland drainage schemes. Bogs are also easily modified or eroded by factors such as trampling, associated with overgrazing or visitor pressure.

Fens (E3) used to be extensive in the UK. However, most of the *rich fens* have been drained, leaving small isolated remnants that are usually surrounded by intensively cultivated land. *Poor fen valley mires* are quite common in heathland areas where they intergrade with wet dwarf shrub heath. However, these are under pressure from the same factors that are destroying heathland. *Basin mires* are often found over deep depressions that were evidently created by glacial activity. Again, these are usually small isolated "islands" in a sea of agriculture. Because they are **geogenous**, fens are highly susceptible to changes in the quantity or quality of groundwater supply, and can therefore be affected by developments located some distance away (see Ch. 10).

12.2.6 Rock exposures and waste

Natural rock-exposures (I1) include the steep slopes and screes of mountain and coastal areas, and *limestone pavement*. The latter is a rare ecosystem type and was afforded special protection under the Wildlife and Countryside Act 1981. *Artificial exposures* (I2) include spoil heaps and refuse tips, which usually present reclamation problems but can have long-term ecological potential, and disused quarries, which can have considerable ecological value.

12.2.7 Miscellaneous

This category includes a heterogeneous group of ecosystem types, many of which have little ecological/conservation value. However, *boundaries* (J2) such as *hedges* and *roadside verges* can be very valuable as refuges for species in an otherwise intensively cultivated landscape, and perhaps also as **wildlife corridors**. In addition, there is increasing interest in the conservation of urban ecosystems, especially in relation to their amenity value (see §11.6.3). Information on urban ecology and conservation can be found in Bornkamm et al. (1982), Goode (1989), the London Ecology Unit (1989), Shimwell (1983), and Sukopp & Hejný (1990).

12.3 Scoping and baseline studies for the flora

12.3.1 Scoping

The majority of ecological assessments concentrate (often entirely) on the flowering plants, partly because these are relatively conspicuous and easy to identify.

In principle, assessment of the flora should include all **vascular plants, bryophytes**, lichens, algae (including stoneworts) and fungi, although the importance of the groups varies in different communities. It is essential to employ experts who are both competent in the identification of the various groups and have some knowledge of their ecology.

Baseline studies for the flora should include information on both plant species and **plant communities**. As indicated in §11.4.2:

- all ecological assessments should include a phase 1 habitat survey in which major habitat/vegetation types are identified and described
- many EIAs will also require more detailed data on the distribution and abundance of important plant species and/or on plant communities, at least for selected sites – and appropriate studies can be divided into phase 2 and phase 3 survey methods (although the distinction between these is not clear cut).

The *Handbook for phase 1 habitat survey* (NCC 1990) suggests that in the context of nature conservation:

- phase 2 survey methods should aim to describe and classify vegetation more precisely than can be achieved by phase 1 habitat survey, e.g. by application of the NVC, for purposes such as assessment of SSSIs (NCC 1989b)
- phase 3 survey methods should aim to provide detailed information on species and communities for site monitoring and management purposes.

Similar arguments can apply in the context of EIA. Thus:

- application of the NVC (§12.3.3) as phase 2 can be useful for establishing the baseline, evaluating sites in relation to the significance of impacts (§11.6.3), and post-development monitoring
- additional (phase 3) studies (§12.3.4) may be justified if it is thought necessary to present evidence on some aspect not adequately covered by phases 1 and 2. This could be to obtain detailed information on the distribution and abundance of selected species, elucidate a complex community pattern, or determine the relationships between species or communities and one or more critical factors – and may be particularly important in relation to post-development monitoring and management.

In principle, phase 3 surveys may include the study of several community attributes, including (a) vegetation structure in terms of **life-form** composition, **vertical structure**, and **age structure**, (b) **biomass** and **net primary production** (NPP), (c) **species composition**, and (d) **species diversity**. The study of community structure can be useful in woodland surveys (Kirby 1988), but is less commonly applied to other communities. It is sometimes suggested that ecological assessments should include measurements of vegetation biomass and NPP, but acquisition of new data on these is probably too costly in terms of time and resources for the majority of EIAs. Application of the NVC requires data on species composition and, in general, this and species diversity are also the most useful community attributes for phase 3 survey in EIA.

12.3.2 Study options

Some of the required information for phase 2 and phase 3 surveys may be available from the sources listed in §11.4.4 and Appendix F. Further information on the ecology of species can be found in Grime et al. (1989) and the *Biological flora of the British Isles* (see BES 1992).

Five important decisions needed in planning a detailed field survey are the selection of (a) sample size, (b) sampling pattern, (c) species abundance measures, (d) relevant **environmental factors**, and (e) methods of data analysis. Available options are explained and discussed in Gilbertson et al. (1985), Goldsmith et al. (1986), Greig-Smith (1983), Kent & Coker (1992), Kershaw & Looney (1985), Krebs (1989), Mueller-Dombois & Ellenberg (1974) and Shimwell (1971), and details of procedures and concepts outlined below can be found in these texts. In addition, guidance is available on conducting surveys in specific habitats such as grasslands (Smith et al. 1985) and woodlands (Kirby 1988).

Sample size can be critical, because ecological data obtained from small samples is generally unreliable. Many ecological assessments are based on inadequate samples, the data from which are sometimes processed by sophisticated analytical procedures. The results may impress an uninformed reader, but would immediately remind a trained ecologist of the adage "garbage in, garbage out". For instance, there is little chance that a few randomly or subjectively placed portable **quadrats** will provide representative data for a site. There is no completely objective way of determining the minimum requirement, and the number of observations taken is usually a compromise between the need for precision and the cost in terms of labour and time. Quadrat sampling results in a proportion of an area being sampled (depending on the size and number of quadrats used) and a percentage-of-area target is sometimes recommended, e.g. to sample 5 per cent of the study area. However, this is rarely achieved, and Greig-Smith (1983) emphasizes that sample accuracy is more dependent on the number of observations. Consequently, it is generally preferable to use many small quadrats than few large quadrats, regardless of the overall area sampled.

The main *sampling pattern* options are outlined in Box 12.1, and illustrated in Figure 12.1. The *species abundance measures* usually employed in the study of plant species and communities are briefly described in Box 12.2.

It is important to assess the influences of major *environmental factors*, and whether they are likely to change in response to impacts from the development. In many cases these will include biotic factors such as grazing and human influences that cannot be easily quantified, but it is usually possible to identify one or more critical abiotic factors (climatic, **edaphic**, or hydrological) that can be measured. In addition, it may be appropriate to consider existing (baseline) pollution levels. All these parameters are relevant to other components of the EIA (see Chs 8, 9 and 10), and early consultation with the appropriate consultants can be highly beneficial in identifying common goals and preventing duplication of

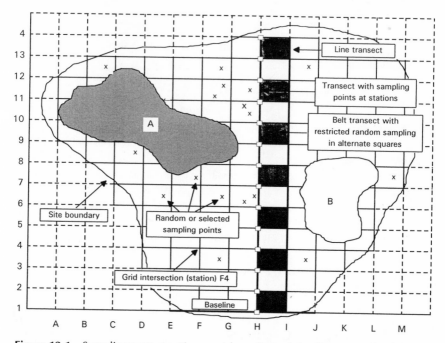

Figure 12.1 Sampling pattern options on a hypothetical site. Squares with solid borders represent a grid system that could be sampled at intersections (stations) or within squares, e.g. by restricted random sampling. To save time, sampling could be limited to a proportion of this, ranging from alternate stations or squares of the whole grid, down to a single transect with sampling points at stations or an interrupted belt transect (as shown). A and B are areas selected for stratified sampling.

effort. However, if it is considered necessary to investigate the relationship between a factor and the community patterns on a site, specific measurements will be needed to obtain compatible data sets, i.e. with equivalent observations. Environmental data are more time-consuming and expensive to collect than floristic data, so it may be necessary to plan a sampling programme in which one environmental measurement can be matched with the mean of several floristic observations.

Appropriate *methods of data analysis* will depend on the nature of the study (see below), but it is important to consider this aspect when planning a sampling programme.

12.3.3 Phase 2 survey

The main phase 2 method recommended by EN is application of the NVC to vegetation samples from the study area(s), preferably collected using the standard NVC sampling procedure. This is the *relevé method* (details in Kent & Coker

234

Box 12.1 Sampling pattern options for plant species and communities

Random sampling (unrestricted)   Observations are made at random points within the site or study area. It is theoretically the most statistically acceptable method, and many ecologists argue that it should be used whenever possible. However, it assumes that species have random distributions, which is rarely if ever true, and it has two serious disadvantages: (a) unless a large sample is taken, the chance observations are unlikely to provide representative data (e.g. they could be clustered in one small area, and (b) the results do not normally permit the detection of community gradients within the sampling area.

Systematic (regular) sampling   Provided that care is taken to avoid regular features (see §11.5.2), this can be the most cost-effective pattern for obtaining representative samples of a site or study area, and of sampling community gradients. Options (illustrated in Fig. 12.1) include:
- *Line transects* which can be useful for reconnaissance or application of the line-intercept method;
- *Interrupted belt transects* which can consist of (a) lines with sampling points (stations) at regular intervals (and intermediate stations if required), or (b) selected (e.g. alternate) squares within which sampling is conducted. They can be particularly useful for sampling community gradients;
- A *grid system* which can cover the site or selected study areas, with observations taken at intersections (stations) or within grid squares. All or a proportion of (e.g. alternate) stations or grid squares may be sampled. A grid can provide the most representative sample of a study area, though precision is affected by the size of the squares which should be as small as practicable.

Restricted random sampling   This is sampling at random locations within areas defined by a systematic pattern, e.g. the squares of a belt transect or grid system. It is considered to be a reasonable compromise between unrestricted random sampling and systematic sampling (Greig-Smith 1983). It does not detect community patterns within the defined areas, e.g. grid squares, which should therefore be small.

Stratified sampling   This involves the selection of study areas, and sampling each of these according to its size. It is particularly useful for large sites where adequate systematic sampling of the whole area is unnecessary or impracticable. The study areas can be defined by features such as fairly discrete vegetation boundaries, land-use boundaries, or tracks, and can be sampled systematically or randomly. Stratified random sampling should not be confused with restricted random sampling.

Selective sampling   This involves the subjective (not random or systematic) selection of sampling points thought to be representative of a community type or to contain a special feature such as a particular species under study. It may be the only practicable method, e.g. when access to sampling points is difficult or when a species only occurs in scattered locations (Goldsmith et al. 1986). However, it is also employed in the *relevé method* in which samples are deliberately obtained within apparently homogeneous vegetation patches that are thought to be representative of a community type. In this case, a requirement is that the area sampled should be large enough to include all the community's species. A popular procedure to ensure this is estimation of *minimal area*, though it is only reliable in truly homogeneous vegetation (see Kent & Coker 1992, Kershaw & Looney 1985). An advantage of the relevé method is that the samples should represent communities amenable to formal classification, e.g. in the NVC; disadvantages are the subjective selection of sampling areas and the deliberate disregard of community gradients.

Box 12.2 Measures of abundance for plant species and communities

Semi-quantitative abundance ratings *Visually estimated measures* made using rating systems such as DAFOR or ACFOR.

DAFOR D = dominant A = abundant F = frequent O = occasional R = rare

ACFOR A = abundant C = common F = frequent O = occasional R = rare

With DAFOR the most abundant species is assumed to be **dominant**; ACFOR may be preferred because it does not require this assumption (although the estimated dominant can be noted). Conversely, it may be difficult to distinguish between C and F in ACFOR. Either system can be supplemented by ratings such as co-dominant and subdominant, and by distributional information such as *widespread* or *local* (e.g. IA = locally abundant). The measures permit rapid survey but are subjective, approximate, and have limited potential for analysis and presentation (although the ratings can be converted to a 5-point numerical scale). Consequently, there is little point in using them if equivalent quantitative data can be obtained.

Number of individuals Numbers are usually counted in **quadrats**, and expressed as *density* (number per unit area) or *population size* (number in site or overall study area). These are appropriate for estimating abundance of selected species, provided that individuals can be readily counted, but generally not applicable in community studies because number of individuals has little meaning when comparing species of widely differing size.

Cover (%) This is the percentage of ground occupied by a perpendicular projection onto it of the aerial parts of a species. Cover can be measured by the following methods:
• *Visual estimation in quadrats* is the most flexible and popular method, but is subjective and prone to observer error, especially in dense or tall vegetation, and tends to under-value species having small scattered individuals. For these reasons, values are usually recorded using incremental cover–abundance scales (see below and Table 12.1).
• *Line-intercept method,* which can be accurate and quite rapid in simple vegetation, but is more difficult and time-consuming in complex vegetation.
• *Point-intercept (point quadrat) method,* which can be the most objective and precise method, but is time consuming and only readily applicable in short vegetation.
Individual observations are quantitative and should be representative; but the time needed with any of the methods can discourage the collection of an adequate overall sample of a study area.

Cover–abundance This aims to avoid underestimating the importance of small species with scattered individuals by using *cover* for species with cover >4% or >5%, but *abundance* in the strict sense (of numbers) for species with cover <4% or <5% (see Table 12.1).

Frequency (%) This is the percentage of observations in a sample that contain the species, i.e. is derived from presence/absence observations (usually in quadrats). Disadvantages are: it is strictly a measure of distribution rather than abundance, and tends to over-represent small species; it fails to discriminate between high density (with many individuals present in quadrats) and density that is just sufficient for at least one individual to be present in a large proportion of quadrats; and frequency values increase with increasing quadrat size, so results from surveys using different-sized quadrats are not strictly comparable. As a general rule, frequency should be measured using small quadrats, and should not be calculated from sets of less than 20 observations. In spite of its limitations, frequency can be the most cost-effective method for obtaining large representative samples in community studies because it is relatively rapid and free from observer error.

1992, Kershaw & Looney 1985, Mueller-Dombois & Ellenberg 1974, Shimwell 1971) the main features of which are:

- samples (relevés) are obtained using *selective sampling* (Box 12.1), with the aim of ensuring that all the observations refer to a given community type
- the species composition of each relevé is determined from observations in quadrats, using a *cover–abundance scale* such as the 10-point Domin scale or the 5-point Braun–Blanquet scale (see Box 12.2 and Table 12.1)
- each community type (vegetation unit) is defined on the basis of its species composition and other diagnostic features, principally its *constancy profile*. This characterizes the unit in terms of the constancy (frequency) with which species occur in relevés, and is unique to the unit. Constancy is expressed on a percentage scale and is divided into classes as shown in Table 12.1. Species with high constancies (IV and V) contribute most to the diagnosis.

Table 12.1   Measures used in the relevé method including versions of the Domin cover–abundance scale and Braun–Blanquet constancy classes used in the NVC (these differ slightly from the original versions).

Domin scale of cover–abundance (%)		Braun–Blanquet scale of cover–abundance (%)		Braun–Blanquet constancy classes and % presence in relevés	
1	< 4	few individuals	+	< 5	few individuals	I	1–20
2	< 4	several individuals	1	< 5	numerous	II	21–40
3	< 4	many individuals	2	5–25		III	41–60
4	4–10		3	26–50		IV	61–80
5	11–25		4	51–75		V	81–100
6	26–33		5	76–100			
7	34–50						
8	51–75						
9	76–90						
10	91–100						

Comparison/matching of individual field-survey samples with NVC communities can be made using simple species lists, but more accurate diagnosis requires quantitative data, preferably Domin scale values. Matching may be achieved by reference to the keys and descriptions in the appropriate volume(s) of *British plant communities* (Rodwell 1991a et seq.), although this can be both time-consuming and difficult. Two alternative computer programs are available that greatly facilitate the process: MATCH (Malloch 1992) and TABLEFIT (Hill 1993). Both programs hold the NVC community data and compare them with a user's data by means of a coefficient that calculates the similarity on a scale of 0–100. MATCH can also provide tables comparing the constancy values and maximum Domin scale values obtained in the samples with those in NVC communities (see Table 12.2).

From a comparative trial of the two packages, Palmer (1992) concluded that both are useful, but should be used with caution and in conjunction with the descriptive text in Rodwell (1991a et seq.), and that indiscriminate use by non-experts could lead to serious misinterpretation. A major reason for the need to apply caution is that no classification can ever hope to be fully comprehensive, because there is an infinite variety of communities and because communities intergrade (§11.4.6). Consequently, there are bound to be samples that cannot be readily allocated to NVC communities, and the temptation to match samples with these at any price should be resisted.

Although the NVC is likely to be the main classificatory system used in the UK, in order to implement the Habitats Directive 92/43/EEC the NCAs will also be required to apply the **CORINE biotope** system, for example to classify SSSIs and NNRs in terms of SACs (see Tables 11.1, 11.2). Many CORINE units are broadly compatible with NVC communities, and TABLEFIT has the facility to cross-reference units in the two systems. However, the CORINE system includes some units that lack vegetation but are important animal habitats (e.g. bare estuarine muds).

Table 12.2 Parts of the results obtained using MATCH to compare samples from Cothill Fen SSSI with NVC mire communities and subcommunities.

(a) Some of the communities and subcommunities most closely matching the samples

Community number	Similarity coefficient	Subcommunity number	Similarity coefficient
M13	65.7	M13c	70.7
M24	46.0	M13b	61.7
M22	44.8	M24b	48.8
M26	40.1	M24a	46.0
M9	37.3	M13a	44.7

(b) Part of the output matching the sample data with NVC community M13: *Schoenus nigricans–Juncus subnodulosus* mire. Every species in the NVC has a number which can be used as an alternative to the species names for data entry. The program provides comparisons between the samples and the NVC community in terms of constancy classes and maximum Domin scale values.

NVC species number	Abbreviated species name	Constancy in samples and (in brackets) in NVC unit	Max. Domin value in samples and (in brackets) in NVC unit
418	Cirs palus	IV (III)	4 (3)
2732	Junc subno	V (V)	9 (7)
876	Moli caeru	V (V)	8 (7)
1046	Pote erect	IV (IV)	6 (4)
1207	Scho nigri	V (V)	8 (8)

12.3.4 Phase 3 survey

As indicated in §12.3.1, most phase 3 surveys will concentrate on detailed studies of the distribution and abundance of one or more selected species, or of selected communities in terms of their species composition and species diversity.

Species diversity is widely employed in the evaluation of sites, but its indiscriminate use can have pitfalls (§11.6.3). Similarly, **species richness** can be used to evaluate hedges in terms of their approximate age. A simple rule of thumb method is that the average number of different tree/shrub species found per 30 m length of hedge is taken to indicate age in 100-year increments, e.g. 5 species = 500 years (Pollard et al. 1979). The estimate is very approximate because of differences in environmental conditions and past management, and it should be backed up by historical evidence where possible. Further information on hedges can be found in Dowdeswell (1987) and Muir & Muir (1987), and a method for surveying and evaluating hedgerows is given in Clements & Tofts (1992).

Time can be saved if the same sampling programme can be used for both phase 2 and phase 3 surveys. However, the relevé method is not usually suitable for phase 3 community surveys, because the subjectively selected sampling areas deliberately avoid community gradients and they may not be representative of the community complex on the site, and in general a systematic, restricted random or stratified sampling pattern is to be preferred (Box 12.2). Conversely, although the data obtained by these sampling patterns can be compared with NVC communities, the matches obtained are relatively poor.

Studies of species composition and species diversity can normally employ a common set of species abundance measurements (Box 12.2), and the data can be processed by *multivariate analysis*, which is the simultaneous analysis of several variables that have the same set of observations, e.g. an abundance value of each species in each observation. Multivariate methods applicable to community studies include *species diversity indices* (such as the Shannon–Wiener index) and procedures to analyze community patterns in terms of species composition using two approaches, classification and ordination (see Kent & Coker 1992, Kershaw & Looney 1985, Krebs 1989).

Classificatory methods seek to identify groups of similar units (e.g. samples or species with similar distributions) that represent communities. These groups can be usefully compared with NVC communities, e.g. using MATCH. However, the communities they represent may be unique (see §11.4.6), so it does not necessarily follow that they can be readily fitted to a formal classification. *Ordination methods* seek to identify community gradients by plotting units (e.g. samples) usually along one to three dimensions, so that similar units lie close together and dissimilar units lie farther apart. Ordination can also be used to analyze community–environment relations. Classificatory and ordination methods can utilize the same data and can provide complementary information.

A variety of computer-assisted multivariate methods are available (see Gauch 1982, Digby & Kempton 1987, Kent & Coker 1992). The computer software can

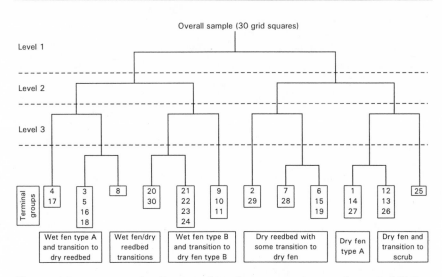

Figure 12.2 TWINSPAN classification of 30 plant community samples at Cothill Fen SSSI, collected by restricted random sampling in grid squares. The dendrogram illustrates the hierarchical nature of the process which progressively sorts the samples (in this case 30) into smaller groups (for mechanism see Kent & Coker 1992) and culminates in the production of terminal groups (in this case 12) of similar samples. Each terminal group can normally be considered to represent a community. The major community types identified by the classification are indicated in the text boxes. Because of community gradients, some samples and groups include transitional communities. This particularly applies to samples 8, 20 and 30, and explains why sample 8 was separated from the other 2 at an early stage (see also Fig. 12.3).

be divided into ecological programs, general multivariate analysis packages, and general statistical packages. Two widely used ecological programs are TWIN-SPAN for classification (Hill 1979a), and DECORANA for ordination (Hill 1979b). These are available separately, or in VESPAN (Malloch 1988) which can be purchased as a joint package with MATCH – with the advantage that a single data set can be entered and used for comparison with NVC communities, classification by TWINSPAN, and ordination by DECORANA. The type of results obtainable with TWINSPAN and DECORANA are illustrated in Figures 12.2 and 12.3. CANOCO (ter Braak 1987, 1988a,b) is similar to DECORANA, but incorporates the facility for analysis of community–environment relations. MVSP (see Appendix B) is an inexpensive multivariate analysis package that contains programs for ordination (including principal components analysis (PCA), and calculation of species diversity indices. Several general statistical packages also contain PCA and programs for classification of units, for instance by cluster analysis.

Selection of methods and packages for use in EIA should depend on accuracy, acceptability and ease of use. TWINSPAN, DECORANA and CANOCO are widely accepted, and are generally considered to be superior to other methods, at least for large data sets. However, "objective tests . . . have never been clearly able to suggest one optimal method" (Kent & Coker 1992), the procedures available in general packages are evidently adequate for small and medium data sets, and advantages of these packages include graphics output (not available in TWINSPAN and DECORANA) and relatively easy data entry.

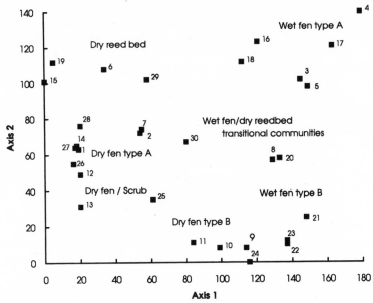

Figure 12.3 DECORANA ordination of the 30 plant community samples collected by restricted random sampling in grid squares at Cothill Fen. The process calculates co-ordinates for each sample in relation to two axes. The distance between samples on the resulting graph reflects their difference in terms of species composition. The communities identified by TWINSPAN can be seen, but the analysis emphasizes the community gradients. These include gradients from wet fen to dry fen or dry reedbed that are related to moisture gradients, and gradients between dry reed bed, dry fen and scrub that are probably influenced by factors such as scrub development, scrub clearance and fire.

12.4 Scoping and baseline studies for the fauna

12.4.1 Scoping

In general, animals are more difficult to sample than plants. Consequently, quantitative animal studies are not attempted in phase 1 surveys (see Table 11.3) and many EISs contain little more than lists of a few species.

It can be argued that the ecological value of a site can be determined largely from a vegetation study, because most resident visiting animals depend on the plants for food and shelter. However, although many animal species are directly

241

associated with specific plants and vegetation types, other factors such as climate affect their distributions. Also, plant populations and communities can be influenced by animals, and it is possible for a development to have no direct impact on vegetation but to affect this indirectly through impacts on the fauna. Finally, there are many rare or endangered animals detectable only by a systematic faunal survey.

The type and quantity of information on the fauna in an EIA are influenced by the nature of the development, the types of habitat likely to be affected, and constraints such as insufficient funding and time. Ideally, information gathered should include an accurate list of the species present, their location/distribution, the nature and extent of their dependence on the site and, if possible, an estimate of their abundance.

The fauna are usually very diverse and often extremely abundant. In Britain there are estimated to be more than 33,000 animal species, of which insects and other invertebrates constitute over 99 per cent. It follows that a faunal survey must be selective. Compilation of animal species lists is a requisite of phase 1 habitat surveys, but, unless a site is already well documented, even this can be a daunting task, and neither existing data nor new surveys are likely to provide a complete list. Studies to obtain data on the abundances and distributions of selected species or groups can be regarded as phase 2 surveys (§11.4.2). Detailed (phase 3) studies of animal populations are likely to be undertaken only in exceptional cases, and studies of whole animal communities are not practicable in EIA.

Unfortunately, ease of study and level of popular interest in particular groups often determine whether they are given adequate consideration. Identification of the animals usually requires the services of several specialists. Some species may be recorded through simple observation, but because of their small size, concealed way of life or mobility, many require special sampling techniques. It is primarily these factors that have led to the terrestrial fauna being largely neglected in many EIAs. Spellerberg & Minshull (1992) examined 45 environmental statements and found that only 47 per cent included information on animals, mostly birds. Because of the large number of amateurs and professionals with an interest in them, and the relative ease with which they can be observed, birds and, to a lesser extent, butterflies often dominate the faunal sections of EISs. However, these constitute less than one per cent of the British fauna and may not necessarily be the best indicators of site quality. Also, concentration on these groups may exclude the study of those species most likely to be adversely affected by a development. Targeting should therefore involve evaluation of species in terms of factors such as their rôles and conservation status (see §11.6.2 and Appendix F).

12.4.2 Study options

It should be possible to obtain some of the information required for a faunal survey from sources given in §11.4.4 and Appendix F, or from organizations interested in particular groups. These include The British Trust for Ornithology (BTO), local bird-watching clubs, the Mammal Society, the Bat Conservation Trust (BCT), and the British Butterfly Conservation Society (addresses in Appendix B). Where existing information is unavailable, inadequate or out of date, it will be necessary to conduct new field work.

Information about animal species present and their numbers can be gathered in a variety of ways. The choice of methods will be influenced by factors such as type of site, nature of the development, time of year, time and manpower available, and the species or groups being studied. In the case of the vertebrates, noting the presence of species on a site is relatively easily undertaken by a competent person, because most species of these groups are easily detected and the numbers of species involved are few. This is in contrast to invertebrates such as insects, of which there may be an abundance of species, many of which are inconspicuous or concealed.

Information on *species abundance*, including population size, can be useful in determining the value of the site for that species and the likely viability of the population in terms of breeding and future numbers. However, most animal species vary in abundance from year to year at any site, and assessment of the likely impact of any change at a particular site is therefore fraught with difficulty. Also, determination of numbers may be difficult, time-consuming and imprecise, so it must be asked whether it is appropriate to enumerate any of the species present for the purpose of an EIA. Where impacts are predicted to be localized, any attention to numeric information should be focused on these areas. Special attention should be paid to sedentary species to avoid impacts in the areas they occupy.

Knowledge of the *distribution* of a species on a site may be important, because many species utilize some areas for feeding and breeding, and have little or no dependence on the rest of the site. On the other hand, a species may be utilizing different parts of the site for different purposes, or may require an extensive area for some activity such as feeding. A full assessment should therefore identify which parts of a site are used for specific activities.

Dependence on a site can be an important consideration, but this may not apply all the time or for the same reason. Some species will be permanently resident, whereas others (such as some birds) may be seasonal visitors, utilizing the site for feeding or breeding and then moving on. Others may be casual or regular visitors, perhaps during migration. If so, they are still dependent on the site, especially when it lies on a regular migration route. It may be possible to study some resident animal species at any time, but some will not be accessible throughout the year, because of seasonal activity patterns (e.g. hibernation) or because they have inaccessible life-cycle stages (e.g. most invertebrates), and migratory species can be sampled only in appropriate seasons. Therefore, effective sampling seasons are restricted for many animal groups, often to the summer months (see Fig. 11.1).

12.4.3 Field survey methods for invertebrates

Sampling insects and other terrestrial invertebrates can be difficult and time-consuming. Even a limited survey will yield many individuals and species. Specimens from a day's intensive sampling may require at least two days for identification, usually involving specialists if identifying to species level.

Techniques need to be carefully selected to meet the aim of a survey which may be to produce:

- a full species list
- a representative list indicative of all the communities present
- a list of notable species
- a classification of invertebrate communities by the use of **indicator species**.

They also need to be appropriate to the study site and its invertebrate fauna. Brooks (1993a) offers guidelines for invertebrate site surveys. There are five basic questions that should be addressed before beginning any survey:

- *Where to sample?* Small areas (micro-habitats) can be utilized by different invertebrate species and even different life stages of the same species. This can range from soil- versus canopy-dwelling species in woodland, through those utilizing different plant species, down to those utilizing different parts of the same plant. Ideally the sampling pattern should reflect the level of micro-habitat diversity, but it must be feasible in terms of both effort and time. A useful starting point is to base sampling locations on previously identified plant community types such as NVC communities (§12.3.3).
- *When to sample?* Species can be easily missed if they are in a concealed phase when the survey is conducted, e.g. soil-dwelling and stem-boring larvae, and the egg phase of many invertebrates. In Britain the optimal sampling period is May–September, when the adult phases of most invertebrates are active (see Fig. 11.1), but, ideally, sampling should be repeated throughout the year. In addition, many invertebrates are active at particular times of day and their activity may be restricted by weather conditions (Brooks 1993a).
- *How many samples?* Single samples are unlikely to be representative, especially where abundance and activity of certain species may change rapidly over short distances and time intervals. Therefore, replication of samples is always desirable.
- *What to sample?* It will be rarely if ever practicable to sample or even list the whole invertebrate fauna for an EIA, so attention should be concen-

trated on notable species, representative species (of the habitat/vegetation type) or indicator species.

- *How to sample?* Sampling techniques for invertebrates can be divided into *observer-dependent methods,* which are carried out by the investigator in the field, and *observer-independent methods,* which employ traps of various types and thus permit sampling in the absence of the investigator (although their efficiency is dependent on appropriate siting). The methods are explained and discussed in Southwood (1978), and main options are outlined in Box 12.3.

12.4.4 Field survey methods for vertebrates

Appropriate methods vary widely for the various vertebrate groups.

Amphibians and reptiles

Survey methods for amphibians are outlined in §13.4.5. Field observations of *reptiles* tend to be fleeting glimpses which, coupled with the variability of most species, can lead to misidentification. Sampling in good weather gives a better chance of finding basking snakes, and slow worms may be found by lifting objects such as corrugated iron sheets. Reptile detection and species identification are also possible from sloughs (shed skins), or the leathery eggs of grass snakes, although these are not often encountered. Reptiles are difficult to survey quantitatively.

Birds

Bibby et al. (1992) is an excellent source of information on bird census techniques. There are many methods, most of which can be very time-consuming and require specialized knowledge. It is usually not possible to count all species equally well and some species are very difficult to detect, let alone count. Even so, an attempt must be made to list those *species present* plus, ideally, their *numbers* and *locations*. General aspects of bird surveys, and features of the four main methods employed, are outlined in Box 12.4. In Britain, the preferred method for estimating numbers and locations of breeding birds is *territory mapping*, mainly because this is the Common Bird Census (CBC) method recommended by the BTO (which means that it is well established and widely used, has standardized rules and instructions, and has been used to produce many published data). However, it can be very time-consuming, is not necessarily accurate (Bibby 1984), and is not suitable for winter visitors and species with weak territorial behaviour, low breeding density, secretive nesting habits or nocturnal activity. Consequently, one of the alternative methods may be preferable for EIA, even though these are less standardized and fewer published studies are available for comparative purposes. Bibby et al. (1992) provides details of methods appropriate for particular species.

Box 12.3   Methods for sampling invertebrates

Observer dependent methods

- *Direct observation and identification*   General searching and recording species found, usually in areas defined as having "interesting" vegetation. Such methods are not normally quantitative, can easily lead to misidentification, only record species that are active when the observations are made, and tend to provide information limited to species that are either common in the study area or extremely conspicuous, e.g. dragonflies (Moore & Corbet 1990).
- *Transect walking*   Involves the observation, identification and enumeration of species on a set route, undertaken within prescribed time and weather conditions; usually restricted to butterflies and day-flying moths (see Pollard 1977, Pollard et al. 1975, Brooks 1993b).
- *Sweep netting*   A hand-held net is swept through vegetation up to 1 m in height. Most invertebrates are swept off the vegetation and can be collected and transferred to a preservative for later identification. The method can be quantitative if a standard number of sweeps are taken, but sweeps in different vegetation types are not directly comparable, because different vegetation types differentially resist the net. Invertebrates occupying the basal parts of vegetation are not normally sampled, and active flying individuals often escape from a net before they are captured. The method is not suitable for thorny or wet vegetation.
- *Swish netting*   Is like sweep netting but is restricted to the air boundary immediately above vegetation. It is especially good at collecting Diptera (flies) and Hymenoptera (bees and wasps).
- *Suction sampling*   A portable vacuum is used to collect invertebrates from the ground layer and/or basal parts of vegetation. It can be an efficient collecting method in dry conditions and where there is little vegetation litter, and can provide quantitative data if a set number of samples are obtained.
- *Soil samples*   Can be taken for identification and enumeration of soil invertebrates. A variety of physical or chemical extraction methods are used to extract the organisms from the soil samples.
- *Beating*   A stout stick is used to knock invertebrates off vegetation onto a sheet from which they are collected. It is usually used to sample the fauna of individual species of tree. With care, the method can be used to obtain quantitative data, but it is not practical in wet conditions.
- *Subsidiary methods*   Most experts have favoured methods for detecting those invertebrate groups with which they are most familiar. Methods include observing flower visitors hand-searching vegetation for plant grazers (especially molluscs), stone turning (especially for beetles, molluscs and millipedes), and investigating litter and dead wood for decomposers.

Observer independent methods

- *Pitfall traps*   A good quantitative method used chiefly for ground-dwelling beetles, which fall into the traps. Pitfalls are placed on a regular grid within selected areas. They usually contain a fluid that kills and preserves the invertebrates caught.
- *Malaise traps*   Flying insects are intercepted by a net and funnelled into a collection vessel by their flight mechanism. The method can collect large numbers of insects, especially Diptera and Hymenoptera (Disney et al. 1982), and is efficient for obtaining quantitative and comparative data. It does not discriminate between insects resident in, or simply flying through, the area.
- *Sticky traps*   Usually consist of a mesh screen on which a viscous oil is applied. They can be used like malaise traps, or placed within vegetation. Fragile invertebrates may become damaged in trying to escape from the trap, and sampled invertebrates have to be removed by a solvent.
- *Water traps*   Rely on the fact that a variety of flying insects (especially flower visitors) are attracted to coloured surfaces. They are simple to use but selective (see Disney et al. 1982, Usher 1990).
- *Light traps*   Night-flying insects are attracted to light sources, especially if these emit ultraviolet wavelengths. They are useful, but (a) they require a power source and are not easily transported, and (b) they may sample species that are not associated with the site but which are flying over it (see Waring 1994 for information on light traps and other types of moth trap).
- *Emergence traps*   Usually consist of a closed mesh canopy placed over vegetation, and a collecting vessel. They are designed to collect most adult flying insects which were in a developmental stage on the vegetation or in the soil when the trap was erected. They can be used quantitatively, because each will have a known basal area, but must be in place for long periods.

Box 12.4   Bird census methods

General aspects   All methods are time consuming, involve extensive site walking, and require expertise, e.g. in the recognition of bird calls/singing. They are affected by (a) seasonal variations, e.g. breeding seasons, seasons in which different species sing, and seasons of migrant visitors, and (b) time of day, e.g. shortly after dawn being generally the best time in relation to bird song. They require repeat sampling, e.g. early morning visits at weekly intervals, variation in routes and directions taken so that particular parts of the site are not always visited at the same time of day, and records of all contacts (sightings or call/song registrations) and the times these are made. Factors affecting census accuracy include:

- birds are easier to find in some habitats than in others, e.g. they are less easily detected in dense scrub and woodland than in open habitats
- some species are less conspicuous and/or noisy than others and therefore may be overlooked
- recognizing different species, counting individuals and determining territories may be difficult where birds occur at high density
- in wet and windy weather birds may be less active and skulk out of sight.

Territory mapping   Can be used to determine densities, locations and territories. It is best conducted during the breeding season when most species are territorial, and territories are often marked by singing, displays and disputes with neighbours. It usually involves walking field boundaries (on agricultural land) or recording within defined plots (e.g. quadrats). It is widely accepted in Britain but requires specialist expertise and is time consuming, which can lead to error, e.g. if small plot sizes are used to save time. Consequently it may have to be restricted to those species which are rare or uncommon on a local, regional or national scale. The territory map should be presented in association with a habitat map, especially if impacts via habitat loss or change are to be predicted.

Line transect method   Involves walking transects of fixed length (e.g. 1km, though this can be sub-divided) and location (arranged randomly or systematically, e.g. as a grid), at a standardised speed, e.g. c.2km per hour. The method can be used to estimate densities, e.g. by counting in relation to a prescribed band each side of the transect, is more rapid than mapping, and can be carried out throughout the year; but its value may be limited in small and/or heterogeneous sites.

Point count method   Involves randomly located points at which observations are made. It can be useful in small sites and heterogeneous vegetation, and permits good analysis of association of bird species (including densities) with given habitat/vegetation types.

Plotless methods   Involve the selection of observation points in relation to specific aims, e.g. at central points in the range of selected species. This can be useful in the study of scarce species that tend to be neglected by the other methods.

Mammals

Some British mammals are protected by legislation. This should be checked before any survey is undertaken (see Corbet & Harris 1991, Morris 1993). For the purposes of an EIA, terrestrial mammals can be divided into three groups: large mammals, small mammals and bats. Each requires different surveying techniques.

Large mammals include hedgehog, mole, rabbits and hares, large rodents (squirrel and dormouse), carnivores (e.g. fox, weasel, stoat, pine marten, badger, otter, polecat and wildcat) and ungulates (e.g. deer). Sources of information on survey methods for some of these are given in Appendix F. For many it is possible to identify species by direct observation or by identification of tracks, spraints (droppings), excavations, feeding damage such as gnawed nuts, and resting locations such as sets, holts or nests (Corbet & Harris 1991). Some can be heard (foxes, deer) and others smelt (foxes). Within the scope of an EIA survey it is rarely possible to determine population sizes, because this requires mark–recapture techniques (see Begon 1979). However, direct counting of communal groups such as badgers can be achieved by recording the number of individuals emerging from a set or den. A major problem with large mammals is that individuals are often wide ranging and they use different parts of a site for different activities. They may also range beyond specific sites, and their use of a site may be seasonal. Consequently, limited periods of recording may miss important species or misrepresent the importance of a site to a species.

Small mammals include the shrews, voles and mice. The presence of shrews may be detected by high-pitched squeaks, and of species such as field voles by turning over objects to expose runs. Hair traps (pieces of sticky tape inside sections of plastic pipe) will capture hairs from small mammals, sometimes making it possible to distinguish shrews, dormice and mice and voles. Identification and enumeration is best performed by live trapping using Longworth traps (see Gurnell & Flowerdew 1990). Small mammals usually have limited home ranges and detection of their presence is generally indicative of breeding in the vicinity of the trapping location.

All British *bats* and their roosts are protected, and any survey technique that may disturb bats in their roost or involves catching and handling bats requires a licence from the appropriate NCA. The Bat Conservation Trust (BCT) must also be contacted to advise on survey methods and appropriate personnel. Bat surveying techniques are specialized and require expertise (see Hutson 1993, NCC 1987). They include the identification of roosts, detection of foraging bats, and identification of flight pathways. Bat roosts may be found above ground in buildings and trees, and below ground in places such as caves, mines and tunnels. Roosts in buildings may be detected by the presence of droppings, insect remains and noise, although a visual search may be necessary to confirm the presence of bats. Observations of flying bats may be made from a fixed point for a specified time or while walking (or sometimes cycling or driving) along a transect.

12.5 Impact prediction

12.5.1 Types of impact

Principles, methods and problems of ecological impact prediction are discussed in §11.5, together with general types of impacts associated with developments.

Little needs to be added in relation to the flora, so this section concentrates on impacts that particularly affect the fauna.

All developments have impacts on fauna within the development area and beyond, although some are less obvious than others. They include:

- habitat loss
- habitat fragmentation and isolation, and associated factors such as move-ment restriction and increased mortality risks, e.g. on roads
- reduction in habitat quality and suitability
- pollution
- disturbance, e.g. during the construction phase of a project or from in-creased human activity.

As indicated in §11.5.3, impact prediction is difficult, because knowledge of the ways in which species and communities respond to impacts is fragmented. In general, the problem increases with the complexity of the impact process (e.g. indirect rather than direct impacts) and of the receptor system. Thus, it is rela-tively easy for single species and it becomes progressively more difficult with species groups and communities. Consequently, professional judgement becomes increasingly important and more subjective as indirect effects and com-munity relationships are considered. One approach with individual species is to seek information on the habitat requirements of species present and then to iden-tify those requirements that will no longer be met following development. How-ever, little is known about the detailed ecological requirements of the majority of animal species (especially invertebrates), and prediction is always hampered by the complex population control mechanisms and biotic interactions that affect any species population.

Habitat loss
As with plants, the impacts of habitat loss by physical destruction are relatively easy to predict, given adequate baseline information. Thus, if data are available on (a) the population size and distribution of a species on a site, and (b) the loca-tion and size of the areas likely to be affected by the project, a predicted loss can be calculated. In addition, if information is available on the local, national and international status of the species and its habitat, then the loss can be estimated proportionately. In the case of British breeding and overwintering birds, pre-dicted losses may be related to critical population sizes, which are given in Andrews & Carter (1993).

Habitat fragmentation and isolation
The main adverse effects of habitat fragmentation are that the resulting remnants become progressively smaller and more isolated from other areas of similar habitat, and both of these factors can have serious ecological implications (see §11.6.3). Small sites may not support viable populations of some species, especially those mobile species having large range requirements. Morris (1993) considers that (a) the main development impacts on mammals are habitat

fragmentation and loss, reduction in habitat quality, and direct mortality caused by road traffic (see below), and (b) habitat fragmentation may be more important than habitat loss, although little is known about the minimum areas required for population survival or even about minimum viable population sizes. Bats may be at risk from fragmentation because being able to move between roosts and feeding sites safely and economically is a major requirement (Hutson 1993). Relatively sedentary, habitat-selective species (e.g. mountain hare, red squirrel) are also particularly vulnerable. In contrast, some species (e.g. badger) appear to benefit from fragmentation, as they are more abundant in countryside in which their woodland habitats are interspersed with fields. Three notable factors causing severance between habitats are roads, fences, and removal of **wildlife corridors**.

Roads act as dispersal barriers causing isolation of populations in adjoining land (Mader 1984), and may also prevent animals from moving between feeding and breeding sites or to new areas if an existing habitat becomes unsuitable. Roads also present the risk of mortality from vehicle strikes, especially when they cross traditional territory or foraging routes (DoT 1993). Groups particularly at risk include: reptiles and amphibians, especially in relation to isolation of amphibian breeding ponds from foraging and hibernation habitats; birds of prey that often hunt along road verges, e.g. barn owls; and mammals with large territories or habitual foraging routes, e.g. otters, deer and badgers. Possibly hundreds of thousands of mammals are killed each year (Arnold 1993, Morris 1993), but information on the true extent of this and, more importantly, the impact on populations, is limited. High mortality may mean few survivors and hence a threat to a small population, and it is probably advisable to assume that populations, especially of species known to be declining in numbers, will be reduced by vehicle strikes if a road is built. However, high mortality may simply reflect a thriving population with individuals dispersing in search of new territories, and evidence from brown hares and hedgehogs indicates that deaths may reflect activity levels (Morris 1993).

Roads can have positive impacts because roadside verges provide habitats for many species. For example, Munguira & Thomas (1992) found that individual sites may support up to 40 per cent of the British butterfly species (with the number increasing in relation to range of breeding habitats and verge width), and butterfly mortality from vehicles was insignificant compared with other causes. Similarly, although kestrel mortality is high on major roads, there is little doubt that motorway verges provide suitable and relatively pesticide-free habitats for this species.

Fences are erected for a variety of reasons, including deliberate exclusion of wild animals (e.g. deer), and as mitigation measures, e.g. as visual or noise barriers, or to protect habitats from disturbance. The degree to which they act as wildlife barriers depends on their location and structure, i.e. solid or open, and with small or large gaps.

Removal of *wildlife corridors* may significantly increase the isolation of animal habitats and populations. Dover (1990) showed that some butterfly species move along hedgelines, possibly because of the shelter from wind and elevated temperature. Some butterflies are very sedentary, so colony expansion is slow and requires habitat continuity (but see Shreeve 1994). Reptiles and amphibians use corridors to move between hibernation and breeding sites. Corridors may also be used by small mammals (e.g. bank vole), and by deer and badgers where there are obstructions such as roads in the area; or they may simply add to the extent of suitable habitat, which is evidently their main value to birds, most of which will readily cross open areas.

Reduction in habitat quality and suitability

In general, animals will suffer if the vegetation on which they depend for food, shelter or habitation is altered, or if there are changes in microclimatic factors such as temperature or humidity (which are often affected by the vegetation). Invertebrates are known to be susceptible to:

- losses of food plants, cover, breeding and hibernating sites
- habitat changes resulting in unsuitable temperature and humidity
- a wide range of pollutants including pesticides.

Fry & Lonsdale (1991) provide information on insect habitat requirements. Birds and mammals are equally habitat dependent, and many need suitable feeding sites (with an adequate food supply) in addition to habitation sites. Increased lighting on roads and at industrial plants can adversely affect insects and birds, which are attracted to lights or suffer disorientation.

Pollution

Any development is bound to generate some pollution, but in many cases it is difficult to isolate the impacts of a specific development from the cumulative effects of others. Similarly, while it is possible to measure pollutant levels in environmental systems such as air, soils and waters (Chs 8, 9, 10), and in plant and animal tissues, it is much more difficult to predict the impacts of these on ecological systems.

Ecotoxicology (the study of pollutants in ecosystems) is a rapidly growing subject (see Beeby 1993, Forbes & Forbes 1993, Moriarty 1988), and is likely to play an increasing role in EIA. However, present knowledge and understanding of responses by species populations and communities to pollutant levels and toxicities is sketchy, and the temptation to draw conclusions from a few simple measurements should be avoided. For example, there is little doubt that **acid deposition** is a major factor causing degradation of terrestrial ecosystems such as forests. However, the assumption that low soil **pH** values or leaf damage on trees are evidence of acid deposition is highly dangerous in the absence of further information such as past data from the study site, data from reference sites, and data on related variables. Similarly, it is known that bioaccumulation of pollutants is a serious problem, especially in top carnivores (carnivores at the top of **food chains**), but the effects of this are rarely well documented. For instance, otters frequently have high levels of polychlorinated biphenyls (PCBs), but the effects on otter populations is not known, and other factors are more likely to be critical. A further problem is that the collection of reliable pollution data is difficult, time-consuming and expensive.

Although the above arguments point to the difficulties of assessing pollution impacts, they should not be taken as an excuse for not estimating, and taking steps to minimize, any potential pollution from a development – rather the reverse. For many terrestrial ecosystems the main threats are from air pollution and consequent soil pollution (Chs 8, 9). The systems most threatened by water pollution (Ch. 10) are the semi-terrestrial wetlands, especially lowland mires and marshes (see also Ch. 13).

Disturbance

Disturbance caused by human activity (e.g. visual, noise, trampling) can be a major factor associated with recreation, traffic, and the construction phase of projects. The risk of impacts often varies seasonally (Fig. 12.4), but vulnerable periods may not be immediately obvious. For example, invertebrates with mobile adult phases (e.g. butterflies and other flying insects) may be most vulnerable when in developmental stages (e.g. eggs, larvae and pupae), because of damage to their foodplants and pupal sites. The majority of invertebrates are also vulnerable to disturbance in winter, because they cannot escape disturbance when dormant. Some groups of molluscs are restricted to (and can be used as indicator species of) long-established semi-natural ecosystems such as ancient woodlands, wetlands and chalk grassland (Willing 1993). These species have poor powers of dispersal, and therefore of recolonization, if a habitat has been disturbed.

Vertebrates that are permanent residents on a site are most sensitive to disturbance during the breeding season, but some species may be also vulnerable in overwintering periods, and the risk to migrant birds is during their visit period. Information on the effects of disturbance to birds is supplied by Hockin et al. (1992) who conclude that, in spite of birds being the most studied animal group, knowledge of disturbance effects is still limited. Disturbance during the breeding season may reduce breeding success, mainly resulting from nest abandonment. It may also reduce population density, as in the case of lapwings and godwits, which showed reduced numbers up to 2 km from roads (van der Zande et al. 1980), and passerines, which have lower numbers in areas of high recreational use or close to car parks (van der Zande & Vos 1984, van der Zande et al. 1984). Recreation and road traffic can affect waterfowl by direct disturbance, and indirectly through alteration of their habitat affecting availability of food and nesting cover. The most vulnerable are species that feed for long periods or in exposed habitats (e.g. wintering geese and ducks), which may be displaced from preferred feeding and roosting areas. Some disturbed habitats have a higher species diversity, mainly of passerines, as a result of more common opportunist species moving in.

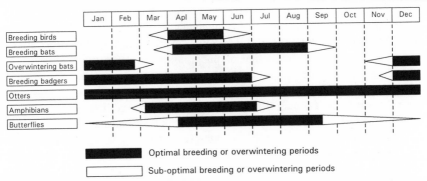

Figure 12.4 Periods during which some animal groups are particularly sensitive to impacts.

Effects of disturbance on bats is discussed by Hutson (1993). All British bats are dependent on buildings or trees for their roost sites and, if a development is likely to affect these, the appropriate NCA must be consulted and allowed time to advise.

12.5.2 Significance of impacts

General principles for estimating the significance of ecological impacts are discussed in §11.6. In the case of individual species, clearly the greatest significance must be attached to protected species and their habitats. However, species that are not protected by law may well be vulnerable at the local and regional level. Animal groups such as birds have sometimes been used as a means of evaluating sites, and scoring systems for amphibians and reptiles have been devised for use in the selection of SSSIs (NCC 1989b). However, in general, evaluation of terrestrial ecosystems for EIA is probably best achieved by means of the criteria discussed in §11.6.3.

12.6 Mitigation

The general principles and methods discussed in §11.7 apply to both the flora and fauna of terrestrial ecosystems, but some aspects of particular relevance to animals may be added. These are listed below.

- Where possible, major construction-phase operations should be avoided during periods when species and groups are particularly vulnerable (Fig. 12.4).
- Fences are often erected around construction sites, sometimes to prevent damage to adjacent wildlife habitats. However, it should be remembered

253

that these involve disturbance and land-take, and may act as barriers to animal movements, especially for mammals, reptiles and amphibians.

- Methods to mitigate against the habitat severance and mortality risk effects of roads include: provision of underpasses for large mammals such as badgers and otters, and small tunnels under roads for amphibians (runs between different locations being protected by fencing).

- Munguira & Thomas (1992) suggest that road verges can be significantly improved for butterflies and burnet moths by reducing the top soil and amount of fertilizer applied, planting native seed mixes and shrubs, creating irregular topography, and making them as wide as possible. However, the success of such measures depends on future management.

- Disturbance, e.g. from traffic and on sites used for amenity purposes, may be reduced by creation of **buffer zones**.

- Deliberate re-introduction of species lost from a site is rarely practicable, and disrupted communities can rarely be re-created. In relation to wildlife corridors, Andrews (1993) stresses that "enhancement of the existing habitat and maintaining the continuity of existing links is more important than establishing new ones". Given the appropriate conditions within a habitat and the presence of nearby populations, most animal species will eventually recolonize a site.

- Enhancement measures have included installation of tubes in artificial banks to encourage kingfisher nesting, and artificial sand martin nests.

12.7 Monitoring

The need for monitoring, and the potential methods and problems involved (discussed in §11.8) apply to both the flora and fauna in all ecological assessments. Animals and annual plants may present particular difficulties, because their populations can vary significantly from year to year, e.g. in response to physical factors such as the weather. This adds to the problem of distinguishing between residual impacts of a development, cumulative impacts from other developments, and natural changes. Consequently it is important to monitor reference sites, to continue monitoring for an extended period, and to employ standardized procedures for repeat sampling throughout the monitoring programme.

CHAPTER 13
Freshwater ecology

Jeremy Biggs, Antony Corfield,
David Walker, Mericia Whitfield,
Penny Williams

13.1 Introduction

This chapter describes the main steps involved in undertaking an EIA for freshwater communities, and highlights measures that can be used to protect aquatic wildlife. The chapter should be read in conjunction with Chapter 10, which provides associated information about physical hydrology and water quality.

13.2 Definitions and concepts

Introduction

The basis of any good EIA is a thorough understanding of the systems that could be impacted. This section aims to highlight the main features of freshwater habitats, including those that are often overlooked. General information about freshwater ecosystems can also be found in standard texts such as Moss (1988), and a good introduction to the natural history of plants and animals in fresh waters is given by Fryer (1991). Although dealing specifically with invertebrates, Kirby (1992a) provides one of the most useful sources of information about the management of freshwater communities, and outlines many conservation principles which are widely applicable.

13.2.2 Rivers and streams

Most natural rivers and streams are characterized by three interconnected zones: the channel, the subterranean hyporheic zone, and the river corridor or flood-plain. Protecting river ecosystems involves maintaining the wide range of habitats associated with all three of these zones.

The *channel* is the most obvious part of a river, and river management is often focused solely on this area. River channels typically support aquatic plants and invertebrates which are restricted to flowing water habitats, but many of the wide variety of species occurring at the channel edges can be found in similar habitats in lakes and ponds (Holmes 1990). Channel habitats, which are often poorly appreciated or ignored, include bare mud and shingle bars, riverside cliffs, and overhung or wooded sections (Shirt 1987, Bratton 1991, Kirby 1992a). River channels are frequently an important fish habitat, so their management can have economic and recreational implications (Crisp 1993).

The *hyporheic zone* is, in essence, a river beneath the river: an area where water flows through the gravels or rocks below the channel base. This zone is often neglected in river assessments. However, in many rivers, particularly those with a gravel bed, it can be a major habitat for invertebrates. Research suggests that the hyporheic zone may be particularly important in temporary streams or headwaters where it can act as a refuge for invertebrates that burrow down into the river bed to escape drought (Ward 1992). In larger rivers, it is also used as a refuge during times of high flows (Marmonier & Dole 1986).

Natural and semi-natural river *floodplains* are rich and complex habitats, which can support a wide range of natural communities including wet **meadow**, wet woodland, **marsh** and **fen**. In Europe, completely natural floodplains are now very uncommon, most of them having been a target for agricultural improvement or urban development. Remaining areas of floodplain, including river-edge strips, will both buffer watercourses from diffuse pollutants and provide habitats for riverside plants and animals. Associated backwaters, abandoned channels and floodplain ponds can add significantly to the variety of open water habitats and may provide flood refuges for fish and other animals.

13.2.3 Ponds

Ponds are small man-made or natural water bodies, less than 2 ha in area (PCG 1993). Many are important wildlife habitats and they can have economic value as fisheries. The best ponds for wildlife generally occur in areas of semi-natural habitat, for example, on heathland, non-intensively managed grasslands or in woodland (Pond Action 1994b). Ponds in these locations often support rich and/ or distinctive communities of plants, invertebrates and amphibians and some contain rare or endangered species (Bratton 1990a, Biggs et al. 1994). High-value ponds typically have abundant stands of submerged and emergent plants, but in some natural habitats, such as woodland, they may be quite bare and muddy, and still support valuable animal communities (Stubbs & Chandler 1978, Biggs et al. 1994). Seasonal ponds are a particularly neglected and internationally threatened habitat (Bratton 1990a), and where they are long established they should always be treated with care.

Ponds in urban or agricultural situations (especially those that receive polluted runoff) or ponds stocked with large numbers of fish or ducks may be of little wildlife interest; however, surveys are always required to confirm this.

13.2.4 Lakes

Like ponds, lakes may be man-made or natural. Lake communities include a wide range of generalist species that can be found in many different freshwater habitats together with more specialized planktonic plants and animals. The composition of these communities broadly reflects the nutrient status of the lakes. Upland lakes tend to have fewer nutrients, and they support distinctive but relatively species-poor communities. These may include rare fish such as the powan (*Coregonus lavaretus*), and rare birds such as divers (*Gavia* spp.) (Fuller 1982). Lowland lakes are often nutrient-rich, and can support a diverse fauna and flora. Unfortunately, many are now culturally **eutrophicated**, reducing their conservation value.

The creation of reservoirs and gravel pits during the 19th and 20th centuries, has substantially increased the number of lakes, particularly in southern Britain. Many of these are now valuable wildlife habitats, supporting uncommon plants and invertebrates, and nationally or internationally important numbers of wildfowl.

13.2.5 Ditches and canals

Ditches and canals are wholly artificial in origin, but in some parts of the country they provide some of the most valuable open-water habitats. Canals support a predominantly still-water community, whereas the flora and fauna of ditches varies from pond-like to stream-like, according to the rate of flow. The best ditch sites can support aquatic plant and invertebrate communities of exceptional nature conservation interest, sometimes including relict fen species. Important ditch systems typically occur in areas with high water tables, particularly in areas of drained **wetland** such as the Somerset Levels. The best sites usually have water that is relatively unpolluted with nutrients, and many are non-intensively managed for the purpose of grazing stock (Newbold et al. 1989).

From a nature conservation point of view, the most important canals are generally those with little or no boat traffic and good water quality. A few of the UK's canals are extremely significant in terms of protecting freshwater plants and animals (Byfield 1990). In intensively managed arable and urban areas, ditches and canals are often degraded by run-off from surrounding land. However, even ditches in intensively managed areas can sometimes support important invertebrate communities (Pond Action 1989, Foster et al. 1989) and they should always be adequately surveyed in EIAs.

13.2.6 Springs and seepages

Despite tending to be small and inconspicuous, springs and seepages may support distinctive invertebrate communities and can be of considerable conservation interest (Kirby 1992a). The best communities usually occur in areas of semi-natural habitat, particularly in areas where seepages and springlines are common.

13.3 Legislative background and interest groups

Specific protection for some rare freshwater plants and animals is given by the Wildlife and Countryside Act, including species listed by the Bonn and Bern Conventions and the Wild Birds Directive (see Ch. 11). Freshwater habitats may also be directly notified as SSSIs and NNRs. However, the proportion of freshwater habitats protected in this way is small; for example, less than 1 per cent of the total length of streams and rivers in England and Wales has been deliberately designated as SSSIs, or indirectly within existing SSSIs (Boon 1991, NRA 1993b).

Freshwater communities are also be indirectly protected by British and EC legislation, which is designed to protect water as a resource. For example, the Fisheries Directive (78/639/EEC) sets water quality standards for waters that have been designated as important fisheries (see **designated waters**), whereas the Environmental Protection Act 1990 establishes regulations for a prescribed list of particularly polluting industrial activities and substances (see Ch. 10).

For developments requiring an EIA, there is a statutory requirement for the planning authority (LA) to consult the Countryside Commission, and the relevant nature conservation authority (NCA). Depending on the development type and possible impacts, the LA may also consult HMIP and the NRA (in England and Wales) or the River Purification Boards (in Scotland) (see Ch. 10 and DoE 1989). For freshwater communities the most important of these consultees are the relevant NCA and the NRA, because both have statutory duties to protect and, where possible, enhance wetland and aquatic wildlife.

Other groups who may have an interest in freshwater ecosystems include a range of quangos, voluntary groups, companies, landowners and individuals (see Turton 1993). Privately and publicly owned water companies are of particular importance, including the water services companies (which supply water and treat sewage), and the statutory water companies (which supply water). The "environmental movement" also has a profound influence on the protection and management of freshwater ecosystems. All the large environmental organizations run campaigns, plan policy and undertake research into fresh waters, and some of the best remaining freshwater habitats are now in the control of such non-governmental organizations as the RSPB or the RSNC. Although mainly concerned with catching fish, anglers and their representative organizations have done much to ensure that fresh waters, especially rivers, have received protection from pollutants.

13.4 Scoping and baseline studies

13.4.1 Scoping

Survey data gathered about freshwater communities for EIAs is mainly intended to answer the question "Is the area/site important or vulnerable?". This typically focuses on assessing wildlife conservation value, usually in terms of **species diversity**, rarity and community type (see Ch. 11). However, it may also involve assessments of fish stocks or gathering data about the abundance, distribution and habitat requirements of uncommon species thought to be present on the site.

In addition, physical, chemical and biological data (such as water depth, nutrient status, tree shade) are needed in order to understand how the freshwater ecosystem functions, and to help predict the possible effects of the development. Many thousands of parameters could be measured for freshwater EIAs, so it is clearly important to choose those most relevant to the study area and the expected impacts (see Ch. 10).

Collection of baseline data needs to be carefully planned so that it can be used both for predicting the impact of a development and as a basis for monitoring post-development impacts. Freshwater systems are dynamic, changing seasonally and often annually, so single surveys made during any one month or year may not be representative. Surveys designed to create a reliable baseline therefore need to be undertaken over more than one season, and ideally over several years (Elliott 1990). Cost and time considerations mean that baseline surveys that fulfil these requirements are not always undertaken. However, inadequate baseline data makes it difficult to (a) describe the state of the system in the absence of a development, (b) predict the effects of a development, or (c) monitor any resulting impacts.

13.4.2 Using existing data

The first stage of most baseline studies is a review of existing data. This can often supply historical data about the site, together with background information such as past and present management regimes. It may also provide data about areas or species of particular value. Organizations that hold information about freshwater habitats and species, or can at least act as a starting point, include the NRA, the NCAs, the Biological Records Centre, county records centres, LAs, county wildlife trusts and the RSNC, the Institute of Freshwater Ecology, the Wildfowl and Wetlands Trust, the British Trust for Ornithology (BTO) and the RSPB. However, with the occasional exception of birds and some wetland plant communities, most data held on freshwater species or habitats will only provide background information. Additional survey work will almost certainly be needed to create adequate baseline descriptions of individual sites.

13.4.3 Gathering new field data

The process of gathering new field data can be divided into three phases, similar to those described in Ch. 11 for ecological surveys in general:

- Phase 1 survey involves brief site visits to *identify freshwater habitats likely to be affected by the development*. Care should be taken to ensure that small and apparently insignificant freshwater habitats have been included, such as springs, seepages and temporary pools. It is also essential that *all* the areas of freshwater habitat that may be impacted by a development are identified. For example, it is often forgotten that impacts on a river running through a development site are likely to extend some distance down stream. Similarly, some developments may bring major changes to regional groundwater levels, affecting freshwater habitats many kilometres away.
- Phase 2 survey provides the baseline for most freshwater EIAs. It mainly involves description and assessment of the *conservation value of freshwater plants and animals* in habitats identified by Phase 1 survey (see §13.4.4 below).
- Phase 3 survey involves more intensive sampling, often to provide *detailed quantitative information about species/communities*. This will mainly be appropriate to larger EIAs or where impacts are likely to be highly disruptive to the environment. However, collection of quantitative data can also be relevant where there are species or communities of particular value that require detailed investigation.

The basic range of survey methods used to undertake Phase 2 surveys are outlined in §13.4.4 below. Information about how to undertake Phase 3 surveys can be gathered from the references given in §13.4.5 and Chapter 12 as appropriate. Techniques commonly used to describe and model the physical and chemical environment, and which are relevant to water resources issues, are discussed in Chapter 10.

13.4.4 Phase 2 survey methods for freshwater ecosystems

There are two ways of determining whether a site supports valuable plant and animal communities: by recording the species present, and by measuring environmental factors believed to indicate that a valuable community may occur. The advantage of *recording species* (e.g. aquatic invertebrate species, mammals, birds, wetland plants, fish) is that it allows direct assessments of the conservation value of the communities, based on the community type, species diversity and rarity. The main disadvantage is that it can be time-consuming and expensive to collect adequate species lists.

Environmental indicators – usually water quality, but sometimes habitat diversity (e.g. NRA 1992b) – are used as indirect measures of the quality of the whole system. They are based on assumptions, e.g. if water quality is high, it is assumed that the conservation value of that ecosystem will also be high, and that special protection measures should be implemented to protect the habitat. The advantages of this approach are:

- it is usually relatively quick and inexpensive, because less time is spent identifying species, and
- it treats the system as a whole, rather than perhaps focusing attention on a few rare species.

The main problems with this approach are: (a) habitats with low water quality may be written off as having little conservation value when valuable species may be present, and (b) it gives no indication about the species and communities present, so it is difficult to target protective or preventative mitigation measures.

In general, surveys that identify species are much more valuable for EIA than indicator data, simply because they provide more information and therefore facilitate more informed decision-making. The basis for species surveys is described below. Some of the main methods of assessing the quality of fresh water using environmental indicators are outlined in Chapter 10.

13.4.5 Species and community-based surveys to assess conservation value

Surveys of species are primarily used to assess the conservation value of communities. In fresh waters, this is normally done by looking at: (a) the range and *number of species* recorded and (b) the presence (and sometimes abundance) of uncommon or *rare species* in all waterbodies likely to be affected by the development. In addition, plant, invertebrate (and sometimes fish) communities are often assessed in terms of *community type,* because this gives more information about their value and about the physicochemical conditions that influence them.

Surveys to assess community type must use the standard survey methods established for each waterbody type (see below). In all cases, the main aim of surveys is to provide data from which to assess whether any species or communities are of local, regional, national, or international significance. In addition, if the development could physically damage a waterbody, it will be important to accurately *locate and map* important habitats or species.

As a basic minimum, most good freshwater EIAs look at the conservation value of the *wetland plant* and *aquatic invertebrate* communities. Amphibians are also routinely surveyed in still waters, and wetland birds in larger lakes, gravel-pit lakes and rivers. Fish are rarely of importance in conservation terms, since most species are widespread and common, but surveys may be relevant because fish have an economic importance and can be very significant in ecosystem function. In addition, a few species are very rare. Surveys of small mammals, bats and semi-terrestrial invertebrates are still relatively uncommon in the freshwater component of an EIA, but may increasingly be included in larger projects.

Wetland plants

Ideally, aquatic and wetland plant surveys should be undertaken in two seasons: in early summer, to catch early flowering species (such as water crowfoots) and in late summer/early autumn, when the majority of wetland plant species are identifiable. The definition of what constitutes an aquatic or wetland plant differs considerably, so use of a standard checklist is essential to allow comparisons of species numbers to be made between sites. A checklist of vascular wetland plants is available from Pond Action (see Appendix B).

A variety of standard methods have been developed for surveying wetland plants in different waterbody types, and it is important that the appropriate methods are followed accurately, so that plant community types can be reliably identified and compared. Surveys of plant communities in lakes should follow Palmer (1989) or Palmer et al. (1992), keeping aquatic and marginal species separate. Surveys of vegetation in ponds (waterbodies up to 2ha) should follow National Pond Survey techniques (Pond Action 1994a). River communities should be surveyed using the methods of Holmes (1983) and need to include **bryophytes**. Ditch surveys should follow Alcock & Palmer (1985). The National Vegetation Classification (NVC) for aquatic communities is still in preparation, but when published could be used as an alternative for classifying communities. However, it will rely heavily on much of the data outlined above. Standard methods for mapping the *location* of plant stands and other habitat features have been developed for rivers (NRA 1992b) and these can be applied to other waterbodies.

A summary list, giving the national and regional status of all vascular wetland plants, is available from Pond Action. Other sources of information on the habitats distribution and conservation status of wetland plant species are listed in Appendix F.

Invertebrates

Invertebrates make up a large proportion of the diversity of most freshwater habitats and often contribute significantly to the conservation value of a site. A definitive list of all aquatic invertebrate species is available from the DOE. The main problem with invertebrate survey work is that it is only possible to record a (usually small) proportion of the species present in any waterbody at any one time. To overcome this, standard survey techniques have been developed for some habitats, which enable sites to be compared.

Ideally invertebrate surveys should allow (a) assessment of the value of the whole site, and (b) assessment of the value of smaller components of that site. For example, in a survey of a gravel-pit lake, samples from different habitats (such as mud or submerged plants) should be kept separate, so that the value of these habitats can be assessed. Samples should be replicated to assess whether perceived differences between habitats are likely to be real.

Surveys of invertebrates in fresh waters have dealt mainly with aquatic species sampled using pond nets. However, there many different dredges, grabs and traps for collecting aquatic invertebrates, which may be appropriate under certain circumstances (see Elliott & Tullett 1983, Southwood 1978). Aquatic invertebrate surveys should be carried out in a minimum of two seasons, and this should include an early spring visit to record mayfly and caddis-fly fauna. Identification to species level should be undertaken with invertebrate groups for which keys are available. These groups are listed in Appendix F, together with references to the relevant keys. Given the emphasis placed on uncommon and rare species, it is essential that identifications of these are confirmed by experts.

Standardized aquatic invertebrate survey methods have been developed for ponds (Pond Action 1994a) and rely on a three-minute hand-net sample from all significant habitats within the pond. River invertebrate communities are most frequently surveyed using methods described by Wright et al. (1984; also a three-minute hand-net method), and this method forms the basis of much routine NRA river invertebrate monitoring. Additional surveys are often conducted for adult dragonflies, either as they emerge or on the wing (Moore & Corbet 1990, Brooks 1993a).

Semi-aquatic and terrestrial invertebrates associated with the margins of waterbodies are an important part of the fauna of many freshwater habitats, and EIA surveys are increasingly likely to include these. The most commonly used survey techniques are described in Chapter 12, and they include pitfall traps for ground dwellers, various aerial traps for flying species (Malaise traps, water traps) and sweep netting (e.g. for flies).

Appendix F lists sources of information on the habitats, distribution and conservation status of aquatic invertebrates.

Amphibians

Most amphibian surveys are undertaken at their breeding sites (usually ponds) during the breeding season (see Fig. 11.1). However, searches of surrounding terrestrial areas, and of ponds during the summer, can provide additional data. The main methods used are:

- pond netting for individuals in the water
- "torching" at night
- bottle trapping, and
- searches for frog, toad and newt egg masses during the breeding season.

The eggs of smooth, and palmate newts cannot be distinguished in the field. However, eggs of great crested newts *can* be identified, and egg searches have proved to be a quick and effective means of locating this species.

Using a combination of survey methods generally proves more effective than one alone e.g. searches for egg masses in spring, followed by summer netting for juveniles and any remaining adults. The methods listed above really provide information only about which amphibian *species* are present, and they cannot give more than a crude idea of population numbers (numbers collected are often "out" by a factor of 10 or more). Collecting accurate population data for any species can be time-consuming and expensive. The most frequently used method involves ring-fencing the breeding site to intercept animals moving to or from the surrounding area. More detail about amphibian survey methods can be found in Swan & Oldham (1989).

Two of Britain's six native species are protected by law (the great crested newt, *Triturus cristatus*, and natterjack toad, *Bufo calamita*), so it is an offence to net or handle them without a licence. It is also illegal to damage their habitat, including the terrestrial areas around the breeding site that they inhabit for most of the year. Great crested newts are relatively widespread in England, and so frequently feature in EIAs.

Data on amphibian distribution is given by Swan and Oldham (1993). The *Herpetofauna worker's guide* (HCI 1993) includes information on amphibian conservation in the UK, and provides a wide range of contacts.

Fish

The majority of Britain's native fish species are common and widespread, but they are of great economic and recreational importance. Most work is therefore concerned with estimating biomass, age structure or species diversity in order to provide data for habitat management and restoration schemes. There are two important exceptions to this: (a) the biology of fish is well enough known for protection of distinctive local races of some species to be considered in conservation planning (e.g. some of the races of brown trout, *Salmo trutta*), and (b) specific surveys and measures may be undertaken to maintain populations of the few rare species (e.g. vendace, *Coregonus albula* and powan, *Coregonus lavaretus*).

There is an extensive literature describing methods for fisheries surveys (e.g. Bagenal 1978). The main techniques are netting, electro-fishing and direct observations of breeding habitats (mainly for salmonids). Radio-tagging and counting at fish passes can be an important part of more sophisticated studies. Recently there has been considerable interest in studies of young (and hence small) fish, which can be surveyed more cheaply than adults. Observations of anglers may give additional information, but are rarely used in scientific assessment of fish populations. Specific measures for the conservation of rare species are reviewed in Maitland & Lyle (1993).

Birds

Birds are one of the few groups of which enough is known about total population sizes to make counts of individual species an important part of EIA. In freshwater habitats, the main areas of concern are likely to be whether there are (a) populations of overwintering waterfowl or waders that exceed the criteria for national or international importance (1 per cent of population), or (b) populations of threatened breeding species. Critical population sizes for British breeding and overwintering birds are given in Andrews & Carter (1993), and significant overwintering populations are likely to be already monitored by the Wildfowl and Wetlands Trust. If population estimates of wetland birds using rivers and canals are required, it may be appropriate to follow the methods of the Birds of Waterways Survey, organized by the BTO. Bibby et al. (1992) summarize a wide range of appropriate techniques for studying bird species and populations, with discussion of the problems of survey design (see also §12.4.4). Sources of information on the distribution and conservation status of birds in Britain are given in Appendix F.

Mammals

Otters (*Lutra lutra*) and small mammals, including bats, may require specific attention in freshwater EIAs. Survey methods for otters are described in NRA (1993a), which should also be used as a starting point for further information about otter distribution patterns and habitat requirements. Small mammals including water shrew (*Neomys fodiens*), harvest mouse (*Micromys minutus*) and water vole (*Arvicola terrestris*) may be directly associated with the margins of waterbodies. Survey techniques for small mammals are considered in Chapter 12, and a summary is given in Gurnell & Flowerdew (1990). Daubenton's bat (*Myotis daubentoni*) is largely reliant on waterbodies, and many other bat species use water opportunistically. All bats are specially protected and a licence is required to handle them. Survey methods are outlined in §12.4.4 and further references are listed in Appendix F.

13.4.6 Analysis of baseline data

Much of the baseline analysis undertaken during EIA involves interpreting information contained in species lists collected using the standard methods outlined above. Interpreting these data commonly involves (a) an assessment of the **species richness** of different sites/samples, noting the presence of uncommon and/or **indicator species**; (b) an assessment of the characteristics of the community type (e.g. uncommon, degraded, or rich in species); and (c) an assessment of the abundance of important species. Each of these three criteria is then described in terms of its local, regional, national or international significance. The literature sources given above and in Appendix F can provide a basis for this assessment, but additional local or regional information about the occurrence of important species/habitats may also be appropriate at this stage (see §13.4.2).

Assessments of waterbody conservation value can sometimes be aided by *numerically* scoring sites according to the richness or the rarity of species they support. Numerical methods are particularly useful where they facilitate comparisons between different sites or waterbodies, or where they are used to combine data about different aspects of conservation value. However, simple numerical scores or indices can also be misleading and may lead to inappropriate conclusions; hence, they should never be used in isolation. Pond Action has developed a numerical score system for assessing the value of wetland plant and aquatic **macroinvertebrate** communities (Pond Action 1994b).

Finally, it is clearly important to interpret wildlife data in the light of other environmental information gathered about the site (e.g. water depth, sediment type, habitat diversity). This is the basis for understanding freshwater communities and it is an essential part of predicting likely impacts from the development.

13.5 Impact prediction

13.5.1 Introduction

Predicting future conditions with the development depends on (a) understanding how the development will change the water environment; and (b) predicting what effect these changes will have upon the existing freshwater flora and fauna. The inevitable difficulty with such predictions is that the detailed effects of most impacts on most freshwater species are poorly known. EIAs therefore generally make broad predictions based on well known tenets, backed up by more detailed work where information is available (see §13.5.7).

Freshwater communities are changed or damaged by five broad types of physical and chemical impact:
- changes in surrounding land use
- changes in water depth
- changes to the flow regime
- reduction of habitat size/complexity
- pollution.

These are discussed below. More detailed information about the physicochemical impacts which might be expected from different development types is given in Chapter 10.

13.5.2 Changes in surrounding land use

Fresh waters are intimately linked to the land surrounding them. Changing the land use around a waterbody may, therefore, considerably influence its wildlife. There are three main interactions.

- Waterbodies surrounded by different habitat types, such as woodlands, heathlands or meadows, frequently also have distinctive aquatic and water's-edge communities. Species in these habitats (e.g. hover-flies, water beetles and amphibians) often need not only water but also specific terrestrial conditions during their life-cycle. Changing either the surrounding habitat type or its management may therefore eliminate these species and change the aquatic community as a whole (e.g. Fry & Lonsdale 1991).
- Fresh waters are sinks for liquids and solids that drain in from the surrounding land. As a result, the quality of the freshwater environment usually reflects the quality of the surrounds. Where waterbodies are bordered by relatively non-intensive land use (e.g. moorland, deciduous woodland), they are often buffered from pollutants. If land use becomes more intensive, then the volume of pollutants such as silt, nutrients, organic wastes and **biocide** sprays draining into fresh waters can rapidly increase (e.g. Ormerod et al. 1993).
- Activities such as land drainage or urban development (see Ch. 10) may also change water-table levels and river flow regimes over large areas, inducing profound alterations in freshwater ecosystems, sometimes at points far away.

13.5.3 Changes in water depth

It is often believed that *stable* water levels are critical for freshwater ecosystems. In fact, some water-level fluctuation is natural in all open water habitats. Typical fluctuations in still-water bodies during the year are often in the order of 0.3 m– 0.5 m, whereas, in rivers and streams, flood levels frequently rise several metres. In addition, not all aquatic habitats are permanently wet. Temporary ponds and streams can be persistent features of natural landscapes and, particularly where they are long established, may contain specialized animals and plants of high conservation interest (Bratton 1991, Foster & Eyre 1992). Damage to freshwater systems occurs when changes go beyond what is normal for the system, particularly if those changes are permanent or erratic. Most community damage is caused by lowering water levels, but deepening can be equally devastating, especially where traditionally temporary water habitats are made permanent.

13.5.4 Changes to the flow regime

Water flow is one of the main factors that distinguish freshwater ecosystems, and is critical to both their functioning and ecology. The effects of flow go far beyond increases in water velocity, because this is inevitably accompanied by changes in other parameters, such as dissolved oxygen concentration, nutrient fluxes and sediment type and volume. In general, still-water habitats such as lakes and ponds accumulate sediments, including organics, heavy metals and such adsorbed nutrients as phosphates, so the ecological effect of any pollutant inputs may intensify with time. Running waters generally export materials (including pollutants) down stream. This means that the effect of pollutants on river and stream communities are more likely to be transitory at any one point, but may also affect far larger areas before becoming degraded or diluted.

Changing the flow rate of a waterbody, be it an increase or decrease, can indirectly damage communities adapted to the prevailing flow, and may irreversibly modify the physical and biological environment. For example, linking a stream and pond to "stop the pond stagnating" will, among other impacts, change the pond's sediment characteristics and increase its infill rate. It will also introduce stream plants and animals which may considerably alter the original pond community. Similarly, creating impervious urban surfaces often causes rapid spatey storm runoff into streams (see Ch. 10). This has been shown to physically modify stream widths, flood regimes and bottom substrates (Walesh 1989) with considerable knock-on effects for the channel and floodplain communities.

13.5.5 Reduction of habitat size and complexity

For any freshwater species there will be a critical minimum area of habitat needed to maintain a viable population. For example, the minimum area of bare gravel substrate needed by nesting little ringed plovers is around 0.2 ha (Andrews & Kinsman 1990). If a development threatens to destroy part of a water body or habitat type, it is important to assess, as far as possible, whether there is sufficient area remaining to retain important species and/or communities on the site.

Habitat damage can be of particular concern for the conservation of aquatic invertebrates (Kirby 1992a). Most invertebrates are very small and many rely on small-scale habitat features. Thus, small areas of bottom sediments such as sand, mud, submerged wood or different complexes of plants may each support very different invertebrate communities (Harper et al. 1992). Habitat damage that destroys any one of these areas may eliminate an entire community. This occurs most obviously where part of a habitat is completely lost to development, but considerable damage can be caused by simplifying habitats. For example, river straightening often gives more uniform flow regimes, water depths, and bank profiles, all of which reduce habitat complexity and associated plant and invertebrate diversity (Brookes 1988).

Many species also live in (or need) more than one part of an aquatic habitat at different stages of their life-cycle. For example, fish fry benefit from backwaters or bays in which they can develop (Schiemer & Waidbacher 1992), and nymphs of pond skaters are known to inhabit plant stands of different density as they grow. Removing any one of these habitats, or blocking the migration route between them, can therefore eliminate those species from the population.

13.5.6 Pollution

The chemistry of natural waters varies considerably, and this contributes to the diversity of species and habitats found in fresh waters. Damage to a freshwater community is most likely to occur when human activities modify the chemical environment to the extent that it strays outside the natural range for that waterbody, e.g. by increasing phosphate to levels that are higher than would be experienced at that time of day/year in the unmodified system. *Physical changes* can also result in chemical impacts to freshwater systems. For example, destroying wet woodland adjacent to a river may, amongst other effects, reduce **denitrification** in the organic soils, and hence increase nitrate inputs to the river channel. The main causes of water pollution, and the effects of pollutants on water quality, are discussed in Chapter 10. Some of the most significant ecological impacts of pollution are outlined in Box 13.1.

13.5.7 Predicting impacts using more detailed information

General principles, such as those outlined above, can give a broad understanding of the impacts likely from a development. However, where *species-level* information is available for plants and animals in a freshwater habitat, then it is usually possible to make much more specific predictions about the impacts on key species or communities. For example, a development that increases the inputs of silt to a stream could cause generally detrimental effects, including the swamping of existing gravel habitats used by aquatic invertebrates and spawning fish. A more detailed desk study of the habitat requirements of the stream's species would then aim to identify if any species were likely to be particularly vulnerable to this damage. Such predictions almost always rely on interpreting existing data from the literature (e.g. establishing habitat preferences, pollution sensitivity, breeding times). Starting points for this information include Grime et al. (1988) for plants, the standard identification keys for invertebrates, Maitland & Campbell (1992) for fish, Swan & Oldham (1993) for amphibians, Cramp & Simmons (1977 et seq.) for birds, and Corbet & Harris (1991) for mammals.

Box 13.1   The effect of pollutants on freshwater flora and fauna

Organic matter (and associated deoxygenation of water)   Decomposition of organic matter by micro-organisms in water can lead to partial or total deoxygenation. Low oxygen levels are particularly damaging to river communities where fish and specialized river invertebrates require consistently high oxygen levels. Lakes may suffer if the bottom waters become highly deoxygenated, so (a) causing loss of bottom-dwelling biota, and (b) exacerbating the effects of eutrophication by promoting the release of phosphorus from the **sediments**. Small still-water bodies have highly variable oxygen levels, and support communities adapted to these conditions. These may also be harmed by organic pollution, but the effects are less easy to detect.

Thermal pollution   Temperatures above the normal range (e.g. near to power stations) can: (a) exacerbate the effects of organic matter pollution by increasing decay rates and hence deoxygenation; (b) disrupt the life cycle timing of native species, (c) cause stress to cold-blooded animals by causing above-normal rates of respiration, and (d) favour an abnormal biota (for the area) which may include exotic and sometimes nuisance species and allow these to become acclimatized to the region. The first three factors may make heated water, particularly that polluted with organic matter, uninhabitable by many forms of aquatic life.

Acidification   Low pH, and the toxic materials (particularly aluminium) brought into solution at lower pHs, are directly injurious to many freshwater animals, and have diverse biological effects including changes in the abundance, biomass and diversity of invertebrates, plants, fish and amphibians. Effects are greatest in the uplands and areas of lowland heaths and bogs – wherever there is high rainfall and/or a prevalence of acidic soils. The most significant commercial effect is the decline of fish populations.

Eutrophication   This results in enhanced plant growth, followed by the decay of this plant material and subsequent oxygen depletion of the water. In many cases enrichment by nitrogen and phosphorus is accompanied by addition of organic wastes which exacerbate the deoxygenation problem. Anoxic (oxygen free) conditions may lead to the release of phosphate from bottom sediments, and hence allow the eutrophication process to become self-perpetuating, irrespective of future decreases in phosphorus inputs to the system. It is primarily a threat to communities of standing waters but may affect slow-flowing and highly regulated rivers. It can result in considerable loss of conservation value, including loss of species diversity, and community changes which allow dominance by a few tolerant plants (particularly algae). Fish community composition may alter, with an initial increase in fish biomass, often followed by high mortality during periods when decay of high standing crops of plants causes deoxygenation. Phosphorus is generally considered to be the principal eutrophicating agent in temperate freshwaters. However, once a system is rich in phosphorus, nitrates may become the main factor controlling aquatic **productivity**, and these conditions tend to promote the growth of undesirable, nitrogen-fixing blue-green algae.

Silt (fine organic and/or inorganic particles)
Silt may contain organic matter and hence have a high biochemical oxygen demand (so causing deoxygenation). It often carries nutrients (particularly phosphate) and adsorbed pollutants (such as micro-organics). Abrasive effects in rivers may kill fish through gill damage. Reduction of light by **suspended sediment** inhibits the growth of **macrophytes**, and may favour algal dominance. On settling, silt may (a) destroy salmon spawning beds, and the habitats of specialized plants and bottom-dwelling invertebrates, and (b) progressively seal waterbodies, isolating them from groundwater flows, and so potentially enhancing eutrophication

Metals, micro-organics and other harmful chemicals   Polluting effects are diverse. Toxicity data are available for very few chemicals, but many have detrimental effects on aquatic life, some at levels considerably below those which cause immediate death, e.g. sublethal levels may enhance the risk of disease, affect reproductive capacity, or alter community structure due to changes in competitive or foraging behaviour. Some toxins may accumulate up the food chain or have synergistic effects (where the effect of a pollutant cocktail is greater than the sum of the parts).

Oils and grease   These deoxygenate water as they are broken down. Oil can blanket the water surface, inhibiting oxygen diffusion, and may directly coat plants and animals, causing injury and death. Oils contain many carcinogens, such as polycyclic aromatics and phenols, which mix with water and poison aquatic life.

13.6 Impact significance

13.6.1 Introduction

Assessing impact significance is one of the most difficult parts of freshwater EIAs. The main problem is that, even where it is obvious that the developments will cause a change in freshwater communities, it can be very difficult to decide whether this change constitutes significant damage. The decision inevitably requires an element of judgement and should therefore draw heavily on expert opinion wherever possible.

The best practice procedure is to look at each potential impact in turn (e.g. nutrient inputs, changes in water level) and assess whether any changes to the ecosystem are likely to lie within the *natural range* of perturbation for that system. This should be considered in both the short and long term, and for all phases of the development, including construction and, if relevant, the decommissioning or restoration phase. As a rule, where impacts are likely to be within the normal range of the system, then the predicted level of change is likely to be acceptable. Where the normal range is exceeded, the significance of impacts will generally depend on (a) how much the system is likely to change from the norm, (b) how *sensitive* it is to damage, and (c) its ecological/conservation value (see also Ch. 11).

13.6.2 Areas of significant impacts

Generally, for any impact, the greater its relative *magnitude*, the greater is its potential for damage. Thus, a large area of physical damage, or the discharge of a large volume of polluting effluent, is likely to affect more individuals and species than a smaller impacted area or volume of effluent. Similarly, an impact is likely to have considerably more effect on a small habitat, such as a stream or pond, than it would on a larger one such as a lake or river, where detrimental impacts would tend to be diluted or ameliorated.

Short-term impacts, such as temporary changes in water depth or turbidity, generally have less significant impacts on aquatic communities than *long-term changes*. This may sometimes also be true of temporary habitat damage, but it depends very much on the habitat. River species, for example, are often very mobile as a result of downstream drift and upstream migration, and if a varied structure can be re-created or redeveloped there may be only temporary damage to aquatic communities. However, survey work is always necessary to ensure that (a) the habitats to be modified do not support unique species or unrecreatable features, and (b) they have good recolonization potential.

For many species and communities there may be *critical limits*; within these limits there is little change, but outside them considerable damage may ensue. Defining where these critical limits are for individual species is generally difficult (but see Hockin et al. 1992 for a review of the effects of disturbance on birds, which are relatively well understood). As stated above, the safest guide is to ensure that impacts lie within the range of what is already natural for the system.

Contamination of freshwater systems by non-mobile elements (such as phosphorus) or non-biodegradable toxins, must be minimized, since their effect may be permanent and effectively *irreversible*. Such impacts are most likely to have significant effects on systems that act as cumulative sinks for sediment, especially lakes and ponds. Although it may be possible to partially repair systems damaged in this way, practical experience suggests that the high cost of treatment makes such work extremely unlikely.

Finally, adhering to *statutory requirements* (such as maintaining water quality standards and protecting scheduled species) is clearly important, since contravening these requirements has been judged to result in a danger to health or in environmental or economic damage. Contravention may, in addition, lead to prosecution.

13.6.3 Sensitivity and ecological value of the community

The likelihood of any impact causing damage to a freshwater community depends on the sensitivity of the existing community. For example, studies have shown that, where an outfall discharges urban run-off into a stream with good water quality and a sensitive community, it can significantly damage that community. The same discharges into an area of low water quality with degraded communities may cause little further deterioration.

Highly sensitive and valuable communities are widely regarded by ecologists as sacrosanct. They include communities that are nationally uncommon and not re-creatable or recoverable, except perhaps in the very long term. Freshwater habitat examples include old pond complexes, such as the pingo systems of Norfolk (PCG 1993), or the relatively unmodified River Spey in Scotland (Gibbons 1993). Communities will also be highly sensitive if they support rare (Red Data Book) species, which require protection in their own right. Care should be taken, however, not to adopt a species-centred approach and just translocate that species or manage for it alone. Often the presence of one rare species may indicate that the habitat and community as a whole supports a suite of threatened species within groups that may not even have been included in the baseline survey.

Sensitive and valuable communities are those which have been comparatively little degraded by human activity, and may be of local or regional value in terms of biodiversity. Often they are characterized by clean water, and they may be buffered by areas of semi-natural habitats. Generally, the higher their quality the lower will be their recovery potential, and the more care needs to be taken to prevent damage.

Less sensitive and valuable communities are usually those that have already been extensively modified or degraded by humans and, having lost most of their sensitive species of plants and animals, they may be relatively insensitive to further degradation. New habitats are also typically regarded as being of low sensitivity, both because they are usually re-creatable and because they may still be changing rapidly. However, new habitats often have areas of bare substrate and sometimes good water quality, so that even these may support uncommon, specialist species (Pond Action 1991).

It is always important to state in EIAs, that the presence of lower-sensitivity communities does not mean that mitigation of development impacts is therefore unnecessary. All ecosystems are interconnected, so that severe degradation in one area can affect others. This can be seen in rivers where new pollution events may have little effect in a reach already badly damaged, but may progressively damage downstream sections and eventually the marine environment. Without mitigation there is also the danger of subtle, persistent environmental damage, even to degraded habitats. Impoverishment of local and regional biodiversity in this way is detrimental in its own right and can reduce the potential for future recovery or restoration of the area. Finally, not all impacts are likely to be damaging to freshwater communities; some may be neutral or beneficial. However, benefits are most likely where change is made to already highly degraded or newly created habitats. Changes to semi-natural freshwater ecosystems are much more likely to cause damage.

13.7 Mitigation

13.7.1 Introduction

The aim of mitigation is to minimize and, where possible, eliminate the damaging effects of developments. As freshwater ecosystems are almost always profoundly influenced by adjacent terrestrial ecosystems, mitigation frequently involves maintaining these areas too. Mitigation is, of course, vital where it prevents damage to sensitive or highly sensitive communities. However, as stated above, it can be important even for damaged communities. The best practice for deciding upon mitigation techniques is to use the precautionary principle: if it is uncertain whether ecosystem damage will occur, then mitigation measures should be implemented. It also has to be recognized that full mitigation is not always possible. If residual detrimental effects are likely to be *severe* and there is the potential for permanent damage to *highly sensitive* or unrecreatable communities, then replanning or relocating the development should be the initial recommendation (Canter et al. 1991).

Examples of the broad range of mitigation measures used to minimize or prevent adverse impacts on freshwater ecosystems are given in Table 13.1. In practice, deciding which method(s) will be most effective in any situation may

Table 13.1 Mitigation measures relating to impacts on freshwater ecosystems.

Impact	Mitigation
Silt	Collect in **silt traps**/sedimentation basins, **french drains**, or siltation ponds/lagoons (proper maintenance is essential). Use vegetated **buffer zones** (30–100m), including wetlands, as filters. Phase major construction periods to avoid wet seasons. Minimize disturbance during construction or operation, e.g. reduce bare areas by zoning, and install fences to protect adjacent areas. Avoid vegetation removal where possible. Revegetate bare areas rapidly, using temporary cover crops or mulches where necessary. Minimize dredging disturbance and erosion associated with bare areas, e.g. grade spoil heaps, and cover with tarpaulins.
Organic matter, nutrients and salt	Reduce silt inputs as above (phosphorus is primarily carried with silt). Reduce nitrogen inputs by minimizing soil disturbance. Encourage formation of wet organic soils (i.e. create wetlands, extensive waterbody margin habitats, and wet woodland) to promote **denitrification**. In sewage treatment use nutrient stripping, tertiary treatments, separation of effluents, storm overflows.
Heavy metals, micro-organics, and other toxic materials	Treat, recycle, re-use, industrial pollutants at source (using effective methods), and monitor effluents. Reduce silt inputs (as above). Reedbeds have been used to remove or manage many industrial and domestic effluents but proper design and maintenance is essential. Buffer zones (30–100m) may give a reprieve from diffuse pollutants but can lead to long-term accumulation and/or release if these are not degradable. Minimize surface drainage from polluted areas. Reduce use where possible (e.g. application of biocides). Test any fill material placed in surface waters during the construction phase. Ensure isolation of waste-storage facilities and landfill sites from surface and groundwater bodies, and monitor for leachates. Discharge vehicle and other wash waters to foul sewers rather than surface water drains. Guard against accidental pollution by: effective safety systems (with back-up systems), security systems to protect against fire or vandalism where potential pollutants are stored or delivered; contingency plans; education/training of personnel.
Oils	Install silt/petrol traps (gully traps) in road or parking areas and ensure a proper maintenance regime. Bund or dike around temporary fuel/oil storage areas during construction phase. Vegetated buffer zones may help to retain petroleum products while they degrade. Guard against accidental pollution (as above).
Acidification	Strip flue gases at power stations. Control the extent of afforestation and modify forestry practices. Avoid use of liming to increase the pH of waterbodies because of adverse effects on other parts of the ecosystem.
Heat	Recirculate and/or use to heat local buildings
Changes in flow regime and aquifer recharge	Appropriate procedures are outlined in Table 10.6. It is difficult to reproduce natural flow conditions using physical structures; so, where possible, mimic natural processes by encouraging infiltration, e.g. use vegetated areas, porous artificial surfaces, or detention basins.
River engineering	Wherever possible, maintain natural river depths, river course, bottom sediments, active floodplain and flood regimes. Use natural materials/methods for bank protection and stabilization, e.g. bankside trees and vegetation fringes instead of concrete or steel reinforcements. Restrict damage during maintenance (e.g. dredging) by working from one bank and retaining vegetated areas.
Physical loss or other damage	Destruction or degradation of long-established semi-natural habitats should be strongly resisted, since current technology and understanding are not sufficient to allow full recreation. Whenever possible, the development should be relocated or rezoned. For other habitats, loss or damage may sometimes be minimized by retaining key areas and protecting specific species' migration routes, shelter and refuge zones. Consider habitat creation or enhancement to ameliorate for loss.
Disturbance of wildlife	Maintain or create buffer zones. During construction phases: restrict working and access areas; limit extent of temporary roads and service areas; physically protect key wildlife areas, including habitat and food areas; and plan activities around critical periods (e.g. breeding and nesting). During operational phases, prevent access to valuable wildlife areas, and provide other focuses to reduce public pressure.

be difficult, since there has been little monitoring of the long-term effectiveness of different mitigation techniques. As a general rule, all point-source pollutants (such as industrial effluents) should be dealt with at source, and ideally removed, recycled, or reused. More diffuse pollution (such as urban runoff) is best dealt with by a combination of measures, possibly including biological techniques (see below). Some mitigation measures will require periodic maintenance, such as changing the filters on oil traps or dredging siltation lagoons. Arrangements therefore need to be set in place to ensure that this is routinely undertaken. Finally it is important to consider whether mitigation measures may themselves have an adverse impact on freshwater habitats. For example, the creation of an on-stream lake to create a landscape feature or intercept sediment may have downstream implications for the flow regime, nutrient cycling and the ecology of the stream.

13.7.2 Pollution control using biological methods

Recently, there has been considerable interest in the use of biological systems to reduce pollutant impacts through biological mitigation techniques, particularly where these are for pollutants that are difficult to mitigate by other means. For instance, structures such as french drains and siltation lagoons are of little use for removing dissolved inorganic materials from motorway runoff (DoT 1993). The main biological systems employed are:

- buffer zones (e.g. along rivers), which usually aim to intercept diffuse pollutants such as agricultural chemicals
- natural and artificial wetlands (especially reedbeds or ponds), which are designed mainly to intercept point-source pollutants (e.g. from roads or in urban situations).

These techniques are generally a positive development, with benefits that go beyond prevention of pollution by, for example, assisting in flood control from storm runoff and creating new wetland habitats. However, they are not a panacea and, although initially effective, they may create residual long-term problems. For example, a reedbed used to intercept road runoff may effectively deal with degradable pollutants such as nitrates, but have only a limited capacity to store non-degradable pollutants such as phosphates and heavy metals. Thus, it may eventually become saturated and then export most of the non-degradable pollutants subsequently received. The nature conservation benefits of biological mitigation techniques have also often been overemphasized and used to justify other forms of damage. For example, the benefits of a creating a pond to intercept urban runoff are exaggerated; ponds filled with polluted water and sediments are unlikely to be good wildlife habitats. Further information on the use and value of biological methods for pollution control can be found in Brookes & Hills (1994), Cooper & Findlater (1991), Hammer (1989), Hardman et al. (1993) and Muscutt et al. (1991a,b).

13.7.3 Habitat restoration, creation, and enhancement

Where a development project is likely to lead to environmental loss or residual damage, compensation may be suggested in the form of the recreation or enhancement of existing habitats or the creation of new habitats. The benefits and perils of compensation and habitat creation are not always obvious, not least because monitoring projects to assess the success of existing schemes are few and far between. Projects that aim to re-create existing waterbodies and wetlands accurately are unlikely to be straightforward. This is particularly true of semi-natural habitats with complexes of open water and wet ground. Little is known about re-creating the complex hydrological and ecological relationships in such habitats, and to date no attempts to do so have fully succeeded. Thus, justifying damage to high-value communities on the basis that the same communities can be re-created elsewhere is not backed up by evidence and should be avoided.

Where damage to existing systems is unavoidable, every effort should be made to compensate for their loss. For example, where a project necessitates rerouting a stream, steps should be taken to ensure that the new section has high environmental and landscape quality. In addition, river works can sometimes be used as an opportunity to repair damage caused by earlier insensitive schemes. Methods of restoring and enhancing channel ecosystems are given in Biggs & Williams (1993), Boon et al. (1992), Brookes (1988), Gordon et al. (1992); and methods for wetlands in general are reviewed in NRC (1992) and Ward et al. (1994). Andrews & Kinsman (1990) describe methods for restoration of gravel pits, and Buckley (1989) reviews enhancement methods for a range of habitat types, including wetlands.

There is currently much scope to improve the quality of habitat creation proposals in EIAs. The most common form of site enhancement is simply digging a pond or lake. It would often be much better to establish a mosaic of habitats, perhaps including pools together with marsh, fen and wet grassland habitats: these not only add complexity, but some (such as wet grassland), are now relatively uncommon habitats in their own right. A practical example of this approach is currently in progress at Pinkhill Meadow in Oxfordshire, where an NRA research project on wetland creation/river corridor enhancement is currently in progress; more information about this project is available from Pond Action (see Appendix B). It is always important that attempts to restore or create aquatic habitats ensure that both the physical and chemical conditions are suitable. For instance, there is little point in trying to restore a river habitat such as a meandering stream channel unless good water quality can be assured (NRC 1992). Similarly, provision should be made to ensure that enhancement or re-creation sites can be adequately managed and maintained after their establishment.

Care should also be taken to ensure that any creation sites are not already of high value, and further, that any management work to enhance the value of an existing site *is* actually an improvement. Temporary pools, damp hollows, wet flushes and well vegetated or shaded ponds are all examples of undervalued habitats that are often and easily damaged by misguided habitat enhancement work (Biggs et al. 1994).

13.7.4 Translocation of rare species

Where species are critical or have legal safeguards, translocation is sometimes undertaken. However, simply transferring rare animals or plants from one existing waterbody to another as a means of mitigation should not be recommended, except as a last resort. This is because (a) one uncommon species in a habitat may well indicate the presence of others, (b) the chances of success are low, and (c) there may be adverse impacts on the community of the newly stocked waterbody. For example, attempts to transfer species such as the great crested newt into existing ponds have often proved unsuccessful, either because the ponds were unsuitable or because they were already at their maximum carrying capacity for that species. Translocation into new, specifically created ponds is more likely to succeed, but may not compensate for the loss of the original ecosystem.

13.8 Monitoring

Mitigation of adverse impacts from developments should not stop at the design stage, and should evolve into long-term protection and monitoring. This is important both to check that mitigation measures are properly functioning, and to ensure that they are adequately preventing damage. Three requirements are essential: (a) baseline data that are good enough to detect detrimental changes caused by a development, (b) funding to carry out the survey work and the monitoring of the mitigation measures, and (c) sufficient contingency funds to enable modifications to mitigation measures to be made, or faults to be rectified, if necessary. Finally, important lessons learnt, or the results produced from well constructed monitoring projects, should be made available to others wherever possible. The ability to undertake effective EIAs and to plan effective mitigation schemes is significantly hindered by lack of information. The combined efforts and experience of those involved in undertaking and monitoring environmental EIAs could do much to change this.

CHAPTER 14
Coastal ecology
Stewart Thompson

14.1 Introduction

The UK has more than 15,000km of coastline and approximately one-third of a million km² of territorial waters (Gubbay 1990). The coastal zone contains a wide variety of valuable wildlife habitats. It is subject to considerable economic and recreational pressures, about which there has been growing concern for many years. Ecologists and conservationists have noted the deterioration of the varied habitats, much of which can be directly linked to the types and levels of development pressure present. Development pressures also have serious social and economic implications, e.g. for the fishing and tourism industries. However, there have been few attempts to assess the ecological implications of coastal developments.

14.2 Definitions and concepts

14.2.1 The coastal zone

Information on the ecology of coastal ecosystems can be found in sources such as Chapman (1976, 1977), Barnes (1979), Jefferies & Davy (1979), Mann (1981), Boaden & Seed (1985), Carter (1993). What precisely is meant by the term *coastal zone* is problematic. In general terms it can be defined as the zone where the sea meets the land; and in relation to the land–sea axis it can be subdivided into three zones: the littoral (or intertidal, or shore) zone, the supralittoral (or maritime) zone, and the sublittoral (or marine) zone. However, the boundaries of these zones are frequently unclear, both ecologically and for the purposes of EIA.

The *littoral zone* is the strip between high and low tide levels. It is often very narrow, as when the slope of the land is steep, but can be quite extensive, as in the case of mudflats. In most littoral zone ecosystems the resident plants and animals are essentially marine (saltmarsh vegetation being an exception). However, they are adapted to the regular pattern of immersion and emersion associated with tidal cycles, which in most locations around Europe consist of two tides per day. The tidal range (rise and fall) varies daily, with a two-weekly cycle of large *spring tides* that advance and retreat much farther than the small neap tides of the alternate weeks. In addition, there are larger seasonal cycles, with the largest spring tides near the spring and autumn equinoxes (in March and September); and it is these extreme high-water spring-tide levels (EHWS), and extreme low-water spring-tide (ELWS) levels, that are used to define the approximate boundaries between the littoral zone and those above and below it (see Fig. 14.1). Although all intertidal organisms must be adapted to the tidal cycles, the immersion/emersion regime varies considerably within the littoral zone. During the year, an organism living near the top of the zone may be submerged for only a few minutes per year, and one living near the bottom may be exposed to air for a similar short period.

Littoral ecosystems vary considerably in substratum type, largely in relation to their location in terms of exposure to wave action and strong currents. Mud accumulates only in estuaries and the sheltered upper reaches of inlets, whereas rocky shores occur where exposure prevents deposition of mud or sand (see Table 14.1). Sand and mud are soft, unstable substrates that do not provide firm

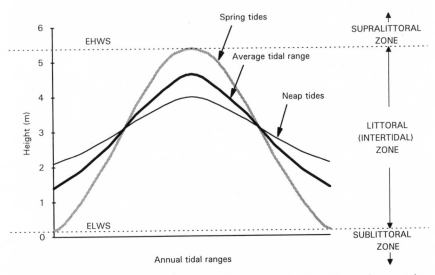

Figure 14.1 Coastal zones in relation to the ranges of spring tides and neap tides (and average tidal range) for a site where the extreme low-water spring tide level (ELWS) is 0.2 m OD and the extreme high-water spring tide level (EHWS) is 5.4 m OD (OD refers to elevation over/above chart datum, which is the designated zero contour).

anchorage for seaweeds or surface-dwelling animals. They are suitable for burrowing shellfish and marine worms, which are protected from wave action and the threat of desiccation during immersion and emersion phases respectively. By contrast, rocky shores provide a firm and generally impenetrable substratum that precludes burrowing, but on which anchorage is both feasible and essential for survival. Therefore, rocky shores are colonized largely by seaweeds and animals that adhere to the rock surfaces. However, severe wave action excludes seaweeds and many animal species from very exposed rocky shores, which are usually dominated by barnacles. Rocky shore ecology is discussed in Lewis (1985) and Moore & Seed (1985).

Table 14.1 Coastal ecosystem types and their locations in relation to exposure to wave action; and common associations between littoral and maritime types.

Littoral ecosystems (essentially marine)	Common associations	Maritime ecosystems (essentially terrestrial)	Locations
Mud flats	. . . grade to . . .	Saltmarshes	Estuaries and sheltered inlets
Sandy shores	. . . backed by . . .	Sand dunes	Fairly sheltered bays
Shingle beaches	. . . backed by . . .	Shingle banks	Fairly exposed coastlines
Rocky shores	. . . backed by . . .	Sea cliffs	Exposed headlands and bays

For the same environmental reasons, different littoral ecosystems are usually associated with specific *maritime systems* of the *supralittoral zone* (Table 14.1 and Appendix E). The vegetation of maritime communities is essentially terrestrial, which is why saltmarshes are usually placed in this category, in spite of their intertidal location. Where there is sufficient shelter, the saltmarsh plants are able to stabilize, and assist in, the accumulation of the muddy substrate; but these plants must be tolerant of saline conditions (Ranwell 1972, Long & Mason 1983, Adam 1990). Saltmarshes are often located in estuaries. These are unique ecosystems in which the mixing of fresh and salt water is a major ecological factor (McLusky 1981, Ketchum 1983, Barnes 1984, Nordstrom 1992).

Sand dunes are usually found where onshore winds blow sand from shores and sand banks that are exposed at low tide. However, they can only develop and be maintained through the stabilizing action of the vegetation, and are very sensitive to disturbance, e.g. from recreational pressure (Ranwell 1972, Ranwell & Boar 1986, Tansley 1939). Shingle beaches, banks and bars are usually found near to coastal erosion sites from which lateral currents drag the material to a location where there is sufficient shelter for it to be deposited. They are often backed by inlets containing saltmarsh or freshwater lagoons. Vegetated shingle beaches are very sensitive to disturbance (e.g. by construction plant) and are slow to recover. Maritime cliffs vary in relation to their geological composition and local landforms. Those composed of soft materials are often subject to quite rapid erosion, as along many stretches of the east coast of England.

In some cases the transition from littoral to supralittoral ecosystems is relatively sharp, but this is not so in estuaries. On their landward side, British maritime ecosystems are usually backed by agricultural land, urbanized areas, or recreational developments such as caravan sites or golf courses. In the absence of these, they would intergrade with terrestrial ecosystems such as forest. Consequently, the landward extension of the coastal zone is less well defined than it might seem at first sight.

Similarly, the lower boundary of the littoral zone is fairly clear, but the seaward extent of the *sublittoral zone* is particularly difficult to define. Ecologically it consists of all subtidal areas, and there is a continuum from the seashore to the open ocean. However, it is reasonable to restrict the coastal zone to the shallow seas of the *continental shelf* that fringes the European land mass. The extent of the continental shelf varies, but it includes the English Channel, the Irish Sea, and most of the North Sea. It also varies in depth, but generally slopes gently to a depth of 200–300 m before the seabed falls steeply to the deep ocean floor. The ecology of the sublittoral zone is discussed in Earle & Erwin (1983), and in marine ecology texts such as Holme & McIntyre (1984).

The sublittoral environment is more constant than the widely fluctuating littoral environment, but can vary appreciably for the following reasons.

- The seabed has a similar range of substrates to that of the littoral zone; these largely control the communities living on or in the substratum, and to a lesser degree those in the water above (which will include organisms that depend on the seafloor for food, shelter or reproduction).
- There is considerable variation in water movement, including turbulence, currents and tidal movement.
- Salinity may be reduced by fresh water, and turbidity increased by suspended sediments, near the mouths of rivers and estuaries, especially in wet weather.
- Water turbidity also tends to be high in areas with a muddy or sandy seabed, especially when the bottom sediments are disturbed during storms.

For the purposes of EIA, the sublittoral zone should clearly include nearshore waters, but the evident pollution of the entire North Sea demonstrates that the impacts of land-based developments can extend far beyond these. A recent inquiry into coastal zone protection concluded that definitions of the coastal zone may vary from area to area and from issue to issue, and that a pragmatic approach must therefore be taken at the appropriate national, regional or local level (House of Commons Environment Committee 1992).

14.2.2 The ecological value of British coastal zone ecosystems

The coast and seas around northwest Europe are one of the most productive wildlife habitats in the world. They are home to a range of flora and fauna often present in numbers that assume international importance. A major factor in their presence is the high nutrient status of the water. This supports large numbers of **primary producers**, which are the mainstay of the large and often complex **food**

webs of marine ecosystems. However, some marine communities have low recruitment and growth rates, and are therefore very prone to disturbance by development.

The British coastal zone is particularly special because of its location. Temperate, warm temperate, and Arctic species can all be found around the shores. The coastline is geologically and topographically varied, is heavily indented, and is subject to a wide range of wave activity and tidal regimes. These features affect a range of environmental factors, including water turbidity (and hence light penetration), silt deposition (and hence the nature of the substratum), and food availability. Consequently, they provide a wide variety of habitats in which different communities can develop, and a mosaic of different littoral and maritime ecosystems are often present within a small stretch of coast.

The UK's coastal areas are particularly important to birds. They provide refuges for many rare species, and British seabird colonies are of global importance. Of the 261 Internationally Important Bird Areas in the UK, 28 qualify because they hold over 1 per cent of the world population of a seabird species, and 61 qualify because they hold over 1 per cent of the EU population total (RSPB 1991).

UK estuaries are an internationally important habitat type. Because of the indented coastline and large tidal ranges, Britain has a higher proportion of estuarine habitat than anywhere else in Europe, just over 2 per cent of the British land–sea area (Davidson et al. 1991). For the same reasons, together with factors such as the warm seas, mild winters, and nutrient inputs from the land, their intertidal mudflats harbour a variety of invertebrate species, often in very large numbers, and they are among the most biologically productive ecosystems in the world (Rothwell & Housden 1990). Consequently, British estuaries provide rich feeding grounds for birds, and are important areas for some species of birds in European and international terms. In particular, they form vital links between the breeding and overwintering grounds of migratory waders and wildfowl.

The ecological consequences of estuarine development are substantial. Any significant loss or alteration to the estuarine habitat means that the diversity and biomass of the invertebrates present will substantially decrease. The knock-on effects of this are felt throughout the food chain, culminating in a loss of food for both birds and fishes. The cumulative implications of estuary development on the UK's resident and migratory biota should therefore be a primary consideration when viewing estuarine development proposals.

14.3 Legislative background and interest groups

The large number of agreements intended to conserve and enhance coastal areas which meet designation criteria reflects the international and national importance of the British coastal zone. The relevant legislation and associated designations of conservation sites are outlined in Tables 11.1 and 11.2. The most important regulations for coastlands are the Ramsar Convention, the EC Wild Birds Directive (79/409/EEC), the EC environmental assessment Directive (CEC 1985), the EC Habitats Directive (92/43/EEC), NNR, MNR and SSSI designations, and the Planning Policy Guidance Note (PPG) on Nature Conservation (DoE 1992g). The ultimate aim must be to comply with the EC Fifth Environmental Action Plan (CEC 1992), which calls for sustainable development of coastal zones in accordance with the carrying capacity of the coastal environments.

Some of the legislation affecting water quality, outlined in Table 10.1, applies to the coastal zone, and the NRA's responsibilities for pollution control extend to coastal waters. In addition, there is some EU and UK legislation for the coastal zone, including two Statutory Instruments (SIS) that supplement the general EIA regulations under SI 1988/1199 (see Table 14.2).

Table 14.2 UK and EU legislation specific to the coastal zone.

Legislation	Aims and aspects
Coastal Protection Act 1949	Powers of coastal protection authorities, control structures below tide level
Salmon and Freshwater Fisheries Act 1975	Regulations for inland fisheries, and for salmon and sea trout within a 6-mile zone
Shellfish Directive 79/923/EEC	Quality required of shellfish waters
Food and Environment Protection Act 1985	Pollution control in coastal waters; dumping at sea; licences for construction works
SI 1988/1336 and SI 1989/424	The Harbour Works (Assessment of Environmental Effects) Regulations 1988, and (no. 2) 1989; implement 85/337/EEC

Much of the UK coastal zone has some form of nature conservation designation, or is owned or managed by conservation bodies such as the RSPB or the Natural Trust. For instance, of the 155 recognized estuaries in the UK, 68 qualify for designation as Ramsar sites or as Special Protection Areas (SPAs) under the EC Wild Birds Directive. However, in spite of the stringent obligations imposed by these designations, they afford little protection (Treweek et al. 1993). Many estuaries recognized as internationally important wildlife sites are still being subjected to development pressures with the blessing of central government.

Coastal zone planning and management are beginning to receive a much higher profile than in the past, as can be seen from the recent publication of PPG 20 on coastal planning (DoE 1992h), which provides a useful framework for a comprehensive coastal management strategy. Of particular note are the recommendations: that the coastal zone be extended landwards and seawards, with its limits defined by the geographical extent of the natural coastal processes and the human activities affecting them; that only those developments that require a coastal location should be approved; that new projects be directed to areas already developed; and that consideration should be given to the offshore impacts of onshore developments.

Perhaps the most important aspect of PPG 20 is the recognition that coastal planning is a strategic issue and therefore relies upon close co-operation between planning authorities and all other parties interested in coastal zone management. However, few guidelines exist which indicate to the consultant ecologist those areas that should be surveyed, what to evaluate, which methodologies to employ, and how best to present the results (not only for the coastal zone but for all major ecosystem types). These problems are exacerbated by the large number of government departments and statutory authorities that have responsibilities in coastal areas. When the other interested parties are added to the list, the scope for confusion and duplication becomes extensive. This suggests that a radical change is needed in the way in which coastlines are managed. A national strategic planning framework for the coast is needed to avoid the piecemeal way in which development proposals are currently dealt with.

14.4 Scoping and baseline studies

14.4.1 Introduction and scoping

Baseline studies form the mainstay of the ecological component of EIA. If these are to be accomplished, ecologists must be consulted at the earliest stages of a planning proposal, and not be brought in when decisions have already been made. The ecologists must be able not only to provide ecological expertise but also to understand the needs of the developer. A brief must be agreed, and adhered to, by all the interested parties. Two underlying factors in this approach are good consultation and sufficient resources; unfortunately, both these receive inadequate treatment under the existing EIA system.

An important scoping task is establishment of the *impact area*. This is frequently difficult for coastal EIA, because of the indeterminate boundaries – especially of the sublittoral zone. A further problem is to estimate the lateral extent of the impact area along the coastline. As with other ecological assessments, the original estimate of the impact area may have to be revised in the light of information that emerges during the study. Much of the ecological interest of the coastal zone is linked to the physical characteristics of the area, and any ecological study must account for this. Consequently, the survey should include both the biota and the physical parameters, such as tidal movement, currents, exposure to wave action, geology and topography.

14.4.2 Study options

It should be possible to compile some of the information required for a coastal assessment from a desk study, using sources such as those listed in §11.4.2.

Ordnance Survey maps and aerial photographs may be particularly useful to establish if there have been substantial changes in topographic features such as coastal erosion. National inventories exist for some ecosystem types, for instance shingle banks (Sneddon & Randall 1993) and saltmarshes (Burd 1989) In addition, information on marine ecosystems can be obtained from the Marine Biological Association, and data on aspects such as tidal movements can be obtained form the Tidal Prediction Service, which publishes tide tables and summary data for stations around Britain.

However, as with other ecological assessments, the existing information is likely to be inadequate and/or out of date. Moreover, in many areas, the coastal ecosystems are virtually undescribed, with good knowledge existing only for resident sea mammal and bird populations. Currently many EIAs make predictions based on outdated fieldwork, on information collected for another element of the overall study (e.g. a landscape study), or on no data at all. This is unsatisfactory, and new surveys should be conducted wherever necessary.

Although fieldwork for some coastal habitats may be difficult, and survey methodologies may not be well developed, these are not a good reasons to exclude new fieldwork. Field surveys are an integral part of ecological assessment, both for identifying important habitats, species and communities, and for quantifying the parameters that enable potential impacts to be assessed (Treweek et al. 1993). If necessary, ecologists must develop survey methodologies that will provide information at a standard that allows good predictions to be made, in spite of the geographical, seasonal and financial constraints with which ecologists conducting fieldwork must contend. The study area surveyed should include as many habitat types and taxonomic groups as possible. Designated sites represent the country's most important wildlife habitats and, although theoretically afforded statutory protection, are clearly not immune from development pressures (see §11.3.1). Therefore, special care should be taken to ensure that good data are obtained for designated sites, and their importance should be clearly stated in the EIS.

14.4.3 Field surveys

Coastal surveys can employ the general strategy of phases/study levels outlined in Chapters 11 and 12. *Phase 1* can aim to obtain a general survey of the littoral and supralittoral zones, using the NCC Phase 1 habitat survey method (Table 11.3) and habitat classification (Appendix E). An alternative method and habitat classification designed specifically for coastal surveys is given by Hiscock (1990). The SEASEARCH Habitat Guide (Earll 1992) provides an equivalent classification for the sublittoral zone, but requires the employment of divers and is therefore expensive and restricted to sites that are safe for diving. A general survey of seabed types and other features may be possible from existing data. *Phase 2* can aim to achieve a more detailed classification of maritime vegetation types,

285

and obtain population estimates of selected animal species. Volume 5 of *British plant communities* (which contains the maritime communities of the NVC) has not yet been published, but should be available soon (Rodwell, in press). *Phase 3* can aim to obtain more detailed information on species populations and communities of selected areas.

The timing of field surveys and (where possible) repeat sampling, are particularly important in a coastal zone assessment, because many of the ecosystems have a high degree of seasonality. Whereas some faunal groups will be found all year round, many fish and bird populations change in relation to breeding and overwintering strategies. In particular, many populations of waders and other migratory seabirds are resident only during the winter. Saltmarsh vegetation grows and flowers relatively late in the summer, and some sand dune animals and annual plants should be sampled earlier than the most suitable period for a general vegetation survey (see Fig. 11.1). Resident shore communities can be sampled at most times of year, but neap tides do not expose the lower shore, and sampling is best conducted during the large spring tide periods in March or September. Consequently, the time available for sampling is strictly limited.

Although there are few guidelines or references available for conducting a coastal-zone EIA, standard procedures are available for most of the habitats and taxonomic groups, and Baker & Wolff (1987) is a very useful source of information. However, coastal zone sampling does present special problems, and many of the methods are both time-consuming and expensive, particularly those required for the sublittoral zone.

The *communities of maritime systems* such as saltmarshes, sand dunes and shingle banks can be studied using standard terrestrial sampling procedures or modifications of these (see Chs 11 and 12). Rocky shores can be sampled by similar procedures, and because the majority of rocky-shore animal species are immobile and visible at low tide, they can be relatively easily sampled (together with the seaweeds) in **quadrats** (see Baker & Crothers 1987). All these communities normally show clear zonations along the land–sea axis, so the use of **transects** along this axis is usually a suitable sampling pattern. Specific survey techniques exist for shingle banks (Sneddon & Randall 1993) and saltmarshes (Burd 1989). The use of **indicator species** has been suggested for saltmarshes (see Dalby 1987), but is likely to have limited application, mainly because the species selected as indicators must have a very narrow range of tolerance to any given change. Detailed surveys of sea cliffs are restricted by the problem of access.

Subsurface macroinvertebrates are an important group in many coastal habitats (including sandy shores and mudflats) because they are at the base of the food chain. They can be surveyed via methods ranging from a simple inspection of the sediment, e.g. to estimate the densities of lugworms from their castes, to a variety of destructive sampling methods which employ the use of corers and grabs to estimate densities and biomass. These methods, which may be appropriate for subsurface saltmarsh animals also, are discussed in Wolff (1987).

One taxonomic group for which there is a relative wealth of information regarding census techniques are the *birds*. Bibby et al. (1992) provide an excellent practical guide synthesizing all aspects of counting birds. Coastal birds are particularly well catered for, with information provided for the counting and monitoring of both seabird colonies and coastal waders. Hockin et al. (1992) give further information in their examination of the effects of disturbance on birds with reference to its importance in ecological assessment. Much of the text is devoted to species and groups found in coastal situations, and the authors provide an extensive set of appendices summarizing the literature on disturbance effects.

The *survey of plankton* is difficult because of their very small size and diversity, and because they are widespread over a large area of the sea. The use of satellites with sensors that respond to chlorophyll fluorescence may provide detailed distribution maps for phytoplankton. However, this is a very expensive and therefore unrealistic method, and most plankton sampling employs the use of equipment such as nets and various of sampling devices that can be filled with seawater samples at prescribed depths. These methods, together with associated techniques for analyzing the samples, are explained in Tett (1987), who suggests that for survey purposes it is convenient to adopt categories based mainly on ecological rather than taxonomic criteria.

Survey techniques for *fish* are numerous and very variable in their level of complexity. Consequently there are many relevant references, including Bagenal (1978), Begon (1979), Blower et al. (1981), Hardisty & Huggins (1975) and Pitcher & Hart (1982). A further useful source of information are the FAO Fisheries Department technical papers. The survey techniques employed are influenced by various characteristics of the fish populations and communities under study, including distribution (both vertical and horizontal), size and mobility, and population and community dynamics (single species shoals, mixed species shoals, cohorts, and seasonal migration and breeding patterns). The techniques can be grouped under two broad headings:

- *observation* – aerial, direct underwater, underwater photography and acoustic surveys;
- *capture* – hand nets, traps, hook and line, set nets, seines, trawls, lift, drop and push nets, and intake screens of cooling water systems (most of which can provide specimens for mark–recapture programmes) (Potts & Reay 1987).

The samples obtained by these methods can be analyzed by procedures that provide information on species abundance, age structure, fish health, dietary requirements and site productivity (see Potts & Reay 1987 for further discussion). All of this information is useful in that consultant ecologists can indicate the relative worth of the site to fish stocks and therefore provide comment as to the potential impact of the proposed development.

Coastal zone mammals can prove difficult to survey, and an ecologist involved in an EIA will have great difficulty in providing information other than casual observations or known presence linked to tracks, e.g. for otter and mink found along the foreshore or on mudflats. Numbers of seals (in common seal and grey seal colonies) are relatively easy to estimate as they are easily recognized and are faithful to particular stretches of coast, especially at pupping time and during the seals' moult, when aerial and boat surveys can be usefully conducted (Thompson & Harwood 1990, Ward et al. 1988). Numbers of cetaceans (whales, dolphins and porpoises) can be estimated by aerial, ship- and land-based sightings (Hammond 1987, Hammond & Thompson 1991, Hiby & Hammond 1989). More precise estimations of coastal mammal populations involves the use of time-consuming and often expensive field techniques such as capture–recapture and radio telemetry, and are therefore unlikely to be considered in EIA.

Further sources of information on the identification, distribution and conservation status of coastal species are given in Appendix F.

14.4.4 Changes without the development

Changes to coastal habitats in the absence of a given development are likely to result from a combination of natural trends and human impacts. Some coastal systems such as sand dunes are intrinsically unstable and dynamic, and can be affected by severe storms. In the short term, however, threats to these systems are more likely to arise from visitor pressure. Potential changes can be assessed from historical records, present management regimes and predictions of future recreational use.

Developments often have the potential to alter the rates of deposition of sediments, which in turn has implications for invertebrates present in the sediment and ultimately for the whole food chain. Therefore, it may be important to establish the rates of accretion, on a saltmarsh for instance. This can be achieved using several different techniques, including the use of: standard levelling techniques, sediment traps (Hargrave & Burns 1979), poles sunk into the sediment with a portion left protruding (Ranwell 1964), or coloured sand over which the newly deposited material can be measured (Coles 1979).

Other changes are likely to arise from impacts of other current and proposed developments of the various types outlined in §14.6, and these should be assessed as far as possible.

14.5 Impact prediction

14.5.1 Introduction

Predictions concerning the likely impacts of the proposed development are fundamental to the EIA process. Each type of development brings with it a suite of potential problems peculiar to that type, and no two development types are the same. For example, the potential impacts of a salmon farm on an inshore sea loch are different from those of a nuclear power station or permeable barrage scheme.

Impact prediction is generally poorly addressed in EIA. Ecologists seem loath to make any concrete statements regarding ecological changes or losses in and around the development site, usually because of inadequate data and/or understanding of the ecosystem in question. However, although this may prevent prediction of the outcomes of changes in a complex community structure, it may still be possible to state that a given species requires a certain set of environmental variables to be present in order to maintain a viable population, and that a given area with these variables present will support a particular size of population. It follows that any proposal that alters the balance of these variables, or the area in which they operate, will have an impact upon that species. Thus, based on the ecologist's judgement, an EIA can give an indication of the ecological consequences in best- and worst-case scenarios.

14.5.2 Types and sources of impact

Because the diversity of the coastal environment is the result of many complex interactions, any development which has the potential to disrupt this fine balance must be viewed with concern. In particular, any development proposal should be rigorously examined in relation to the following potential impacts:

- removal and fragmentation of coastal habitats
- alteration or removal of tidal activity
- alteration of turbidity levels (and hence light attenuation)
- increasing or decreasing rates of sedimentation
- alteration of water temperature
- increasing water and sediment pollution by the regular discharge of effluents
- increasing the risk of pollution related incidents/accidental spillage.

As in all ecosystems, primary impacts inevitably lead to indirect and cumulative impacts, as illustrated by the accumulation of pollutants in marine food chains (see Brouwer et al. 1990, Davies & McKie 1987, NERC 1983, and Walker 1990)

Many types of development threaten coastlines generally and estuaries in particular. A recent survey (Rothwell & Housden 1990) concluded that of the 123 UK estuaries surveyed, 80 were under some degree of threat, with 30 in imminent danger of permanent damage. Some of the impacts are irreversible and seriously reduce the extent of intertidal habitat available. A large range of threats was identified, including:

- land-take and reclamation
- pollution from urban development and heavy industry
- barrage schemes

- recreational pressures (including marinas)
- coastal defences
- marine fish farming.

Land-take and resultant loss of intertidal and maritime habitats is a major problem associated with a variety of development pressures, including industrial and
urban expansion, recreation and tourism, waste disposal, agriculture, and coastal
defence. *Reclamation* has a long history and is a continuing threat. About one
third of all British intertidal estuarine habitat and about half the saltmarsh area
have been reclaimed since Roman times (Thornton & Kite 1990), and about
32,000 ha of fenland on the Wash have been claimed for agriculture during the
same period (Doody & Barnett 1987). The intertidal area of the Tees estuary has
been reduced by around 90 per cent in the past 100 years (Rothwell & Housden
1990). Moreover, the pressure has increased rather than declined in recent years:
for instance, in 1989 at least 50 UK estuaries were subject to one or more proposals involving land claim (Davidson et al. 1991), and this has resulted in extensive losses, especially on the south and east coasts of England where the largest
tracts were once found. Estuarine habitat loss has also been widespread in other
countries: for example, in France 40 per cent of Brittany's coastal wetlands have
been lost over the past 20 years, and 54 per cent of all US wetlands have been
destroyed since colonial times (Maltby 1988). Further problems are linked to the
fragmentation of the marsh habitat that occurs behind the sea walls, as drainage
patterns are established that aim to minimize winter floods and lower summer
water levels (RSPB 1991). Land-take is further discussed in Therivel et al.
(1992).

Habitat fragmentation is an increasing problem. Many of the UK's primary
wildlife sites have become fragmented as a result of increased coastal development effectively removing areas of scarce habitat. An additional impact of
fragmentation is the formation of barriers to animal movement between the
already scarce habitats. The increased isolation of these remaining fragments
effectively reduces the levels of interaction between populations and hence gene
pools. It is therefore important to (a) identify the extent of direct habitat loss on
both the landward and seaward side of the proposed development, and wherever
possible to emphasize the need to minimize or avoid fragmentation impacts; and
(b) suggest measures for how this might be achieved at both the construction and
operation phases.

Heavy industry (including the chemical industry) has a long association with
coastal areas. One reason for this has been free access to a method of ridding the
industry of unwanted products by simply emptying them into the sea or river.
Many of these developments are very large, such as oil terminals and their associated infrastructure, and power stations. The levels of industrial discharge are
now under tighter control than in the past, but many water bodies still bear the
scars of years of uncontrolled discharge, and several pollution incidents have
occurred in prime estuary sites.

Barrage schemes are particularly controversial. Barrages fall into two basic categories: permeable and impermeable. Impermeable barrages are primarily intended for recreational purposes to produce static water for water sports or to provide aesthetically pleasing views for waterside developments (Therivel et al. 1992). The consequences of these impermeable barrages are not well known, yet proposals are under consideration for the Taff–Ely estuary and the Truro River. Much more widely publicized are the permeable barrages intended to generate tidal power, such as the Mersey and Severn proposals. By 1990, 22 estuarine sites had been subject to preliminary investigation for barrage construction (Rothwell & Housden 1990).

Recreation pressures on estuaries are increasing, as more people with more leisure time are pursuing such water-borne pursuits as small vessel sailing, wind-surfing and jet-skiing. Visitor pressure on sensitive maritime systems such as sand dunes can cause serious erosion problems (Ranwell & Boar 1986). The growth of *marinas* is of particular concern; by 1990, 154 marinas existed in UK estuaries, with a further 78 proposed (Davidson et al. 1991). These developments affect wildlife because of the disturbance they cause. Birds in particular are prone to disturbance, as they are heavily reliant upon undisturbed feeding sites (Therivel et al. 1992). The problem is exacerbated by the concentration of these facilities along the popular stretches of coast in the south and southeast of England. Although the growth of recreation pressures is difficult to forecast accurately, demand is likely to increase by 40–50 per cent by the year 2000 (Sidaway 1991). Strategic planning by central government for marinas is needed on both a regional and national basis.

Coastal defence works may have major impacts upon the flora and fauna of the site. Although guidance on such developments exists (DoE et al. 1982), it is not satisfactory for EIA, since it was produced prior to the EIA regulations. A recent report from the NRA to the National Audit Office stated that one sixth of the UK's sea walls, groynes and embankments had fewer than five years before they would need some form of renovation, with the proportion rising to a third in the southern region (Pearce 1993). Replacement of these artificial defences may result in the loss of valuable wildlife habitats, some of which (in the case of saltmarsh) are already disappearing at a rate of more than a metre a year. Currently, there is a move towards the concept of a "managed retreat" whereby salt-marshes, which act as natural coastal defence systems, are encouraged to form by removing existing sea walls, allowing a new marsh to form behind the old wall. The ecological benefits of these managed retreats are potentially large, and EN's coastal group aims to generate 250ha of saltmarsh per year by the approach.

Marine fish farming relies completely upon the availability of coastal sites. Most UK salmon farms are located in the sheltered waters of inshore sea lochs. These farms have a high potential to lower water quality in and around the cages in which the salmon are reared, as they have a heavy reliance upon chemicals to control pest outbreaks. Further pollution results from the high loadings of organic carbon and nitrogenous compounds in faecal material and uneaten food (Thompson et al., in prep.). Consequently, the environmental quality of the sea lochs may be eroded, thus reducing their ability to support viable populations of characteristic wild species. Additional concerns include the disturbance caused by fish farm operational activities, the excessive use of wild stocks of fish to feed the captive fish, and the effects on the genetic constitutions of wild salmon as they breed with captive bred stock. Many of the problems associated with salmon farming could be avoided if a system of EIA were in place that gave adequate consideration to the resident biota, with particular emphasis on the cumulative and synergistic impacts associated with the industry's heavy reliance upon chemicals, and its concentration in a limited geographic location. Despite the requirements of the 1988 Environmental Assessment (Salmon Farming in Marine Waters) Regulations, only one out of the 327 leases granted for salmon farms had been subject to an EIA at the time of writing.

Eutrophication can be problematic at sites that fall within agricultural catchment areas or sewage outfalls, as both of these activities may increase the nutrient loadings (mainly nitrates and phosphates) to levels that result in algal blooms and associated oxygen deficit (see Ch. 13). In the short term, this can result in the death of the aquatic flora and fauna, and in the long term may lead to changes in species composition, diversity and abundance. The problem can be exacerbated by development types that effectively slow down or remove the flushing regime along the relevant area of coastline. Barrage schemes in particular fall within this category.

Groundwater contamination from either saline intrusion or polluted materials such as contaminated runoff or disposal of bilge waters, is an important issue when considering coastal development. Many coastal communities rely on groundwater for their supply of drinking water, and the natural mixing of salt water and fresh water is crucial to the viability brackish water flora and fauna. It is therefore necessary to consider the implications of any development that has the potential to pollute the groundwater supply or to disturb the relationship between marine and terrestrial aquifers (Carter 1993). Any development that abstracts groundwater may result in saline intrusion, as may the removal or alteration of certain habitat types such as sand dunes, because these maintain the water table at an elevated level.

Dredging takes place in coastal situations for several reasons: improving or creating navigable waterways, harbour and marina creation, to provide fill for shoreline protection, to create new areas of land, and to provide a source of washed aggregate (Carter 1993). There are many problems linked to the dredging operation. For example, it results in (a) physical damage to the site, which has serious implications for the biota in that their required habitat is removed or altered, and (b) deepening of inshore waters, increasing shoreface slopes and allowing larger waves to break closer to the shore (Carter 1993). This increased wave activity results in shoreline erosion, and ultimately removes valuable habitat. The removal of the dredged material creates turbid conditions that result in lower light levels in and around the extraction site. This again causes problems to those flora and fauna requiring clear water conditions. Another problem is linked to the settling of the suspended sediment, which often occurs at some distance away from the extraction site and tends to smother areas that otherwise provide refuges for some benthic species. A final potential impact is the release of toxins and nutrients that normally remain locked up in the sediment, these have the potential to poison wildlife and to initiate algal blooms.

The significance of impacts on coastal species and ecosystems can be assessed by the methods outlined in Chapter 11, but again the sublittoral zone may present particular problems.

14.6 Mitigation

Although the need to mitigate potential adverse impacts is generally acknowledged, proposed mitigation measures are often vague, with little and often no reference to specific ecological impacts. Ecologists involved in the EIA of coastal developments should have appropriate mitigation measures as one of their main objectives. They should provide detailed prescriptions for the proposed measures, indicate how they would actually be put in place, and propose how they might be modified if unforeseen post-project ecological impacts manifest themselves. The last point is particularly relevant to coastal zone developments, as relatively little is known about their impacts because of their (often) large scale, the dynamic nature of the habitats they affect, and a lack of monitoring.

The use of simple classifications (e.g. of the sensitivity of habitats and/or species to impacts) can assist in formulating mitigation measures. Saltmarsh vegetation often acts as an "oil trap" for stranded spills, and in some locations marshes may be subjected to several successive oilings within a few years. A series of field experiments in areas composed of differing saltmarsh vegetation has resulted in a classification of British saltmarsh plants into five groups which reflect their susceptibility to oil spillage damage and also their ability to recover (Baker 1979). This classification is useful, in that field surveys conducted before the siting of an oil terminal can identify sites supporting saltmarsh vegetation that is the most likely to recover from accidental spillage. It should be stressed that the classification is very simple and must not be viewed in isolation, as many other factors need to be considered when positioning an oil terminal. However, an advantage of such simple classifications is that they can provide definite advice to decision-makers regarding the nature of the proposed development site.

Work undertaken in the Netherlands has classified all the major coastal dune landscape types according to where they occur, and the communities associated with each type (number of characteristic vascular plants, number of characteristic bird species, etc.). A matrix has been produced which demonstrates potential human influences in coastal areas, arranged according to the ecosystem component they potentially affect (see van der Maarel 1979). This provides a relatively complete picture of a threatened coastal habitat prone to development pressure. A matrix similar to this should be produced in the UK, as it would appear that the information is available (e.g. see Doody 1985).

Problems linked to eutrophication can be mitigated by various techniques that depend upon the source of the problem. Sewage disposal predominantly involves the use of a pipe outfall, and many of the potential problems posed can be overcome by good planning; this includes predicting the levels of sewage the stretch of coastline can accept in relation to such factors as water circulation (tides and currents), which is a major buffer to environmental stress imposed by sewage disposal (Carter 1993).

There are several ways of mitigating problems associated with groundwater supplies; again they depend upon the nature of the problem. It is important to maintain and enhance natural features such as sand dunes, in order to maximize their rôle in maintaining water tables, and no development that removes this habitat should take place. Another way of minimizing, or sometimes removing, problems linked to groundwater contamination by saline intrusion, leachates and other noxious substances, is to employ the use of artificial recharge of the aquifer. This can be achieved by importing fresh water from outside the catchment or by re-routing streams or storm runoff into infiltration pits, which reduce evapotranspiration (Carter 1993). However, care is needed to ensure that such measures do not generate other impacts on the freshwater systems involved. In certain circumstances it may be that simple operations such as raising the height of the sea wall to protect the aquifer from flooding will prevent any problems linked to seasonal inundation.

The impacts of dredging activities can be mitigated by the use of a carefully planned extraction programme, and controlled techniques. Impacts can be reduced by confining operations to ebb tide periods. Careful site surveillance will indicate those areas that have high nutrient and toxin loadings, and the area to be dredged should seek to avoid these areas. Similarly, a knowledge of tidal movement and sediment loadings can be used to predict where turbidity and the settling out of the suspended solids is likely to have serious impacts on the biota.

14.7 Monitoring

Given the importance of the coast, it is essential that a monitoring system be in place to alert the interested parties to any development-linked ecological problem. Ecological monitoring should take place in and around the area of the proposed development. Monitoring is an important element of EIA, yet few schemes even propose post-development monitoring, let alone undertake it. The success of mitigation measures can be established only if the area in question is subject to adequate, repeated, surveys; if a monitoring programme is in place to measure residual impacts; and, most importantly, if a commitment to both these elements has been secured from the development proponents.

Monitoring should be undertaken by experts from several environmental fields, and in consultation with statutory/regulatory bodies such as the relevant NCA, MAFF (or country equivalent), and HMIP, and NGOs such as the RSPB or the National Trust. This will allow ecologists undertaking future EIAs to consult monitoring information about those elements of previous developments that have had positive and negative impacts upon the coastal zone ecology. Finally, monitoring gives ecologists a tool by which they can gauge their own predictive ability, which in turn indicates where field methodologies and techniques are either adequate or inappropriate, and therefore in need of refinement.

14.8 Recommendations and conclusions

Recommendations regarding coastal zones fall into two broad categories: interim recommendations to limit or remove the environmental and ecological damage of specific coast-based developments in the immediate future, and long-range recommendations concerning strategic environmental assessment, and management of the coastline to ameliorate any future development-linked impacts.

In the short term, LAs should require EIAs for coastal projects to give greater consideration to indirect and cumulative ecological impacts. Currently, the vast majority of project-specific EIAs carried out for the coastal zone fail to address the high levels of cumulative impacts that coastal developments generate. Attention should be paid to the UK's resident and migratory biota in all components of the coastal zone. The impact of development proposals upon these should be considered together, not individually. Central government should bring under planning control those developments currently outside the scope EIA (e.g. agriculture, marine aggregate extraction) and additions or changes to existing techniques of coastal defence. Jurisdiction for these developments should be handed over to the LA (rather than to the DoE), who would then be able to insist that the ecology of a proposed development site be given adequate consideration. The developer must demonstrate a commitment to this by allocating resources for the appropriate fieldwork and post-development monitoring.

In the longer term, LAs should consider, in their structure plans, limiting the development of coastal projects that affect key habitat types or nationally important populations. Ideally they should consider a phased removal or ban of all development within such areas, assisting their relocation to less ecologically sensitive areas. Central government should adhere to the recommendations of the North Sea Ministers Conference of 1990, which advocated the precautionary principle for policies and activities affecting the marine environment. This approach acknowledges the current lack of knowledge about the biology and ecology of marine systems.

Reports from both government and non-government organizations confirm that coastal management practices are poor, and that there is a need for co-ordination of the interested parties, and guidelines as to how integration of interests might be achieved. The coastal zone is an outstanding habitat type in the UK, and EIA provides a means of providing checks on development activities that undermine its ecological worth. Adoption of a system of strategic environmental assessment would lead to the removal of piecemeal development and would provide an arena in which to bring together conflicting and overlapping interests along the coastal zone.

CHAPTER 15

Interactions between impacts

Riki Therivel & Peter Morris

15.1 Introduction

As the previous chapters have shown, all of a project's impacts interact with other impacts. This is related to the fact that ecosystem components interact. Plants, animals and microbes are all affected by and affect environmental systems and other organisms. Human populations are components of these interactive systems and depend on their health and viability. In turn, social and economic conditions influence factors such as noise, traffic, waste generation and land use, and all such factors affect the natural environment. Thus, it is impossible to consider any one impact of a project in isolation.

EC Directive 85/337 and the subsequent UK regulations on EIA require, in addition to an assessment of a range of listed impacts (see Ch. 1 and Appendix A), an assessment of the interaction between these impacts, and of indirect and cumulative impacts as well as direct impacts. These requirements represent the culmination of the "Heineken" approach mentioned in Chapter 2: they reach those parts that not even socio-economic analyses reach, ensuring that the EIA is comprehensive. However, few EISs to date have explicitly addressed these types of impacts. A study by Jones et al. (1991) showed that only 14 in a sample of 100 EISs considered impact interactions. Even where interactions are considered, it is generally not as the result of a formal scoping process, but rather as part of the analysis of other impacts. Indirect and cumulative impacts are also rarely considered in EIAs, despite the clear links between certain development types and certain forms of cumulative pollution, e.g. acid rain and greenhouse gases. Undoubtedly, the fragmented way in which many EIAs are carried out – with separate reports written by separate individuals about separate impacts and only later brought together into a joint document – fosters this approach.

This chapter aims to help redress this imbalance by suggesting approaches towards the assessment of impact interactions. Impact interactions are very complex phenomena, and a comprehensive assessment of them would need to be correspondingly complex, in terms of both the range of impacts it considered and the depth in which it considered them. Such assessment methods are not yet agreed upon and, even if they were, would be beyond the scope of this book, and probably of most EIAs. As such, this chapter focuses on the most cost- and time-efficient methods. After an introduction to the relevant definitions in §15.2, it discusses the need to ensure that individual impact assessments are based on the same assumptions and project proposal in §15.3. Interactions between a project's impacts and between impacts of the proposed project and other projects are discussed in §15.4 and §15.5. §15.6 suggests techniques for predicting impact interactions, and §15.7 discusses how to integrate and summarize a project's impacts.

15.2 Definitions and concepts

There is no one agreed terminology for describing impacts, but they can be broadly classified as (a) *direct*, primary or first-order impacts which are caused by the project and occur in the same time and place; (b) *indirect*, secondary or higher-order impacts which are triggered by the project but affect the environmental component under consideration as knock-on effects between sub-components or via other components; or (c) *cumulative* impacts caused by the sum of the project's impacts on the environmental component, and/or the project's impacts when added to those of other past, present or future projects. In turn, cumulative impacts can be described as: (c1) additive, aggregate, or "nibbling", namely the simple sum of all the impacts; (c2) synergistic, where impacts interact to produce an impact greater than the sum of the individual impacts; or (c3) neutralizing or antagonistic, where impacts counteract each other, reducing the overall impact. An example of additive impacts could be the employment opportunities created by several developments, which can often (given assumptions about, for example, skill levels and unemployment) be determined by adding the jobs created by each individual development. An example of synergistic impacts is the interaction of SO_2 from coal-fired power stations with NO_x from vehicle emissions, which affect the growth and health of plants. Neutralizing impacts could include the "neutralization" of a loss of terrestrial biodiversity through gravel extraction by an increase in aquatic biodiversity.

Interactions between impacts can mean both interactions between the impacts of one project and the interaction between the impacts of the proposed project and those of other projects. An example of the first type is given in Chapter 3, where the employment generated by a project can cause changes in accommodation, services and the area's economic structure. An example of the latter type is given in Chapter 5, where future traffic generated by other developments may limit the ability of a road network to accommodate that generated by the proposed development. The interactions between a specific project's impacts are nothing more than its indirect and cumulative impacts as well as its direct impacts. The interaction between several projects' impacts are cumulative impacts. As such, the definitions overlap to an extent.

In addition, a distinction may need to be made (as in Ch. 11) between impacts and *impact factors*. For instance acid rain can be considered to be an impact factor where it affects soils or plants, or an impact caused by atmospheric pollutants such as SO_2 or NO_x. In turn the atmospheric pollutants are impact factors affecting acid rain, or impacts resulting from the combustion of fossil fuels.

15.3 Ensuring consistency and accuracy

The details of the proposed project's siting, layout and design, and consequently of its impacts and agreed mitigation measures, will evolve over the course of project planning. Several versions of the project may be proposed and circulated to the experts carrying out the assessment of individual impacts, and in turn a wide range of modifications, amendments and additional measures may be recommended. When assessing the interaction between impacts, the EIA coordinator will need to ensure that all parties are basing their assessments on the same, most up-to-date version of the proposed project.

It is particularly important to ensure that all the impacts of the mitigation measures outlined in separate analyses are fully agreed upon and assessed. For instance, plantings which reduce visual impacts may have beneficial side-effects for noise and wildlife, or have negative ecological side-effects or intrude on archaeological remains; proposals to maintain or enhance the diversity of wildlife habitats or reduce noise may in turn affect the landscape; and new roads or road improvements that channel a development's traffic around a built-up area may also affect that area's economy. When mitigation measures are proposed and adopted, they become part of the most recent version of the project proposal (see Fig. 1.1), and the environmental impacts of this version need to be reassessed.

Research by Frost (1993) suggests that about half of projects described in EISs are changed substantially between the time they are approved and when they are built. In such cases, either a new EIS, or an alteration to the existing EIS which explains how the impacts of the new proposal would differ from those of the old one, should be prepared.

15.4 Identifying interactions between a project's impacts

No one method exists for identifying a project's impact interactions, but the joint and iterative use of several of the following methods should ensure that no significant interaction is overlooked.

Structured techniques for identifying impact interactions include (a) cause–effect models such as networks, flow diagrams or matrices; (b) map overlays; (c) statistical data analysis or simulations based, for example, on aerial photos or field data; and (d) computer models and GIS systems based on the above. Networks, flow diagrams and simulation models intrinsically consider indirect and cumulative as well as direct impacts. The simplest matrix method is a table that lists all environmental components on one axis (component A), and again on the second axis (component B). If component A affects component B, this is identified in the appropriate cell and discussed further in the text. Matrix multiplication, or simply tracing impacts through several consecutive tables, allows more complex impact interactions to be identified (Wathern 1984). Maps of the areas affected by individual types of impact can be overlain to show the areas most affected by the project. Glasson et al. (1994) review these methods, and Appendix D summarizes the capabilities and limitations of GIS.

The *experts* involved in preparing the EIA will often be familiar with likely sources of indirect and cumulative impacts on the environmental component they are assessing, as well as the impacts of the mitigation measures they propose. The EIA co-ordinator should encourage the experts to clearly state these impacts, and communicate this information to the other relevant experts.

The co-ordinator also needs to determine whether any *environmental components other than those listed in the EC Directive* or UK regulations need to be considered. For instance, many projects affect an area's noise, agriculture, forestry, recreation, services, retail, traffic and/or access. None of these issues is listed in the EC Directive, yet they could all be described as interactions between other impacts or impact components. For instance, the interaction of impacts on soils, water, air and human beings may be addressed in the agricultural section of an EIS:

Relevant legislation and standards
- MAFF Agricultural Land Classification
- PPG7, The Countryside and the Rural Economy

Baseline data
- land use (% land allocated to various uses/crops)
- land quality (% land of each agricultural grade)
- land ownership, size of holding, access
- soil types
- hydrology, drainage, irrigation, water quality
- climate, including microclimate

Possible impacts
- loss of land, long-term land use restrictions, severance, loss of access, segmentation of farms (temporary, permanent), loss of livelihood
- disturbance or loss (temporary, permanent) of crops and/or production
- disturbance of soil resources, e.g. change in soil condition, erosion, compaction
- change in drainage and/or water supplies

- risk of flooding, drought, wetness, siltation, bank instability
- disturbance or loss of hedges, shelterbelts, boundaries
- transmission of plant or animals diseases, need for special consideration of organic farms

Possible mitigation

- timing of operations to avoid, e.g. planting or harvesting times
- avoiding high-grade land
- separate stripping and storage of different forms of soil, handling of soil only when the soil has an adequate moisture content, careful handling and reinstatement of soils, preventing long-term storage of soils
- planting or seeding to avoid erosion
- avoidance of hedges and "sensitive" crops (e.g. orchards, fruit bushes)
- restoring field drainage
- careful timing of severance (e.g. to avoid hay-making), providing alternative access
- compensating for loss of crop or income
- preventing disease by avoiding bringing plant/animal matter from one area to another

A conceptual shift away from the range of impacts caused by the project's individual components, and towards the *total impacts on an environmental component*, should be sought. This is exemplified in Figure 15.1.

Finally, *local residents' perceptions* should be determined and addressed. For instance, the noise nuisance caused by a development tends to be perceived as being worse when the development is also visible, and vice versa. Reducing

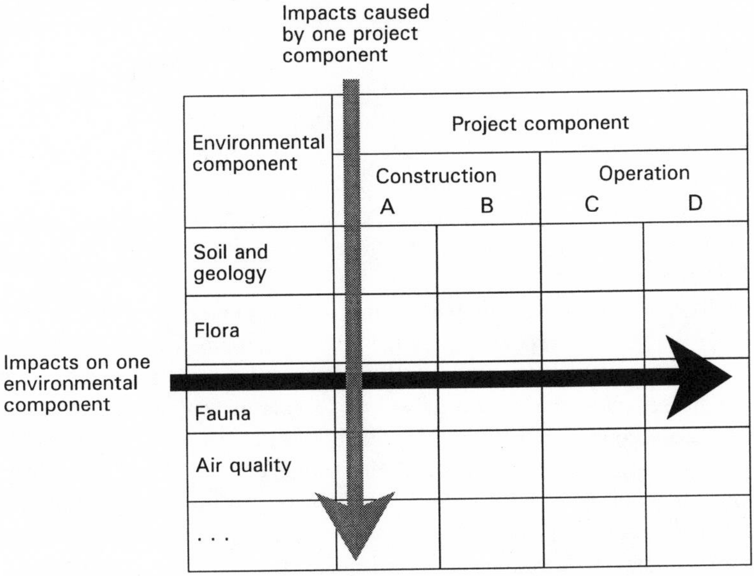

Figure 15.1 Focusing on total impacts on an environmental component.

visual disturbance may thus also reduce the perceived noise impacts, even if the actual noise levels remain the same.

15.5 Identifying interactions between the impacts of the proposed project and other projects

Projects are not planned, built, operated and decommissioned in isolation, but within a regional, national and international process of change, which includes other projects, programmes, plans and policies. Strategic environmental assessment, the assessment of the environmental impacts of more than one project, is not yet a legal requirement in the EU, but may well become one soon (see Glasson et al. 1994). However, already in project EIA some projects are so inextricably related to other projects, or their impacts are so clearly linked, that an assessment of these interactions should be carried out.

An example of a project that generates other major infrastructure development would be a gas-fired power station which may require the construction of a new pipeline and gas reception/processing facility to receive the gas, and transmission lines to carry the resulting electricity. An example of a development that is part of a broader programme of related developments would be a project to widen a section of road as part of a programme of road management and construction. Although these projects may have different planning and permission processes, and may require EIA under different regulations, they should be considered together as one large project, since one could not proceed without the other(s). As such, they should be addressed in a single EIA, or at least their cumulative impacts should be addressed in each of their EIAs.

Other projects are "growth-inducing", namely not themselves dependent on the construction of other projects, but necessary precursors to other projects. For instance, a new motorway may induce the construction of motorway service stations, hypermarkets, or new towns; or the infrastructure provided for one project may make a site more attractive, or may present economies of scale, for further development. EIAs should acknowledge the possibility of these further developments.

In still other cases the developments may not be so clearly linked, or a given developer will not be responsible for the other developments, but their impacts may still interact. For instance the closure of one footpath as a result of a development project may still leave a network of rights of way accessible, but also closing off a second or third footpath as a result of another project may remove all access to the network. The projects' EIAs should clearly address this situation. Information about other developments can be gathered from, for example:

- development briefs
- local authority development plans, especially given the implications of the new Section 54A (which requires developments to be in accordance with

the development plan unless material considerations dictate otherwise)
- environmental appraisals of development plans
- waste disposal and minerals plans
- transport policies and plans
- traffic growth forecasts and the national roads programme
- construction programmes by large companies, e.g. Nuclear Electric, British Coal, British Rail and its successors
- employment forecasts
- industrial directories
- other EISs
- local planning authorities, and
- local research and academic institutions.

In some cases, this stage may suggest that impacts of other small developments not usually requiring EIA may need to be assessed, or that project subcomponents require EIA.

Identifying the impacts of multiple projects is not straightforward and it requires a clear definition of the time, area and range of projects to be considered. Perhaps the simplest method is to list all the relevant projects on one axis of a matrix, and environmental components on the other, and summarize each project's impacts on the environmental component in the relevant cell. Considerable further literature exists on the subject, although it is often quite theoretical and not always easy to obtain. The proceedings of a US/Canadian conference on cumulative impacts (Beanlands et al. 1986) give a good background. Various US government departments have developed relevant methodologies (e.g. USDHUD 1981, Bain et al. 1986, USACE 1988), and a report by the USEPA (1992) on methods for cumulative impact assessment lists almost 350 further references. Research preceding and developing from the enactment of the Canadian Environmental Assessment Act in 1992 has also spawned several excellent publications (e.g. CEAA 1993, Davies 1992, Lane & Wallace 1988). A special edition of *Environmental Management* (Bedford & Preston 1988) focused on cumulative impacts on wetlands.

In the UK, local authorities have had to review the environmental impacts of their development plans as a result of PPG 12 (DoE 1992a). Methods of carrying out such appraisals are given in DoE (1993c) and Therivel (1994). Initial suggestions for how more general strategic environmental assessment could be implemented are given in Therivel et al. (1992).

The broadening of an EIA's remit to encompass other projects may allow trade-offs to be made between impacts and between projects. For instance, an environmentally beneficial "shadow project" may be proposed to neutralize the negative impacts of a development project. An example of this is the mitigation measures proposed as compensation for the loss of waterfowl feeding grounds caused by the construction of the Cardiff Barrage, namely the "creation" of a new feeding ground further up the coastline. However, this has raised concerns about both the scheme's effectiveness in providing alternative feeding grounds, and its aesthetics in terms of landscape authenticity.

15.6 Predicting and integrating impact interactions

Methods for predicting the *magnitude* of impact interactions can in many cases be based on methods used for individual environmental components. In other cases, however, completely new indicators may need to be considered, using novel techniques. For instance, Chapter 11 shows that ecosystems as a whole cannot be described merely through their subcomponents (e.g. water quality, flora, fauna), but may require composite indicators such as biodiversity. Impacts on human health and accident risk require still different methodologies such as risk analysis, which are not discussed here.

Predictions of impact interactions involve a greater degree of uncertainty than predictions of impacts on specific environmental components. Often only broad trends and issues of concern can be identified. Where the impact interactions have an additive or synergistic impact, this will compound any trends in the direct impact. Neutralizing impacts, in contrast, will slow down or reverse such a trend.

However, techniques exist for reducing or clarifying the uncertainty. The impacts of similar existing projects can be monitored, and extrapolated to the proposed project and likely future projects. Worst-case scenarios and the precautionary principle can be used to ensure that the project(s) cause the least amount of environmental disruption possible. Decision trees can be drawn to determine the likelihood of events happening. Scenarios and sensitivity analyses can be used to test the robustness of predictions. Impacts can be described using ranges to take uncertainty into account. All possible causes of uncertainty can be listed in the EIA, so that decision-makers can account for this uncertainty. These methods are discussed further in e.g. De Jongh (1988) and Glasson et al. (1994).

The key impacts of the proposed project will have been initially identified during the EIA scoping process. The assessment of impact interactions allows the EIA co-ordinator to integrate and reassess the relative *significance* of all of the project's impacts. The co-ordinator should consider (a) the entire project(s), including all agreed mitigation measures, and (b) the project's entire package of direct, indirect, and cumulative impacts and their interrelationships, using various scenarios concerning the likely effectiveness of the mitigation measures.

The significance of the project's total impacts should first be determined in relation to any thresholds (e.g. legal standards, agreed emission standards or environmental quality standards) identified for individual environmental components. Links to other consent procedures that affect impact interactions should be highlighted, particularly those for waste licenses and for Integrated Pollution Control under the Environmental Protection Act 1990. Where possible, cumulative impacts should be tested for sustainability, namely whether their scale is similar to what might be experienced naturally. Impacts should also be addressed in terms of their distributional effects. One sector of the community may be much more affected by multiple development projects, or by the concentration of all of a project's impacts, than other sectors. A table that lists receptors (e.g.

residences, sites of ecological importance, or groups of people such as cyclists, children, wheelchair users) on one axis, and the main impacts of a project on the other axis, can identify which groups are most strongly affected by the project. A more equitable distribution of impacts may be sought as a result, or strongly (negatively) affected groups may be compensated in some way. Glasson et al. (1994) discuss this in greater depth.

These analyses should result in (a) a broad ranking of the project's various impacts in terms of their significance; (b) a statement concerning the project's relative contribution to overall environmental change regarding, for instance, the percentage of the change due to the project, how significant this is, and how the project compares with other projects which have similar functions; (c) a non-technical summary of the impacts; and (d) a statement concerning any assumptions, uncertainty, or further data needed in the analysis.

15.7 Conclusions

Despite the lack of established methodology for addressing impact interactions, these impacts undoubtedly exist and may be a major component of an EIA. Compared to the methods described in previous chapters, those presented here are still relatively underdeveloped, and have not often been used in practice. However, current trends towards more holistic, strategic and cumulative EIAs (Glasson et al. 1994) mean that these methods will be increasingly used in the future.

APPENDIX A
EIA/EIS contents

This appendix presents verbatim extracts from the Town and Country Planning (Assessment of Environmental Effects) Regulations 1988 and the DOE's guidebook *Environmental assessment: a guide to procedures*.

A.1 Schedule 3 of the *Town and Country Planning (Assessment of Environmental Effects) Regulations 1988*

1. An environmental statement comprises a document or series of documents providing for the purpose of assessing the likely impact upon the environment of the development proposed to be carried out, the information specified in paragraph 2
   . . .
2. The specified information is –
   (a) a description of the development proposed, comprising information about the site and the design and size or scale of the development;
   (b) the data necessary to identify and assess the main effects which that development is likely to have on the environment;
   (c) a description of the likely significant effects, direct and indirect, on the environment of the development, explained by reference to its possible impacts on – human beings; flora; fauna; soil; water; air; climate; the landscape; the interaction between any of the foregoing; material assets; the cultural heritage;
   (d) where significant adverse effects are identified with respect to any of the foregoing, a description of the measures envisaged in order to avoid, reduce or remedy those effects; and
   (e) a summary in non-technical language of the information specified above.
3. An environmental statement may include, by way of explanation or amplification of any specified information, further information on any of the following matters:
   (a) the physical characteristics of the proposed development, and the land-use requirements during the construction and operational phases;
   (b) the main characteristics of the production processes proposed, including the nature and quantity of the materials to be used;
   (c) the estimated type and quantity of expected residues and emissions (including pollutants of water, air or soil, noise, vibration, light, heat and radiation) resulting from the proposed development when in operation;
   (d) (in outline) the main alternatives (if any) studied by the applicant, appellant or authority and an indication of the main reasons for choosing the development proposed, taking into account the environmental effects;

307

(e) the likely significant direct and indirect effects on the environment of the development proposed which may result from (i) the use of natural resources; (ii) the emission of pollutants, the creation of nuisances, and the elimination of waste;

(f) the forecasting methods used to assess any effects on the environment about which information is given under subparagraph (e); and

(g) any difficulties, such as technical deficiencies or lack of know-how, encountered in compiling any specified information.

In paragraph (e), "effects" includes secondary, cumulative, short, medium and long term, permanent, temporary, positive and negative effects.

4. Where further information is included in an environmental statement pursuant to paragraph 3, a non-technical summary of that information shall also be provided.

A.2 Non-mandatory EIA checklist from *Environmental assessment: a guide to the procedures* (DOE 1989)

This checklist is intended as a guide to the subjects that need to be considered in the course of preparing an environmental statement. It is unlikely that all the items will be relevant to any one project . . . The environmental effects of a development during its construction and commissioning phases should be considered separately from the effects arising whilst it is operational. Where the operational life of a development is expected to be limited, the effects of decommissioning or reinstating the land should also be considered separately.

Section 1

Information describing the project

1.1 Purpose and physical characteristics of the project, including details of proposed access and transport arrangements, and of numbers to be employed and where they will come from.

1.2 Land use requirements and other physical features of the project:
   a during construction;
   b when operational;
   c after use has ceased (where appropriate).

1.3 Production processes and operational features of the project:
   a type and quantities of raw materials, energy and other resources consumed;
   b residues and emissions by type, quantity, composition and strength including:
   i discharges to water; ii emissions to air; iii noise; iv vibration; v light; vi heat; vii radiation; viii deposits/residues to land and soil; ix others.

1.4 Main alternative sites and processes considered, where appropriate, and reasons for final choice.

Section 2

Information describing the site and its environment

Physical features

2.1 Population – proximity and numbers

2.2 Flora and fauna (including both habitats and species) – in particular, protected species and their habitats.

2.3 Soil; agricultural quality, geology and geomorphology.

2.4 Water; aquifers, water courses, shoreline, including the type, quantity, composition and strength or any existing discharges.

2.5 Air; climatic factors, air quality, etc.

2.6 Architectural and historic heritage, archaeological sites and features, and other material assets.

2.7 Landscape and topography.

2.8 Recreational uses.

2.9 Any other relevant environmental features.

The policy framework

2.10 Where applicable, the information considered under this section should include all relevant statutory designations such as national nature reserves, sites of special scientific interest, national parks, areas of outstanding natural beauty, heritage coasts, regional parks, country parks, national forest parks and designated areas, local nature reserves, areas affected by tree preservation orders, water protection zones, nitrate sensitive areas, conservation areas, listed buildings, scheduled ancient monuments, and designated areas of archaeological importance. It should also include references to structure, unitary and local plan policies applying to the site and surrounding area which are relevant to the proposed development.

2.11 Reference should also be made to international designations, e.g. those under the EC "Wild Birds" Directive, the World Heritage Convention, the UNEP Man and Biosphere Programme and the Ramsar Convention.

Section 3

Assessment of effects

(Including direct and indirect, secondary, cumulative, short, medium and long-term, permanent and temporary, positive and negative effects of the project.)

Effects on human beings, buildings and man-made features

3.1 Change in population arising from the development, and consequential environment effects.

3.2 Visual effects of the development on the surrounding area and landscape.

3.3 Levels and effects of emissions from the development during normal operation.

3.4 Levels and effects of noise from the development.

3.5 Effects of the development on local roads and transport.

3.6 Effects of the development on buildings, the architectural and historic heritage, archaeological features, and other human artefacts, e.g. through pollutants, visual intrusion, vibration.

Effects on flora, fauna and geology

3.7 Loss of, and damage to, habitats and plant and animal species.

3.8 Loss of, and damage to, geological, palaeontological and physiographic features.

3.9 Other ecological consequences.

Effects on land

3.10 Physical effects of the development, e.g. change in local topography, effect of earth-moving on stability, soil erosion, etc.

3.11 Effects of chemical emissions and deposits on soil of site and surrounding land.

3.12 Land-use/resource effects:
   a quality and quantity of agricultural land to be taken;
   b sterilisation of mineral resources;
   c other alternative uses of the site, including the 'do nothing' option;
   d effect on surrounding land uses including agriculture;
   e waste disposal.

Effects on water

3.13 Effects of development on drainage pattern in the area.

3.14 Changes to other hydrographic characteristics, e.g. ground water level, water courses, flow of underground water.

3.15 Effects on coastal or estuarine hydrology.

3.16 Effects of pollutants, waste, etc. on water quality.

Effects on air and climate

3.17 Level and concentration of chemical emissions and their environmental effects.

3.18 Particulate matter.

3.19 Offensive odours.

3.20 Any other climatic effects.

Other indirect and secondary effects associated with the project

3.21 Effects from traffic (road, rail, air, water) related to the development.

3.22 Effects arising from the extraction and consumption of materials, water, energy or other resources by the development.

3.23 Effects of other development associated with the project, e.g. new roads, sewers, housing, power lines, pipelines, telecommunications, etc.

3.24 Effects of association of the development with other existing or proposed development.

3.25 Secondary effects resulting from the interaction of separate direct effects listed above.

Section 4

Mitigating measures

4.1 Where significant adverse effects are identified, a description of the measures to be taken to avoid, reduce or remedy those effects, e.g.:

   a site planning;

   b technical measures, e.g.: i process selection; ii recycling; iii pollution control and treatment; iv containment (e.g. bunding of storage vessels).

   c aesthetic and ecological measures, e.g.: i mounding; ii design, colour, etc.; iii landscaping; iv tree plantings; v measures to preserve particular habitats or create alternative habitats; vi recording of archaeological sites; vii measures to safeguard historic building or sites.

4.2 Assessment of the likely effectiveness of mitigating measures.

Section 5

Risks of accidents and hazardous development

5.1 Risks of accidents as such are not covered in the Directive on EA or, consequently, in the implementing Regulations. However, when the proposed development involves materials that could be harmful to the environment (including people) in the event of an accident, the environmental statement should include an indication of the preventive measures that will be adopted so that such an occurrence is not likely to have a significant effect. This could, where appropriate, include reference to compliance with the Health and Safety at Work Act 1974 and its relevant statutory provisions such as the Control of Industrial Major Accident Hazards Regulations 1984.

5.2 There are separate arrangements in force relating to the keeping or use of hazardous substances and the Health and Safety Executive provides local planning authorities with expert advice about risk assessment on any planning application involving a hazardous installation.

5.3 Nevertheless, it is desirable that wherever possible the risk of accident and the general environmental effects of developments should be considered together, and developers and planning authorities should bear this in mind.

APPENDIX B
Useful addresses

For further addresses, see Donn & Wade (1994), DOE (1992), DOT (1993), Essex POA (1992) and Kent CC (1991)

Ancient Monuments Society, St Andrew-by-the-Wardrobe, Queen Victoria St, London EC4 5DE

Atmospheric Research and Information Centre, Dept of Environmental and Geographical Studies, Manchester Polytechnic, Chester Street, Manchester M1 5GD

Bat Conservation Trust, c/o The London Ecology Centre, 45 Shelton Street, Covent Garden, London WC2H 9HJ

Biological Records Centre, see Environmental Information Centre

British Butterfly Conservation Society, Conservation Office, PO Box 444, Dorchester, Dorset, DT2 7YT.

British Geological Survey, Information Services Group, Keyworth, Nottingham NG12 5GG

British Rail Property Board, Great Northern House, 79–81 Euston Road, London NW1

British Tourist Authority, Thames Tower, Blacks Road, Hammersmith, London W6 9EL

British Trust for Conservation Volunteers, 36 St Mary's Street, Wallingford, Oxfordshire OX10 0EU

British Trust for Ornithology, National Centre for Ornithology, The Nunnery, Thetford, Norfolk IP24 2PU

British Waterways, Willow Grange, Church Road, Watford WD1 3QA

Byways and Bridleways Trust, The Granary, Charlcutt, Calne, Wiltshire SN11 9HL

Cadw (Welsh Historic Monuments), Brunel House, 2 Fitzalan Road, Cardiff CF2 1UY

Cambridge Environmental Research Consultants, 3D King's Parade, Cambridge, CB2 1SJ

Chartered Institute of Public Finance & Accountancy, Statistical Information Service, 3 Robert Street, London WC2N 6BH

Chelmer Population & Housing Model, Population & Housing Research Group, Anglia Polytechnic University, Victoria Road South, Chelmsford, Essex CM1 1LL

Civic Trust, 17 Carlton House Terrace, London SW1Y 5AW

Civil Aviation Authority, Aviation House, Gatwick Airport South, West Sussex RH6 0YK

Clwyd/Powys Archaeological Trust, 7A Church Street, Welshpool

Commission of the European Union, Directorate General for Environment, Nuclear

Safety and Civil Protection (DGXI), Rue de la Loi 200, B-1049, Brussels, Belgium

Commission for the New Towns, Glen House, Stag Place, London SW1E 5AJ

Computer programs and packages:

MATCH and VESPAN – Dr A. J. C. Malloch, Unit of Vegetation Science, Division of Biological Sciences, Lancaster University, Lancaster LA1 4YQ

MVSP – Kovach Computing Services, 85 Nant-y-Felin, Pentraeth, Anglesey LL75 8UY, Wales

TABLEFIT – TABLEFIT Programs, Institute of Terrestrial Ecology Monks Wood, Abbots Ripton, Huntingdon PE17 2LS

Council for British Archaeology, Bowes Morrell House, 111 Walmgate, York YO1 2UA

Council for the Protection of Rural England, Warwick House, 215 Buckingham Palace Road, London SW1W 0PP

Council for Scottish Archaeology, c/o National Museums of Scotland, Queen St, Edinburgh EH2 1JD

Countryside Commission, John Dower House, Crescent Place, Cheltenham GL50 3RA

Countryside Council for Wales, Plas Penrhos, Ffordd Penrhos, Bangor, Gwynedd LL57 2LQ

Department of Agriculture for Northern Ireland, Dundonald House, Upper Newtownards Road, Belfast BT4 3SB

Department of the Environment, 2 Marsham Street, London SW1P 3EB (also regional offices)

Department of the Environment, Air Quality Division, Romney House, 43 Marsham Street, London, SW1P 3PY

Department of the Environment for Northern Ireland, Environment Service, Calvert House, 23 Castle Place, Belfast BT1 1FY

Department of National Heritage, 2 Marsham Street, London SW1P 3EB

Department of Trade and Industry, Environment Unit, 151 Buckingham Palace Road, London SW1W 9SS

Department of Transport, 2 Marsham Street, London, SW1P 3EB (also regional offices)

Dyfed Archaeological Trust, The Old Palace, Abergwili, Carmarthen

Energy Efficiency Office, 1 Palace Street, London SW1E 5HE

English Heritage, Fortress House, 23 Saville Row, London W1X 1AB

English Nature, Northminster House, Northminster, Peterborough, Cambridgeshire PE1 1UA

Environmental Information Centre, Biological Records Centre, Institute of Terrestrial Ecology, Monks Wood, Abbots Ripton, Huntingdon PE17 2LS

European Community Directives are published in the *Official Journal of the European Community* (Brussels: European Commission). A CD ROM database of European Law is held on the JUSTIS CD-ROM, but this excludes Annexes of Directives.

Food and Agriculture Organisation, Environment and Energy, Via delle Terme di Caracalla, 00100 Rome, Italy

Forestry Commission, 231 Corstorphine Road, Edinburgh EH12 7AT

General Register Office (Scotland), Ladywell House, Ladywell Road, Edinburgh EH12 7TF

Georgian Group, 37 Spital Square, London E1 6DY

Glamorgan/Gwent Archaeological Trust, Ferrywise Warehouse, Bath Lane, South Dock, Swansea

Groundwork Foundation, 85/87 Cornwall Street, Birmingham B3 3BY

Gwynedd Archaeological Trust, College Road, Bangor, Gwynedd LL57 2D9

Health and Safety Executive, Baynards House, 1 Chepstow Place, Westbourne Grove, London W2 4TF

Her Majesty's Inspectorate of Pollution, Romney House, 43 Marsham Street, London SW1P 3PY

Her Majesty's Stationery Office, PO Box 276, London SW8 5DT

Historic Buildings and Monuments Commission (see English Heritage)

Historic Scotland (see Scottish Office)

Indic-Airviro, S–60186 Norrkoping, Sweden

Institute of Environmental Assessment, Gregory Croft House, Fen Road, East Kirkby, Lincolnshire PE23 4DB

Institute of Field Archaeologists, Minerals Engineering Building, University of Birmingham, PO Box 363, Birmingham B15 2TT

Institute of Freshwater Ecology, The Ferry House, Far Sawrey, Ambleside, Cumbria LA22 0LP

Institute of Hydrology, Maclean Building, Crowmarsh Gifford, Wallingford, Oxfordshire OX10 3BB

Institute of Terrestrial Ecology, Monks Wood, Abbots Ripton, Huntingdon, Cambridgeshire PE17 2LS

International Institute for Environment and Development, 3 Endsleigh Street, London WC1H 0DD

International Union for the Conservation of Nature (IUCN), Avenue de Mont Blanc, CH-1196, Geneva, Switzerland

JMP Consultants Ltd, 172 Tottenham Court Road, London W1P 9LG

Joint Nature Conservation Committee, Monkstone House, City Road, Peterborough, Cambridgeshire PE1 1JY

Local authorities (UK): see various directories

The Mammal Society, Conservation Office, Zoology Department, Woodland Road, Bristol BS8 1UG

Meteorological Office, Penderel House, 284 High Holborn, London SC1V 7HX

Meteorological Office Air Pollution Consultancy Group, Meteorological Office, Johnson House, London Road, Bracknell, Berkshire RG12 2SY

Ministry of Agriculture, Fisheries and Food, Environmental Protection Division, Nobel House, 17 Smith Square, London SW1P

Ministry of Defence, Main Building, Whitehall, London SW1A 1HE

National On-Line Manpower Information System, Unit 3P, Mountjoy Research Centre,

University of Durham, Durham DH1 3SW
National Power plc, Sudbury House, 15 Newgate Street, London EC1A 7AV
National Rivers Authority, 30–34 Albert Embankment, London SE1 7TL
National Rivers Authority (Thames Region), Howard House, 10–11 Albert Embankment, London SE1 7TG
National Rivers Authority (Welsh Region), Rivers House/Plas-yr-Afon, St Mellors Business Park, St Mellons, Cardiff CF3 0EG
National Society for Clean Air and Environmental Protection, 136 North Street, Brighton BN1 1RG
National Trust, 36 Queen Anne's Gate, London SW1H 9AS
Nature Conservancy Council for England (see English Nature)
NETCEN (National Environmental Technology Centre), AEA Technology, Culham, Abbingdon, Oxfordshire OX14 3DB
Nuclear Electric plc, Barnett Way, Barnwood, Gloucester GL4 7RS

Office of Population Censuses & Surveys, St Catherine's House, 10 Kingsway, London WC2B 6JP
Office of Population Censuses & Surveys, Segensworth Road, Titchfield, Fareham, Hampshire PO15 5RR
Open Spaces Society, 25A Bell Street, Henley on Thames, Oxon RG9 2BA
Organisation for Economic Cooperation and Development, 2 rue Andre Pascal, 75775 Paris Cedex 16, France

Pond Action, School of Biological and Molecular Sciences, Oxford Brookes University, Oxford OX3 0BP
Powergen plc, 55 New Broad Street, London EC2M 1JJ

Radiocommunications Agency, Waterloo Bridge House, Waterloo Road, London SE1 8VA
Ramblers Association, 9 Frederica Road, London E4 7AL
Royal Commission of Ancient and Historical Monuments of Scotland, John Sinclair House, 16 Bernard Terrace, Edinburgh EH8 9NX
Royal Commission on the Historical Monuments of England (see English Heritage)
Royal Society for Nature Conservation, The Green, Witham Park, Waterside South, Lincoln LN5 7JR
Royal Society for the Protection of Birds, The Lodge, Sandy, Bedfordshire SG19 2DL
Royal Society for the Protection of Birds: Northern Ireland, Belvoir Park Forest, Belvoir, Belfast
Rural Development Commission, 11 Cowley Street, London SW1P 3NA

Scottish Civic Trust, 24 George Square, Glasgow G2 1EF
Scottish Natural Heritage, 12 Hope Terrace, Edinburgh EH9 2AS
Scottish Office, Environment Department, St Andrews House, Edinburgh EH1 3DD
Scottish River Purification Boards: (different boards for Tay (in Perth), Highland (Dingwall), Tweed (Galashiels), Forth (Edinburgh), Clyde (East Kilbride), North East (Aberdeen), and Solway (Dumfries))
Scottish Wildlife Trust, Crammond House, Kirk Crammond, Edinburgh EH4 6NS

315

SERPLAN (London and South East Regional Planning Conference), 14 Buckingham Gate, London SW1E 6LB
Society for the Protection of Ancient Buildings, 37 Spital Square, London E1 6DY
Soil Survey and Land Research Centre, Silsoe Campus, Silsoe, Bedford MK45 4DT
Soil Survey of England and Wales, Rothamsted Experimental Station, Harpenden, Herts AL5 2JQ
Sports Council, 16 Upper Woburn Place, London WC1H 0QP

Theatres Trust, 22 Charring Cross Road, London SC2H 0HR
Transport Research Laboratory (TRL, formerly TRRL), Old Wokingham Road, Crowthorne, Berkshire RG11 6AU

Ulster Wildlife Trust, 3 New Line, Crossgar, Downpatrick, County Down BT30 9EP
UK statutes: a convenient guide is Halsbury's Laws of England
United Nations Economic Commission for Europe, Palais des Nations, CH 1211 Geneva 10, Switzerland
United Nations Educational Scientific and Cultural Organisation, 7 Place de Fontenoy, 75700 Paris, France
United Nations Environment Programme, Post Box 30552, Nairobi, Kenya
United States Department of Commerce, Computer Products, National Technical Information Service, Springfield, VA 22161, USA
United States Environmental Protection Agency, Office of Air Quality Planning & Standards, Research Triangle Park, NC 27711, USA

Victorian Society, 1 Priory Gardens, Bedford Park, London S4 1TT

Welsh Office, Cathays Park, Cardiff CF1 3NQ
West Midlands Joint Data, Solihull, Midlands
World Bank, 66 Avenue d'Iena, 75116 Paris, France
World Conservation Union (see IUCN)
World Health Organisation, 20 avenue Appia, 1211 Geneva 27, Switzerland

APPENDIX C
The rational method for runoff prediction

A key feature is the use of a single, dimensionless *runoff coefficient* (*C*) which is defined as the ratio of runoff rate to rainfall intensity; if half the rainfall becomes surface runoff, $C = 0.5$ (Hudson 1981). The coefficient has been found to depend largely on infiltration rates and hence on soil type, topography, vegetation cover, and the presence of impervious surfaces. Published values based on these factors are available, as in Tables C.1 and C.2.

Table C.1 Values of the runoff coefficient (*C*) for combinations of topography and other rural or urban factors

(a) Combinations of topography, vegetation type and soil texture

Vegetation and topography	Soil texture		
	Open sandy loam	Clay and silt loam	Heavy clay
Woodland			
Flat (0–5% slope)	0.10	0.30	0.40
Rolling (5–10% slope)	0.25	0.35	0.50
Hilly (10–30% slope)	0.30	0.50	0.60
Pasture			
Flat	0.10	0.30	0.40
Rolling	0.16	0.36	0.55
Hilly	0.22	0.42	0.60
Cultivated			
Flat	0.30	0.50	0.60
Rolling	0.40	0.60	0.70
Hilly	0.52	0.72	0.82

(b) Combinations of topography and extent of impervious urban surfaces

Urban area topography	30% of surface impervious	50% of surface impervious	70% of surface impervious
Flat	0.40	0.55	0.65
Rolling	0.50	0.65	0.80

Source: Hudson (1981).

If, as is likely, a catchment contains areas with different *C* values, or a proposed development includes different land uses, a weighted average can be calculated and applied to the whole system, as in Example 1.

Table C.2 Values of the runoff coefficient (C) for urban areas.

Land uses			Surfaces		
Commercial	Town centre	0.7–0.95	Roads	Asphalt and concrete	0.7–0.95
	Suburban	0.5–0.7		Brick	0.7–0.85
Industrial	Light	0.5–0.8	Drives and		0.75–0.85
	Heavy	0.6–0.9	walks		
			Roofs		0.75–0.95
Residential	Single-family	0.3–0.5	Lawns on	Flat (≤2% slope)	0.13–0.17
	Multi-units, detached	0.4–0.6	heavy soil	Average (2–7% slope)	0.18–0.22
	Multi-units, attached	0.6–0.75		Steep (≥7% slope)	0.25–0.35
	Suburban	0.25–0.4			
Other	Railyards	0.2–0.35	Lawns on	Flat (≤2% slope)	0.05–0.10
	Playgrounds	0.2–0.35	sandy	Average (2–7% slope	0.10–0.15
	Parks	0.1–0.25	soil	Steep (≥7% slope)	0.15–0.20
	Unimproved	0.1–0.3			

Sources: Dunne & Leopold (1978).

Example 1 A 500 ha catchment contains 150 ha of flat pasture on heavy clay, 250 ha of rolling cultivated land on clay/silt loam, and 100 ha of rolling urban area with 70% of surface impervious. From the data in Table C.1 the average runoff coefficient is

$$C = (0.3 \times 0.4) + (0.5 \times 0.6) + (0.2 \times 0.8) = 0.58 \tag{C.1}$$

Maps and aerial photographs can be used to estimate the area of the catchment, and of relevant components within this. The appropriate values of C and catchment area are incorporated in the following formula.

$$Q = 0.0028\,CiA \tag{C.2}$$

where:

Q = the design peak runoff rate in $m^3\ s^{-1}$ (cumecs)

C = the runoff coefficient

i = rainfall intensity in $mm\,h^{-1}$ for a duration equal to the time of concentration

A = the catchment area in ha

The *design peak runoff rate* is the maximum flow rate associated with a given **design storm**. The *time of concentration* is the time taken for runoff to flow from the farthest point in the catchment (in terms of flow time, not necessarily distance) to the point of concern. It is assumed to be the shortest time for the whole catchment to contribute to flow at the point of concern, and thus determines the time of peak flow at this point. It is also used to select an appropriate rainfall intensity. For example, if the time of concentration is 30 min, and the highest rainfall likely to occur in a 30-min period in a 50 year interval (the 50-year, 30-min design storm) is 25 mm, then this is the value of i selected. There are several methods for calculating the time of concentration. A simple, widely used method (Kirpich 1940) uses the formula

$$T = 0.02L^{0.77}S^{-0.383} \tag{C.3}$$

where:

T = time of concentration in minutes

L = maximum length of flow in m

S = average stream (catchment) gradient, or the difference in elevation between the point of concern and the most remote point divided by the length, L.

Some values derived by this method are shown in Table C.3.

Table C.3 Time of concentration in catchments calculated from maximum length of flow and catchment gradient (method of Kirpich 1940).

Maximum length of flow (m)	Catchment gradient (%)						
	0.05	0.1	0.5	1.0	2.0	5.0	10.0
	Time of concentration (min)						
100	12	9	5	4	3	2	1.5
200	20	16	8	6	5	4	3.3
500	44	34	17	13	10	8	6.6
1000	75	58	30	23	18	13	10
2000	130	100	50	40	31	22	15
3000	175	134	67	55	42	30	22
4000	216	165	92	70	54	38	30
5000	250	195	105	82	65	45	35

Source: Hudson (1981).

Example 2   A 500 ha site catchment has an average runoff coefficient of 0.58 (as in Eq. C.1), and a furthest point (in time) that consists of a length of flow of 1000 m over which there is a difference in elevation of 5 m (i.e. a 0.5% gradient). From Equation C.3 or Table C.3, the time of concentration is 30 min, and if the 50-year, 30-min design storm is 25 mm, the design peak runoff is

$$Q = 0.0028\,CiA = 0.0028 \times 0.58 \times 25 \times 500 = 20.3\,\text{cumecs} \qquad (C.4)$$

To assess an increase in runoff associated with a proposed development, appropriate present and predicted values of C can be selected.

Example 3   A 500 ha unimproved catchment is to be developed into a light industrial park. Using (a) the maximum present and predicted values of C in Table C.2, i.e. 0.3 (unimproved) and 0.80 (developed), and (b) the same design storm as in Example 2, the present and predicted design peak runoff rates (Q_0 and Q_1 respectively) are

$$Q_0 = 0.0028C_0iA = 0.0028 \times 0.3 \times 25 \times 500 = 10.5\,\text{cumecs} \qquad (C.5)$$

$$Q_1 = 0.0028\,C_1iA = 0.0028 \times 0.8 \times 25 \times 500 = 28.0\,\text{cumecs} \qquad (C.6)$$

APPENDIX D
Geographical information systems
Agustin Rodriguez-Bachiller

Geographical information systems (GIS) are computer-based databases that include spatial references for the different variables stored, so that maps of such variables can be displayed, combined and analyzed with relative speed and ease. Developing this speed and ease of display and combination of maps has taken over 20 years of development of computer technology, and still today the computational problems of storing, retrieving and displaying maps and the improvements in solving these problems is what drives the GIS technology, with the analytic aspect of GIS lagging behind. Today, there are about 1,000 commercial firms engaged in GIS in the world, covering a market of about $2 billion per year, with a rate of growth estimated at about 25–35% each year, suggesting a forecast (perhaps not entirely justified) of about $8 billion for the year 2000.

GIS are a combination of a computerized-cartography system (that stores map-data) and a database-management system (or DBMS, that stores attribute-data, an attribute being a characteristic of a map-feature, like land use of an area or slope of a road). They share the issues and problems that these two types of system have, namely issues of:

- data capture and storage
- data manipulation and analysis
- presentation of results
- use and value.

D.1 Data capture and storage

The technology for GIS map-data capture is quite varied, and changing rapidly. *Primary data capture* (from reality) techniques include: (a) ground surveys based on sampling; (b) remote-sensing based on classifying "pixels" on a satellite photograph: the Landsat (US) series are available since 1972 with a frequency of 16 days and 20–30 m resolution; Spot (France) pictures are available since 1986, every 2.5 days, with 10–20 m resolution; and (c) the hand-held Global Positioning System that can register instantly the coordinates of a point with accuracies of less than 1 m, probably the most important hardware advance of recent times.

Secondary data capture (from maps) techniques include: (a) digitizing on a magnetic table the "caricature of a line", a line broken down into points and straight segments between them; (b) using automatic scanners, with accuracies of up to 500 dots per inch (20 dots per mm), which are still expensive and require a lot of storage space in the computer; and (c) overlay-digitizing (or "heads-up" screen-digitizing), which combines scanning and digitizing facilities: a map is scanned and displayed on the screen, and vector-features are superimposed by digitization using the "mouse".

Data (raw facts) become information when interpreted by conceptual data-models. The type of *map-data model* used is one of the clearest dividing lines between different types of GIS: (a) the traditional spaghetti models, which represent geometric objects as unrelated series of straight lines and points; (b) regular tessellation models or "raster" models (see Fig. D.1) which are stored using more or less abbreviated versions of a matrix-file, where the different cells (rasters) are stored by rows and columns in order, with their value. Such maps can store only one value or attribute in each cell. Well known raster-based GIS are GRASS (free, by the US Army), IDRISI (very cheap, by Clark University, Worcester, Massachusetts, USA), and SPANS (powerful, £6,000, by TYDAC); (c) irregular-tessellation "vector" models (see Fig. D.2) represent geometric features (points, lines and polygons) as well as their topological relationships (links, crossings, overlaps, adjacencies) with files containing the map-dimensions, and files containing the feature-attributes. Vector-data can be stored by "layers" (each containing one feature) or by "objects" (the latest approach, still in the research stage), with the attention being on individual objects, their properties, and their memberships in different "classes". A typical vector-based topographic map needs about 10Mb of storage. Well known vector-based GIS are ARC-INFO by ESRI (about £30,000 for workstations or £10,000 for microcomputers), and TIGRIS by INTERGRAPH; (d) integrated GIS capable of combining vector and raster data are being researched now as the obvious future development of this industry.

A	A	A	A	C	C	C	C
A	A	A	A	C	C	C	C
A	A	A	B	B	C	C	C
A	A	B	B	B	B	B	B
A	B	B	B	B	B	B	B

Figure D.1   Raster.

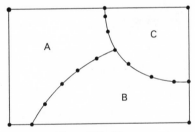

Figure D.2   Vector.

The most common *attribute-data model* used in the GIS database is the entity–relationship model: entities have several attributes, each with a value, and relationships between entity-sets can be seen as tables of rows (entities) and columns (attributes). An alternative DBMS model gaining popularity at the moment is the object-orientated model: objects are classified into classes and subclasses, with the possibility of inheritance of properties between them.

D.2   Data manipulation and analysis

Despite the complicated array of technical operations that can be performed with a full GIS, in terms of spatial analysis the tasks that can be performed easily by such systems are quite limited, and can be summarized in a fairly typical short list:

- calculating areas, and sometimes also volumes under a certain altitude (like the water-content of a lake);

- calculating straight-line distances and, in some systems, distances along networks;
- identifying viewing areas from a point;
- identifying the nearest points to selected features;
- using distances to construct buffer-zones around some features, with which to include or exclude parts of other maps (buffering);
- interpolating attribute-values between those recorded for a given set of points;
- drawing contour-lines using interpolated values between points; and
- superimposing maps to produce combined maps (map-overlay) of simple maps or of some of the above (see Fig. D.3).

The limited range of analytical tasks has been one of the classic areas of criticism of GIS. Stan Openshaw of Leeds University argues that, of the 1000 commands of a typical GIS, none relates to true spatial analysis, and functions like those listed above really correspond to data description more than to real spatial analysis, which should include near-

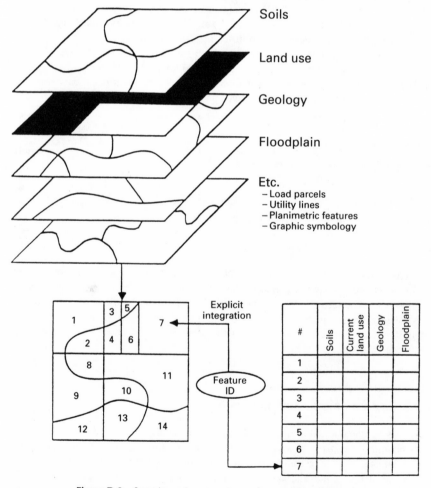

Figure D.3 Superimposing maps to produce combined maps.

est neighbour analysis or quadrat methods for point-features; network and graph analysis, fractal analysis and edge detection for line-features; and shape measurement, spatial auto-correlation, spatial regression, multivariate analysis, cluster analysis and regionalization, spatial interaction modelling and location modelling for area features.

D.3 Presentation of results

The output of GIS is probably the best developed and most appealing aspect of these systems. Output can be produced for a range of devices (the computer screen, pen-based or electrostatic plotters, laser printers), and can be classified by its dimensional level:

- 2-D displays (maps) are most common, with the associated problem of map-generalization: the level of simplification in the representation appropriate to the scale and purpose of the map (below 1:10,000 generalization is unnecessary, but as the scale is decreased and covers a wider area less detail is possible and some gen-eralization must take place);
- so-called "2.5-D" digital elevation models (DEM) or digital terrain models (DTM) which use the z-dimension over an x–y plane. These can be used to calculate slope and orientation values; calculate convexity and concavity; derive specific features like drainage channels and networks, basins and hills; display perspective and "drape" other land-use maps over them; map the area visible from certain points; and estimate relief-shadows and cast-shadows to help give the impression of three-dimensionality.
- 3-D models which link GIS to the computer-aided design (CAD) tradition, currently the object of considerable research, can be used in geological studies, for oil explo-ration, and to represent the shape and size of deposits or reservoirs.

The current trend in GIS output (especially on the computer screen) is towards inter-active multimedia output that combines maps, photographs, sound, animated drawings (for instance to simulate movement), and video recordings as part of the emerging approach of "hypermedia", in which each of these media represent a "zooming in or out" of one another.

D.4 Use and value

Implicit in the development of GIS is the notion that their rôle is to support decision-making. GIS are being considered more and more in combination with other decision-related technologies: (a) expert systems and artificial intelligence techniques to improve the cartographic capabilities of GIS on the one hand, or their decision-making capabilities (with natural query languages and the like) on the other; and (b) spatial decision support systems that put GIS in the context of unstructured problems for which no predetermined solution-plan exists, approached through "what if" questions, and with an implicit learn-ing element.

A full GIS for a large organization can be quite expensive. Initial costs can range from hundreds of thousands of pounds up to millions, with over 50% of the cost in GIS opera-tion being data-related and 20–30% being staff-related. The main benefits of GIS seem to be associated with long-term cost-savings in map-making activities, and with developing unforeseen uses for the system that improve the overall performance of the organization. The cash-flow of GIS-related experiences in large organizations suggest initial higher costs (about twice) than manual mapping, with GIS costs dropping below the manual alternative after about 7 years. GIS also raises legal concerns: issues of access (data ownership, privacy, political control of information), and issues of liability derived from errors or from GIS misuse.

Areas of GIS applications are varied:

* development of base-cartography
* Land Information Systems (LIS) or land inventories, the oldest form of GIS in many countries
* utilities-management for networked or piped utilities, Automated Mapping and Facility Management (AM/FM) in utility companies
* navigation systems for cars, ships and others with the Global Positioning System
* trainers and simulators; and
* Census-related use, like the US street-referenced TIGER files.

GIS applications in EIA are also varied:

* terrain maps for slope and drainage analysis
* land-resource information systems for land management
* soil information systems for soil studies
* geoscientific modelling of geological formations
* disaster planning related to geographically localized catastrophes
* analysis of irrigation suitability
* monitoring of development
* contamination and pollution monitoring
* flood studies
* linking environmental databases, like the European attempt CORINE; and
* constructing global databases for environmental modelling.

D.5 Further reading

The literature on GIS is vast. Fortunately, there is a benchmark publication by Maguire et al. (1991) that is an excellent (but rather expensive at £120) reference source on GIS research up to 1990. For a good introduction to all GIS issues see Antenucci et al. (1991). On environment-related GIS see Goodchild et al. (1993), which contains papers given at a conference in the USA in 1991. The best-known journal on GIS is the International Journal of Geographical Information Systems. Useful magazines containing up-to-date information on products and gossip are *Mapping Awareness* for the UK, and *GIS World* for the USA, and now also *GIS Europe* for other countries.

Conferences on GIS include research/development conferences like the GIS/LIS Conference in the USA, the Canada GIS Conference, or the European EGIS Conference. Application/diffusion conferences include the yearly URISA (Urban and Regional Information Systems Association) Conference in the USA and the UDMS (Urban Data Management Symposium) in Europe (every two years).

APPENDIX E

Outline of the NCC Habitat Classification (NCC 1990)

Categories		Definitions and features
A Woodland and scrub		
A1 Woodland		Vegetation dominated by trees, > 5 m high when mature, forming a canopy.
A1.1	Broadleaved	> 10% broadleaved deciduous trees (e.g. oaks, beech, ash) in the canopy.
A1.1.1	Semi-natural	Old woodlands and plantations with semi-natural shrub and ground flora.
A1.1.2	Plantation	> 30% of the canopy evidently planted, usually recent.
A1.2	Coniferous	> 10% conifers in the canopy.
A1.2.1	Semi-natural	As A.1.1.1, usually dominated by pines.
A1.2.2	Plantation	As A.1.1.2, usually commercial plantations of larches, firs and spruces.
A1.3	Mixed	10–90% of either broadleaved or conifer species in the canopy.
A2 Scrub		**Seral** or **climax** vegetation dominated by native shrubs (usually < 5 m tall), e.g. common gorse (but not dwarf gorses), broom, juniper, hawthorn, bramble, blackthorn, dog rose, grey willow (and all willow **carr** < 5 m tall). May be dense (A.2.1) or scattered (A.2.2).
A3 Scattered trees		Tree cover < 30%, e.g. on pasture/parkland, heath, bog, limestone pavement.
A4 Felled woodland		Only used when future land use is uncertain.
B Grassland and marsh		Herbaceous vegetation usually dominated by **graminoids**.
B1 Acid grassland		Often unenclosed, e.g. hill grazing; soils acid (pH < 5.5); usually species-poor.
B1.1	Unimproved[1]	Typically abundant species include sheep's fescue, brown bent, wavy hairgrass, mat grass, hard rush, sheep's sorrel, and heath bedstraw.
B1.2	Semi-improved	Usually less species-rich (especially in **forbs**).

1 Unimproved grasslands are not heavily grazed, drained or treated with herbicides; semi-improved grasslands show some influence of one or more of these factors; and improved grasslands are markedly affected.

Categories		Definitions and features
B2 Neutral grassland		Usually enclosed and relatively intensively treated; soils neutral (pH 5.5–7.0).
B2.1	Unimproved	Usually species-rich in grasses and forbs, e.g. old **meadows**.
B2.2	semi-improved	Usually less species-rich (especially in forbs).
B3 Calcareous grassland		Often unenclosed; soils calcareous (pH usually > 7.0).
B3.1	unimproved	Can be: little grazed and hence tall with large grasses, e.g. upright brome; or grazed and hence short with smaller grasses, e.g. sheep's fescue, and a variety of forbs.
B3.2	semi-improved	Usually less species rich, especially in forbs.
B4 Improved grassland		Usually: > 50% of agricultural species (e.g. perennial rye grass, clovers); species-poor in forbs; with generally bright green, even sward.
B5 Marsh / marshy grass-land		On wet but predominantly mineral soils (or peat < 0.5 m deep); usually grass-dominated but with high proportions of rushes, sedges and **hydrophilous** forbs, e.g. meadow sweet.
C Tall herb and fern		
C1 Bracken		Areas dominated by bracken (C1.1) or with scattered bracken patches (C1.2)
C2 Species-rich ledges		Upland ledges with large forbs (e.g. wild angelica) and ferns.
C3 Other		Stands of tall herbs such as rosebay willow-herb and nettle.
D Heathland		
D1 Dry dwarf shrub heath		Mainly on acid soils; > 25% cover by **ericoids** (e.g. heathers) or dwarf gorses
D2 Wet dwarf shrub heath		As D1 but with purple moor grass, sedges and **sphagna**; intergrades with **mire**.
D3 Lichen/bryophyte heath		Montane & some lowland areas; < 30% **vascular plant** cover.
E Mire		Peatland; water table at or just below ground level; peat normally > 0.5 m deep.
E1 Bog		**Ombrogenous**; **oligotrophic** and acid; vegetation mainly Sphagna, ericoids and cotton sedges.
E1.6.1	Blanket bog	In uplands; generally covers land surface except steep slopes.
E1.6.2	Raised bog	In flood plains, often developed over glacial lakes.
E1.7	Wet modified	Sphagna replaced by purple moor grass or deer grass, or bare patches.
E1.8	Dry modified	Sphagna replaced by other mosses and lichens.
E2 Flush/spring		Where groundwater seeps to the surface, usually on thin peat.
E3 Fen		**Geogenous**; can be divided into (a) *Rich fen* – **eutrophic**, pH > 5, species-rich, with tall plants, e.g. blunt-flowered rush and black bog-rush, and (b) *Poor fen* – **oligotrophic**, pH < 5, species-poor and *Sphagnum*-rich.

Categories			Definitions and features
E3.1		Valley mire	In valleys; **soligenous**; Rich fen or Poor fen depending on local geology.
E3.2		Basin mire	In depressions; **topogenous**; often Poor fen due to lack of nutrient inflow.
F Swamp, marginal and inundation			Have standing water (water table above ground level) permanently or for most of the year.
F1 Swamp			Transitional between open water and land; dominated by tall, emergent plants, e.g. reeds and other large grasses, sedges, and rushes.
F2	F2.1	Marginal	Narrow strips of emergent vegetation at margins of lowland water-courses.ʼ
	F2.2	Inundation	Periodically submerged, e.g. on river gravels, lake and reservoir margins etc.
G Open water			Beyond the limits of emergent vegetation; may contain **phytoplankton** and submerged, free floating or floating-leaved **macrophytes**.
G1 Standing water			
	G1.1	Eutrophic	Water often turbid/green (due to algae); pH > 7; substrate usually organic mud.
	G1.2	Mesotrophic	Medium nutrient status; pH approximately neutral or slightly below.
	G1.3	Oligotrophic	Water clear; pH < 7; plankton sparse; substrate rocky, sandy or peaty.
	G1.4	Dystrophic	Water peat-stained; pH low (3.5–5.5), plankton sparse.
	G1.5	Marl/tufa	Water clear, usually eutrophic or mesotrophic, highly calcareous.
	G1.6	Brackish	Usually coastal; water somewhat saline.
G2 Running water			Rivers and streams; can be subdivided (like standing water) into six categories (G2.1–G2.6) in terms of water quality.
H Coastlands			
H1 Intertidal (seashores)			Subdivided mainly according to substrate, e.g. H1.1 – mud/sand, H1.3 – rocky
H2 Saltmarsh			On mud in estuaries or sheltered bays; vegetation dominated by cord grass, or more species-rich with sea poa and forbs such as sea lavender.
H3 Shingle above high tide			Open vegetation with scattered vascular plants and lichens.
H4 Rock above high tide			Mainly lichen communities with scattered vascular plants in crevices.
H5 Strandline vegetation			Open community in front of sand dunes; small plants such as sea sandwort.
H6 Sand dune			Wind-blown sand stabilized by vegetation; dune systems normally show zonation of dune age and associated vegetation along sea–land axis.
	H6.4	Dune slack	Moist/wet depressions between dunes, with marshy vegetation.
	H6.5	Grassland	On consolidated, flat dunes; dominated by grasses, e.g. creeping fescue.

Categories	Definitions and features
H6.6 Dune heath	On consolidated dunes; dominated by ericoids, especially heather.
H6.7 Dune scrub	As inland scrub, but often dominated by coastal species, e.g. sea buckthorn.
H6.8 Open dune	Includes: Fore dunes – with sea couch and lyme grass; Yellow dunes – dominated by marram grass; Grey dunes – with mosses, lichens and forbs.
H8 Maritime cliff and slope	Hard cliff (H8.1) and soft cliff (H8.2) with < 10% vascular plant cover; crevice and ledge vegetation (H8.3); coastal grassland (H8.4) and heathland (H8.5).
I Rock exposure & waste	All natural and artificially exposed surfaces with < 10% vegetation cover.
I1 Natural exposures	Includes inland cliff (I1.1), scree (I1.2) and limestone pavement (I1.3).
I2 Artificial exposures	Includes quarry (I2.1), spoil (I2.2), mine (I2.3), and refuse-tip (I2.4).
J Miscellaneous	
J1 Cultivated/disturbed land	
J1.1 Arable	Includes cropland, horticultural land and recently re-seeded grass/clover **leys**.
J1.2 Amenity	Includes intensively managed grassland, e.g. lawn, golf course fairway, etc.
J1.3 Ephemeral	Short patchy vegetation on freely drained, usually thin, soils of derelict land.
J1.4 Introduced shrub	Dominated by non-native shrubs, e.g. *Rhododendron, Cotoneaster*.
J2 Boundaries	Includes: intact hedge (J2.1), defunct hedge (with gaps) (J2.2), species-rich hedge (J2.4), wall (J2.5), dry ditch (J2.6) and earth bank (J2.8).
J3 Built-up areas	Includes caravan site (J3.4), sea wall (artificial material) (J3.5), and buildings (J3.6).
J4 Bare ground	Any bare substrate not included elsewhere in the classification.
J5 Other habitat	Any habitat not covered by the classification.

APPENDIX F
Sources of information on species and taxonomic groups

Key: T = terrestrial   F= freshwater   M = marine and intertidal
✓ = some information   ✓✓ = major emphasis

Species and groups	Major habitats	Identification guides	Distribution (including atlases)	Conservation status (incl. red data books)	Habitat requirements	Survey methods	References
Protected species (all groups)	T/F/M	✓		✓✓	✓✓		Whitten (1990)
Vascular plants – general	T/F/M	✓✓	✓	✓	✓	[1]	Clapham et al. (1987), Stace (1991), Tutin et al. (1964 et seq.)
			✓✓	✓	✓		Grime & Lloyd (1973), Perring & Walters (1990)
			✓	✓✓	✓		Curtis & McGough (1988), Lucas & Synge (1978), Perring & Farrell (1983), Stewart et al. (in press)
			✓	✓	✓✓		Grime et al. (1989), BES (1992)
Trees, shrubs and forbs	T/F	✓✓	✓		✓		Mitchell (1974), Rose (1981)
Graminoids (grasses, rushes, sedges)	T/F	✓✓	✓	✓	✓		Hubbard (1984), Rose (1989)
Pteridophytes (ferns, etc.)	T/F	✓✓	✓	✓	✓		Jahns (1983), Merryweather & Hill (1992), Rose (1989)
Nonvascular plants – general	T/F		✓	✓✓	✓		Hodgetts (1992)
Bryophytes – general	T/F	✓✓	✓	✓	✓		Smith (1978, 1990), Watson (1981)
			✓✓	✓	✓		Hill et al. (1991, 1992, 1994)
				✓✓			Stewart & Church (in press)
Sphagna	T/F	✓✓	✓	✓	✓		Daniels & Eddy (1990)
Algae – marine	M	✓✓	✓				Hiscock (1979, 1986)
Stoneworts	F		✓	✓✓	✓		Stewart & Church (1992)

Species and groups	Major habitats	Identification guides	Distribution (including atlases)	Conservation status (incl. red data books)	Habitat requirements	Survey methods	References
Lichens	T/F	✓✓	✓	✓	✓		Dobson (1992), Duncan (1970), Purvis et al. (1992)
				✓✓			Stewart & Church (in prep.)
Freshwater plants	F	✓✓	✓	✓	✓		Fitter & Manuel (1986), Haslam et al. (1975)
			✓✓				Croft et al. (1991)
			✓✓	✓✓			Palmer (1989), Palmer & Newbold (1983), Palmer et al. (1992)
			✓	✓✓	✓		Holmes (1990), Pond Action (1994b)
Vertebrates – general	T/F/M					✓✓	Buckland et al. (1993), Southwood (1978)
Mammals – general	T/F/M	✓	✓	✓	✓✓	✓	Corbet & Harris (1991)
		✓✓	✓	✓		✓	Arnold (1993), Velander (1983), Watson & Hewson (1973)
			✓✓	✓✓	✓✓	✓	Cresswell et al. (1990), Easterbee et al. (1991)
		✓		✓✓	✓		Morris (1993), Whilde (1993)
			✓✓	✓		✓✓	Staines & Ratcliffe (1987)
					✓	✓✓	Angerbjorn (1983), Buckland (1992), Gurnell & Flowerdew (1990), Harris et al. (1989), Hoodleff & Morris (1993), Mead-Briggs & Woods (1973), Putman (1984), Twigg (1975a,b)
Bats	T			✓	✓	✓	Hutson (1993)
			✓	✓	✓	✓✓	Mitchell-Jones (1987)
				✓	✓	✓✓	Kapteyn (1991)
Freshwater mammals, e.g. otter and water vole	F/T	✓✓	✓		✓	✓	Andrews & Crawford (1986), Chapman & Chapman (1982), Crawford et al. (1979), Green & Green (1987), Lenton et al. (1980), Strachan et al. (1990)
		✓✓	✓✓				Strachan & Jefferies (1993)
						✓✓	Kruuk & Conroy (1987), Kruuk et al. (1986), Woodroffe et al. (1990)
Marine mammals (whales, dolphins, seals)	M	✓			✓	✓✓	Hammond (1987), Hammond & Thompson (1991), Hiby & Hammond (1989), Thompson & Harwood (1990), Ward et al. (1988),

Species and groups	Major habitats	Identification guides	Distribution (including atlases)	Conservation status (incl. red data books)	Habitat requirements	Survey methods	References
Birds	T/F/M	✓✓	✓	✓	✓✓		Cramp & Simmons (1977--)
		✓✓	✓	✓			Lack (1986), Wingfiled Gibbons et al. (1994)
			✓	✓✓	✓		Andrews & Carter (1993), Batten et al. (1990), Lloyd et al. (1991), Whilde (1993)
				✓		✓✓	Bibby (1984), Bibby et al. (1992), Brown & Shepherd (1993), BTO (1982, 1983, 1992)
Reptiles and amphibians	T/F	✓✓	✓		✓		Smith (1951), Arnold & Burton (1978)
			✓✓	✓✓	✓✓		NCC (1983), Hilton-Brown & Oldham (1991), Whilde (1993)
			✓✓	✓	✓		Corbett (1989), Swan & Oldham (1993)
		✓	✓	✓	✓	✓✓	BHS (1990), HCI (1993), JNCC (1994)
Fish – general	F/M	✓✓	✓		✓		Maitland (1972), Maitland & Campbell (1992), Potts & Reay (1987)
				✓✓			Maitland & Lyle (1993), Whilde (1993)
Salmonids	F/M		✓		✓✓		Crisp (1993)
Invertebrates – general	T/F	✓✓					Tilling (1987)
			✓✓	✓✓	✓		Ball (1986), Collins & Wells (1987), Wells et al. (1983)
					✓✓		Kirby (1992a)
					✓	✓✓	Brooks (1993a), Southwood (1978)
Aquatic macro-invertebrates – general	F	✓✓	✓		✓		Fitter & Manuel (1986)
					✓	✓✓	DoE (1979), Elliott & Tullett (1983), Pond Action (1994a)
Insects – general	T/F		✓	✓✓	✓		Shirt (1987)
			✓		✓✓	✓	Fry & Lonsdale (1991)
Aquatic insects – general	F				✓✓		Ward (1992)
Ephemeroptera (mayflies)[2]	F/T	✓✓					Elliott et al. (1988)
			✓	✓✓	✓		Bratton (1990b)
Odonata (damselflies & dragonflies)[2]	F/T	✓✓	✓		✓		Askew (1988), Hammond (1983)
			✓✓				Merritt et al. (in press)
						✓✓	Brooks (1993b), Moore & Corbet (1990)
Plecoptera (stoneflies)[2]	F/T	✓✓					Hynes (1977)
			✓	✓✓	✓		Bratton (1990b)
Orthoptera (crickets & grasshoppers)	T	✓✓	✓✓	✓	✓✓		Marshall & Haes (1988)

Species and groups	Major habitats	Identification guides	Distribution (including atlases)	Conservation status (incl. red data books)	Habitat requirements	Survey methods	References
Hemiptera (true bugs) – general[2]	T/F		✓	✓✓	✓		Kirby (1992b)
Hemiptera (water bugs)[2]	F	✓✓	✓		✓		Savage (1989)
Megaloptera (alderflies)[2]	F/T	✓✓	✓		✓		Elliott (1977), Elliott et al. (1979)
Neuroptera (spongeflies)[2]	F/T	✓✓	✓		✓		Elliott (1977)
			✓	✓✓	✓		Kirby (1991)
Lepidoptera (butterflies & moths)	T	✓✓	✓✓	✓	✓✓		Emmet & Heath (1979–), Goater (1986),
			✓	✓✓	✓		Heath (1981), Parsons (1993)
			✓	✓	✓✓	✓	BUTT (1986)
						✓✓	Pollard (1977), Thomas (1983), Waring (1994)
Trichoptera (caddisflies)[2]	F/T	✓✓	✓				Edington & Hildrew (1981), Wallace et al. (1990)
			✓	✓✓	✓		Wallace (1991)
Diptera (true flies)	T/F	✓✓	✓		✓		Stubbs & Chandler (1978), Unwin (1981)
			✓	✓✓	✓		Falk (1991a)
						✓✓	Disney et al. (1982)
Hymenoptera (bees, wasps, ants)	T	✓✓	✓		✓	✓	Prys-Jones & Corbet (1991), Wilmer (1985)
			✓	✓✓	✓		Falk (1991b)
Coleoptera (beetles) – general	T/F		✓	✓✓	✓		Hyman & Parsons (1992, 1994)
Coleoptera (water beetles)[2]	F	✓✓					Foster (1981–), Friday (1988)
Invertebrates other than insects	T/F/M		✓	✓✓	✓		Bratton (1991)
Tricladida (flatworms)[2]	F	✓✓	✓				Reynoldson (1978)
Hirudinea (leeches)[2]	F	✓✓	✓				Elliott & Mann (1979)
Mollusca – non-marine, general	T/F		✓✓				Kerney (1976)
Land Gastropoda (snails & slugs)	T	✓✓	✓		✓		Kerney & Cameron (1979)
			✓	✓✓	✓		Willing (1993)
Freshwater Gastropoda[2]	F	✓✓	✓		✓		Macan (1977)

Species and groups	Major habitats	Identification guides	Distribution (including atlases)	Conservation status (incl. red data books)	Habitat requirements	Survey methods	References
Freshwater Bivalvia (mussels)[2]	F	✓✓	✓		✓		Ellis (1978)
Malacostraca (shrimps, slaters, crayfish)[2]	F/M	✓✓	✓ ✓✓	✓✓	✓ ✓		Gledhill et al. (1993) Goddard & Hogger (1986), Holdich (1991)
Isopoda (woodlice, etc.)	T/F	✓✓			✓		Hopkin (1991)
Araneae (spiders)[2]	T/F	✓✓	✓ ✓	✓ ✓✓	✓ ✓		Fitter & Manuel (1986), Jones-Walters (1989), Roberts (1993) Merrett (1990)

1. Terrestrial plant species and groups are normally surveyed using the standard vegetation sampling methods (with modifications for aquatic habitats) referred to in Chapters 11–14.

2. Groups for which identification to species level is required in freshwater surveys.

Glossary

abundance See **species abundance**.

acid deposition Deposition on vegetation, soils and waters of acid substances such as sulphuric acid; can be divided into dry deposition (gravitational settling of particulates) and wet deposition/ acid precipitation (sulphur dioxide and nitrogen oxides scavenged by precipitation).

age structure The structure of a plant or animal community in terms of the proportions of individuals of different stages of maturity, e.g. seedlings, saplings and adults of trees in a woodland; can provide an estimation of regenerative success of different species and hence permit prediction of likely changes in **species composition**.

air pollutants Substances or energy (e.g. waste heat) in the atmosphere in such quantities and of such duration likely to cause: harm to plants or animals (including people); damage to materials (e.g. fabrics) and structures (e.g. buildings); changes in weather and climate; or interference with the enjoyment of life or property (e.g. as a result of effects of odours or noise).

air quality standard The concentration of a pollutant over a specified period above which adverse effects on health or the environment may occur, and which should not be exceeded. Health-based standards (often called primary standards) are usually legally enforceable; environment-based standards (often called secondary standards) may be long-term objectives that are not legally binding.

alkalinity (a) the state when the **pH** of a solution is >7; (b) more strictly the concentration of carbonates in water (its carbonate hardness) and hence its ability to resist (buffer) changes in pH, in which terms it is possible for water with pH <7.0 to have high alkalinity and for water with pH >7 to have low alkalinity. Values are often quoted in $mg\,l^{-1}$ calcium carbonate, but are better quoted in milli-equivalents of acid per litre, i.e. the amount of acid needed to change the pH.

alluvial soil A soil that has accumulated by deposition of water-borne sediments, e.g. in a river floodplain.

animal community An assemblage of animal species populations living together in a given location at a given time; usually associated with a particular vegetation type or complex; part of the total **community** of the location.

anthropogenic Generated and maintained, or at least strongly influenced, by human activities.

aquifer A rock stratum that contains groundwater and allows this to flow through. Depending on its geological composition, an aquifer will have a given porosity (and hence water-holding capacity) and permeability (which affects the potential rate of groundwater flow).

biodiversity A comprehensive term for the degree of nature's variety including the number and frequency of **ecosystems**, species and genes in a given assemblage (McNeely 1988); it can refer to global, regional or local systems.

biomass (of vegetation) The mass of plant material (and the energy this contains) present in a community at a given time; usually measured as dry weight per unit area, e.g. dry $g\,m^{-2}$.

biocide A lethal chemical, e.g. herbicides (designed to kill plants such as weeds) and pesticides (designed to kill animal pests).

biota All the organisms in a **community** or area.

bog A type of **mire**; traditionally distinguished from **fen** by being **oligotrophic**, acid, and **species-poor**; now usually distinguished in terms of water supply, i.e. **ombrogenous** rather than **geogenous**, although their dependence on rainwater means that bogs are always oligotrophic, acid, and species-poor (see Appendix E).

335

bryophytes Mosses and liverworts.

buffer zones Vegetated strips of land that are intended to screen **ecosystems** from impacts such as pollution or disturbance, and/or to reduce the area restrictions of protected sites; can be located (a) adjacent to developments or components of these, usually with the aim of filtering out pollutants, (b) around protected sites, with aims such as providing additional **habitat** for some animals, protecting the site from pollutants and disturbance, and perhaps encouraging expansion through species dispersal, or (c) within sites, usually with the aim or permitting their use for both amenity purposes and conservation.

capacity The traffic flows that a road or junction can accommodate; used as a starting point for the design of roads and junctions, but does not specify ultimate traffic flows; discussed in various DOT documents, which are reviewed in Simpson & Baker (1987).

carr A semi-terrestrial **wetland** dominated by **hydrophilous** trees such as willows and alders; usually a **seral community** that replaces herbaceous wetland communities and is eventually replaced by forest.

climate (at a given place) The totality of the **weather** experienced; not simply average weather, since climate includes extremes or deviations from the mean state of the atmosphere, e.g. fogs, frosts, and storms; the behaviour of the atmosphere over periods of weeks, months, seasons, years and decades, i.e. the integration of weather over long periods; usually characterized using long-term records, e.g. 30 years.

climax community/vegetation In general terms the **community/vegetation** type that develops as a result of **succession**, and is then sustained under a given **climate**, e.g. broadleaved deciduous forest in Britain; but it may be strongly influenced by other factors, e.g. many current communities are **anthropogenic** climaxes. Even climax communities change slowly in response to factors such as long-term changes in soil conditions.

colluvial soil A soil that has accumulated at the base of a steep slope as a result of the downward movement of material under gravity.

community (biotic) An assemblage of plant, animal and microbial species populations living in a given location at a given time; involves interrelationships between the species and is considered to constitute a functional system.

community gradient A continuum of **community** change, e.g. in terms of **species composition**, in response to environmental gradients such as changing soil conditions from one area to another; a gradual change from one community to another. Intergradation is considered by many ecologists to be the overriding pattern of natural communities, recognized community types being convenient but artificial abstractions.

controlled waters Defined by the NRA to include all groundwaters, rivers, streams, canals, lakes and reservoirs, the only exclusions being landlocked water bodies that do not drain into other controlled waters (e.g. clay-lined ponds) (see also **designated waters**).

CORINE biotope A site defined in CEC (1991a) as "an area of land or a body of water which forms an ecological unit of Community significance for nature conservation" in terms of the presence of threatened species (listed CEC 1991b), the presence of sensitive/threatened **habitat** types, and the site's richness for a taxonomic group or a collection of habitats. To qualify as an important site at the Community level, a biotope must satisfy specific conditions relating to the rarity of threatened species or habitats. However, these conditions are likely to be modified to conform with the criteria set out in the Habitats Directive (92/43/EEC) and thus enhance the use of the CORINE biotopes database as a database of SACs. CORINE biotopes are categorized using an hierarchical habitat classification (set out in CEC 1991b) in which major habitat types are characterized by a combination of physical and vegetation features (as in the NCC habitat classification), and subdivisions are characterized in terms of the plant communities (as in the NVC, but using European classifications such as that in Ellenberg 1988) except where a biotope lacking vegetation is an important habitat for animals, e.g. tidal mudflats.

denitrification Process of nitrogen removal from soils; occurs naturally in waterlogged soils by the action of denitrifying bacteria that utilize nitrate and release nitrogen gas (which is returned to the atmosphere).

design storm A rainstorm of given magnitude and probability; usually derived from existing rainfall data as the maximum rainfall likely to occur (during a given period, such as 1 hour) within a specified recurrence interval (return period) that is measured in years; e.g. a 50-year, 1-hour design storm is the probable maximum rainfall in a 1-hour period within any 50-year interval. Design storms are used for various purposes, including the design of flood-protection structures, and they may be selected in relation to the type of planned structure; e.g. recurrence intervals of 10 years are often considered adequate for storm sewers or vegetated flood-control structures, but intervals of 100 years may be stipulated for large structures such as dams, or where towns are threatened by major rivers.

designated waters Water bodies or sections of a river designated by the relevant water authority under one or more EU Directives, e.g. a river with an important salmon fishery could be designated under the Fisheries Directive and would have to comply with the water quality standards set in this directive.

determinand A catch-all term for any chemical or associated parameter that is measured, e.g. biological oxygen demand (BOD), nitrate, aluminium, turbidity, etc.

dominant species The species having most influence on others (subordinate species) in a community; in **plant communities**, usually the most abundant species. Two or more species can be co-dominant.

ecosystem An interactive system consisting of a **community** and the environment in which it exists.

edaphic Pertaining to soil or **peat**.

emissions inventory List of the location and type of pollutant sources in the area under study, together with the amount of pollutant discharged in a specified period.

emission standard The maximum amount or concentration of a pollutant allowed to be emitted from a specified source.

environmental factors All environmental variables that are known to affect organisms; can be divided into *abiotic factors,* which involve physical and chemical environmental components (e.g. water, temperature, light, oxygen, nutrients, **pH**, and toxins), and *biotic factors,* which involve interactions between organisms (e.g. competition, predation, parasitism and mutually beneficial relationships such as pollination).

ericoids Shrubby plants mainly belonging to the heath family (Ericaceae), e.g. heather, heaths, bilberries, cranberry; normally the main constituents of heathland vegetation (including lowland heaths and heather **moorlands**).

eutrophic Nutrient-rich (referring to soils, **peats**, waters, or whole **ecosystems**).

eutrophication Process or trend of nutrient enrichment, especially by nitrogen and phosphorus; may occur naturally, but usually refers to **anthropogenic** nutrient enrichment which often leads to ecosystem degradation through excessive nutrient loading.

evaporative pull Process by which water is drawn upwards in plants and soils as a result of evaporation from the leaves or soil surface.

evapotranspiration (ET) Total evaporation from a terrestrial **ecosystem**, including evaporation from soils and surface waters, **transpiration** from vegetation, and interception loss (although the latter is often quantified separately).

fen A type of **mire**; traditionally distinguished from **bog** by being **eutrophic, alkaline**, and **species-rich**; now usually distinguished in terms of water supply, i.e. **geogenous** rather than **ombrogenous**, and subdivided into (a) *rich fens* and *poor fens* on the basis of nutrient status and species-richness, and (b) *valley mires* and *basin mires* on the basis of topography and associated degree of water flow (see Appendix E).

food chain A major route of energy flow (in food) through a **community** from green plants (the **primary producers**) to other trophic levels such as herbivores and carnivores (in the *grazing food* *chain*), or bacteria, fungi, earthworms, etc.(in the *decomposer food chain*).

food web   The various pathways (network) of energy flows between the species populations of a **community**.

forbs   Broadleaved herbaceous flowering plants, usually having conspicuous flowers; often called herbs or flowers.

french drain   A trench over a drainage line, backfilled with layers of material (coarse at the bottom and grading to fine-grained at the top) to act as a sediment filter; usually with a vegetated surface (see Schwab et al. 1993).

geogenous   Referring to **mires** that are dependent largely on mineral groundwater rather than directly on rainfall, with **pH** and nutrient status therefore strongly influenced by the local geological conditions; used to distinguish between **fens** (which are geogenous) and **bogs** (which are **ombrogenous**). (See Appendix E)

graminoids   Grasses and grass-like plants (rushes and sedges).

groundwater   Water in a saturated zone beneath the Earth's surface (see **aquifer**).

growth forms   See **life forms**.

habitat   A place where an organism lives; a type of environment inhabited by particular species and/or **communities**; often chracterized by dominant plant forms, physical characters, or a combination of these, e.g. forest, grassland, marsh and stream habitats.

higher plants   See vascular plants.

hydrological   Concerning water in the environment, including surface water (lakes, rivers, etc.), subsurface water (soil water and groundwater) and atmospheric water.

hydrophilous   "water loving"; tolerant of wet conditions.

indicator species   Species indicative of (a) some environmental or historical influence (e.g. lichens can be atmospheric pollution indicators, and woodland ground-flora species can be indicative of ancient woodland), or (b) a **community** or **habitat** type (e.g. some species can be used to classify invertebrate communities, or are indicative of particular habitats).

isochrone   A map showing equal travel times to a given site.

key species (or **keystone species**)   A species on which several other species, or the functioning of an **ecosystem**, may depend.

labile organics   Organic compounds such as carbohydrates and proteins that are easily degraded in the aquatic environment, as opposed to non-labile (or refractory) compounds (such as many of those in wood) that are resistant to decay.

land cover   Combination of land use and vegetation cover.

landform   The shape and form of the land surface; a combination of slope and elevation.

leakages (economic)   The flows of money out of a national, regional or local economy, following from an initial injection of money into that economy. The most significant leakages are for taxation (direct and indirect), savings and improved goods and services.

ley   Arable land temporarily under grass.

life forms (growth forms)   Plant types in terms of morphology (body form) and function rather than taxonomy, e.g. (a) herbaceous (subdivided into **graminoids** and **forbs**) or woody (subdivided into trees or shrubs, with further subdivisions on the basis of leaf forms – broadleaved, needle-leaved etc.), or (b) categorized according to the position of the perennating (e.g. overwintering) buds in relation to ground level.

macroinvertebrates   Invertebrate animals large enough to be seen by eye, or can be captured using a sieve of mesh 0.5–1.0mm.

macrophytes   Plants large enough to be seen by eye.

marsh   Semi-terrestrial **wetland** on predominantly mineral soil; occurs where poor drainage results in periodic waterlogging, e.g. in river floodplains (see Fig. 10.2), but where this is not chronic enough to permit the development of **mire**; the vegetation is usually dominated by grasses, but has high proportions of sedges, rushes, and **hydrophilous forbs** (see Appendix E).

meadow Grassland maintained primarily for the production of hay; usually found on poorly drained land (low-lying and/or on heavy clay soil) where the traditional management practice is to take one late hay crop (usually in July) and then introduce grazing stock for a few months in the autumn (see **pasture**).

mire (peatland) A semi-terrestrial **wetland ecosystem** with a water table normally at or near ground level, and a substratum consisting predominantly of **peat**. The two main types of mire are **bog** and **fen**.

moorlands Upland **community/ecosystem** complexes on acid, **oligotrophic peat**; usually overly nutrient-poor rock, but where the peat is thick enough to isolate plant roots from the mineral substratum, can occur over limestone. Can be subdivided according to the dominant plant species, e.g. heather moor, cotton-sedge moor, *Molinia* moor, and bilberry moor (see also Appendix E).

multiplier A measure of the scale of the increase in income or employment in a local, regional or national economy, resulting from an initial injection of an amount of money into that economy.

net primary production The surplus energy when that needed by plants is subtracted from the amount they assimilate by photosynthesis (gross primary production); hence the energy available for other trophic levels in communities (see **food chain**); usually expressed as the increase in plant **biomass** during a growing season or year (see **primary production**)

non-labile organics See **labile organics**.

oligotrophic Nutrient-poor (referring to soils, **peats**, waters, or whole **ecosystems**).

ombrogenous Referring to **mires** that are maintained entirely by rainwater, and are thus restricted to high-rainfall areas; used to distinguish between **bogs** (which are ombrogenous) and **fens** (which are **geogenous**).

pasture Grassland maintained primarily for and by grazing, and on which grazing stock is kept for a large part of the year (see **meadow**).

peat Partly decayed organic (mainly plant) material that accumulates where lack of oxygen, associated with near-permanent waterlogging, inhibits the activity of microbial decomposer organisms.

peatland See **mire**.

pH Scale of 0–14 defining the acidity/alkalinity of solutions including those in soils and water bodies; 0 = extremely acid, 14 = extremely alkaline, and 7 = neutral (although soils and waters with pHs between c. 6.5 and c. 7.5 are often referred to as neutral). Although generally accepted, this use of the term **alkalinity** is not strictly correct.

phytoplankton Microscopic, free-floating plants; normally the main **primary producers** in open waters.

plant community An assemblage of plant species populations living together and constituting the **vegetation** in a given location at a given time; part of the total **community** of the location.

potable water Water intended for human consumption.

primary producers Green plants (including **phytoplankton**) that utilize light energy to synthesize organic compounds (by photosynthesis) and form the basis of **food chains**, because they produce more organic material than they require.

primary production The process of light-energy assimilation by **primary producers**, and hence of energy fixation by a **plant community**; can be divided into gross primary production (the total amount of energy assimilated) and **net primary production**; can be expressed in terms of energy or, since this is incorporated in the plants' organic compounds, in terms of **biomass**; often expressed as the amount fixed per unit area per year (e.g. $gm^{-2}yr^{-1}$) which is strictly a measure of rate and hence of **productivity**.

productivity The rate at which a community fixes and utilizes energy.

quadrat Strictly a four-sided (usually square) sampling area, but can loosely include other sampling units such as circles; widely used in the study of plant species and communities, e.g. for recording species presence, density or cover; can refer to (a) *portable quadrats* (m^2 or smaller) in which individual observations are made, e.g. with restricted random sampling, (b) *larger plots* ($2m \times 2m$ or larger) in which cover-abundance values are estimated, e.g. with the relevé method (see Rodwell 1991a et seq.), or (c) *study areas/plots* (usually $10m \times 10m$ or larger) within which sampling is conducted, e.g. with portable quadrats.

runoff Usually defined as channel flow (streamflow) from a site or catchment; but variously used by different authors, e.g. to include (a) overland flow (and hence all surface flow), (b) overland flow *and* interflow, or (c) all flows (including groundwater flow).

sediment Organic or inorganic material that has precipitated from water to accumulate on the floor of a waterbody; commonly consists of fine particulate matter (**silt**) that was previously suspended in the water, but can include (a) coarser materials including sands and gravels, and (b) material such as calcium carbonate that was present in the water as dissolved chemicals and has precipitated through chemical reaction; sometimes incorrectly used to include particulate matter that is still suspended in water (suspended sediment/solids).

sedimentation basin See **silt trap**.

seral community A stage in ecological **succession** (a sere is a given type of succession).

silt Fine particulate organic and inorganic material; strictly confined to material with an average particle size intermediate between those of sands and clays, but often taken to include all material finer than sands.

silt trap A hard-lined stilling well/basin with inflow and outflow pipes for drainage water; designed to slow the flow sufficiently for collection of suspended solids by sedimentation.

soil moisture deficit (SMD) State when the soil moisture content is below *field capacity* (the moisture content of a soil when water percolating downwards under gravity has drained out); usually expressed in millimetres (rainfall equivalent) to indicate the amount of rain needed to cancel the deficit. As SMDs increase, the availability of soil water to plants decreases and they tend to wilt.

soligenous mires Peatlands with water flow and associated nutrient input, with the result that they tend to be more **eutrophic** than **topogenous mires**.

species abundance Strictly the number of individuals of a species in an area, often expressed as density; in community studies, e.g. of **species composition**, more usually (and usefully) the amount of the species as a function of both numbers and size, estimated in terms of frequency, cover or biomass (see Box 12.2).

species composition In simple terms the species comprising a **community** as indicated by a species list; more usually includes a measure of the importance of each species in terms of its abundance (see **species importance** and **species abundance**).

species diversity The structure of a **community** in terms of the number of species present and the relative abundance/importance of each; consists of two components, **species richness** and *evenness* (which is the degree to which the sum of all abundances is distributed between the constituent species). Two communities with the same number of species can have different species diversities, e.g. a community with one or two very abundant species and low amounts of all other species has low evenness and lower species diversity than one with the same number of species but greater evenness.

species importance Strictly a species' rôle in the community (see **dominance** and **key species**), but usually estimated in terms of its **abundance** or *relative abundance*, i.e. its abundance expressed as a percentage of the sum of the abundance values of all species present in the community; can also be a composite value derived from two or more abundance measures.

species-rich/species-poor Can refer to high or low **species richness** in terms of species number only, but usually implies high or low **species diversity** also.

species richness Can refer simply to the number of species in a **community** or area, but is also a component of **species diversity**, when it is defined as the ratio of the number of species to the sum of (a) all individuals or (b) all species' importance values.

sphagna Species belonging to the genus *Sphagnum* (bog moss); occur mainly on wet, acid substrates and are normally abundant in bogs and poor fens.

stomatal closure Partial or complete closing of *stomata* (pores in the leaves and other non-woody parts of plants), usually as a mechanism to reduce **transpiration**.

succession (ecological) The natural process of community change that culminates in the development of the **climax community/vegetation** of the area.

swamp A **wetland ecosystem** with a water surface above ground level permanently or for part of the year, but shallow enough to allow the growth of emergent plants such as reeds, which form the dominant vegetation (see Appendix E).

topogenous mires Peatlands with impeded drainage and hence little water flow or associated nutrient inflow, with the result that they tend to be less **eutrophic** than **soligenous mires**.

transect A line or belt that traverses a study area, usually along a pre-determined compass bearing or one selected to pass through selected **habitats** or **communities**; normally straight, but can have segments with different bearings; particularly useful for studying **community gradients**; can be permanently marked for monitoring purposes.

transmission loss The ability of a panel to resist transmission of energy/noise from one side to the other.

transpiration Evaporative loss of water from plants; normally the largest component of **evapotranspiration** from vegetated **ecosystems**.

vascular plants (often called higher plants) Pteridophytes (ferns and horsetails), gymnosperms (mainly conifers), and angiosperms (flowering plants), all of which transport water and nutrients in a specialized structural system (vascular system) that is not present in simple plants such as **bryophytes**.

vegetation The plant cover and hence **plant communities** in an area, usually classified as specific types, e.g. woodland, grassland.

vertical structure The structure of vegetation in terms of its stratification into horizontal layers, e.g. tree (canopy), shrub (understorey), field/herb and ground/moss layers in woodland.

vice-county An area used for recording the distribution of fauna and flora in the UK. Many vice-county boundaries are similar to those of the administrative counties (see Donn & Wade 1994).

visual amenity The popularity of an area, site or view, in terms of visual perception.

visual envelope The extent of visibility to and from a point or site.

water budget The balance between water inputs to and outputs from a system, affecting the amount of water stored in the system at a given time.

weather (in a given place) The condition of the atmosphere at a given time with respect to the various elements, e.g. temperature, sunshine, wind, precipitation; refers to the behaviour of the atmosphere over a few hours or at most over a few days (see **climate**).

wetland Any **ecosystem** in which there is surface water or the substratum is waterlogged for at least part of the year. Freshwater wetlands include *open waters* (streams, rivers, ponds, lakes, reservoirs, etc.), **swamps**, and *semi-terrestrial wetlands* (**mires**, **marsh** and **carr**). Brackish and saline wetlands include estuaries, maritime saltmarshes and shores (see Appendix E).

wildlife corridors Linear vegetated features, such as a hedgerows or roadside verges, that are linked to other wildlife areas and may act as interconnecting routes for the movement of animals between different areas needed during life, or facilitate dispersal of animals and plants by providing access to new or replacement sites. They may also increase the overall extent of **habitat** for animals with large range requirements (see Andrews 1993), and in urban and agricultural areas may constitute the main remaining wildlife habitats.

zone of intrusion See **visual envelope**.

References

For abbreviations and acronyms, see pp. 17–20

ACAO (Association of County Archaeological Officers) 1992. *Sites and monuments record: policies for access and charging.* Chelmsford: Essex County Council (Planning Department).

Adam, P. 1990. *Saltmarsh ecology.* Cambridge: Cambridge University Press.

Alcock, M. R. & M. A. Palmer 1985. *A standard method for the survey of ditch vegetation.* CST Report 37. Peterborough: Nature Conservancy Council.

Allen, S. E., H. M. Grimshaw, A. P. Rowland 1986. Chemical analysis. In *Methods in plant ecology* (2nd edn), P. D. Moore & S. B. Chapman (eds), 285–344. Oxford: Blackwell Scientific.

Anderson, M. G. & T. P. Burt (eds) 1985. *Hydrological forecasting.* Chichester: John Wiley.

Andrews, E. & A. K. Crawford 1986. *Otter survey of Wales 1984–85.* London: Vincent Wildlife Trust.

Andrews, J. 1993. The reality and management of wildlife corridors. *British Wildlife* **5**, 1–7.

— & S. Carter (eds) 1993. *Britain's birds in 1990–91: the conservation and monitoring review.* BTO (Thetford) and JNCC (Peterborough).

— & D. Kinsman 1990. *Gravel pit restoration for wildlife: a practical manual.* Sandy: Royal Society for the Protection of Birds.

Angerbjorn, A. 1983. The reliability of pellet counts as density estimates of mountain hares. *Finnish Game Research* **41**, 433–48.

Antenucci, J. C., K. Brown, P. L. Droswell, M. J. Kevany 1991. *Geographic information systems: a guide to the technology.* New York: Van Nostrand.

Appleby, C. 1991. Monitoring at the county level. In *Monitoring for conservation and ecology*, F. B. Goldsmith (ed.), 145–78. London: Chapman & Hall.

Appleton, J. 1975. Landscape evaluation: the theoretical vacuum. *Institute of British Geographers, Transactions* **66**, 120–3.

— 1990. *The symbolism of habitat: an interpretation of landscape in the arts.* Seattle: University of Washington Press.

Appleyard, D., K. Lynch, J. R. Meyer 1964. *The view from the road.* Cambridge, Mass.: MIT Press.

Arnold, H. R. 1993. *Atlas of mammals in Britain* (ITE Research Publication 6). London: HMSO.

Arnold, E. W. & J. A. Burton 1978. *A field guide to the reptiles and amphibians of Britain and Europe.* London: Collins.

Arup Environmental 1993. *Redhill Aerodrome Environmental Statement.* London: Ove Arup & Partners.

ASH Consulting Group 1993a. *South East tranquil areas.* London: Council for the Protection of Rural England/Countryside Commission.

— 1993b. *Ecological identity areas: preliminary ecological assessment at the landscape scale.* Memorandum and plan sample, ASH Consulting Group.

Askew, R. R. 1988. *The dragonflies of Europe.* Colchester: Harley.

Avery, B. W. 1990. *Soils of the British Isles.* Wallingford: CAB International.

Ayres, Q. C. & D. Coates 1939. *Land drainage and reclamation.* New York: McGraw-Hill.

Bagenal, T. B. (ed.) 1978. *Methods for the assessment of fish production in freshwaters.* Oxford: Blackwell Scientific.

343

Bain, M. B., J. S. Irving, R. D. Olsen, E. A. Stull, G. W. Witmer 1986. *Cumulative impact assessment: evaluating the environmental effects of multiple human developments*. Washington DC: US Department of Energy.

Baker, J. M. 1979. Responses of salt marsh vegetation to oil spills. In *Ecological processes in coastal environments*, R. L. Jefferies & A. J. Davy (eds), 529–42. Oxford: Blackwell Scientific.

— & J. H. Crothers 1987. Intertidal rock in biological surveys of estuaries and coasts. In *Biological surveys of estuaries and coasts*, J. M. Baker & W. J. Wolff (eds), 157–97. Cambridge: Cambridge University Press.

— & W. J. Wolff (eds) 1987. *Biological surveys of estuaries and coasts*. Cambridge: Cambridge University Press.

Ball, D. F. 1986. Site and soils. In *Methods in plant ecology* (2nd edn), P. D. Moore & S. B. Chapman (eds), 215–84. Oxford: Blackwell Scientific.

Ball, S. & S. Bell 1991. *Environmental law*. London: Blackstone.

Ball, S. G. 1986. *Terrestrial and freshwater invertebrates with Red Data Book, notable or habitat indicator status* (Site register report (CTD Report 637) 66). Peterborough: Nature Conservancy Council.

Bannister, P. 1986. Water relations and stress. In *Methods in plant ecology,* 2nd edn, P. D. Moore & S. B. Chapman (eds), 79–143. Oxford: Blackwell Scientific.

Barnes, R. S. K. 1979. *Coasts and estuaries*. London: Rainbird/Hodder & Stoughton.

— 1984. *Estuarine biology*, 2nd edn. London: Edward Arnold.

Barrowcliffe, R. 1993. *The practical use of dispersion models to predict air quality impacts*. Paper presented at the IBC Technical Services Conference on Environmental Emissions: Monitoring Impacts and Remediation. London: Environmental Resources Management.

Batey, P., M. Madden, G. Scholefield 1993. Socio-economic impact assessment of large-scale projects using input–output analysis: a case study of an airport. *Regional Studies* **27**(3), 179–92.

Batten, L. A., C. J. Bibby, P. Clement, G. D. Elliott, R. F. Porter 1990. *Red Data Birds in Britain*. London: T. & A. Poyser.

Beanlands, G. E. & P. N. Duinker 1984. An ecological framework for environmental impact assessment. *Journal of Environmental Management* **18**, 267–77.

— W. J. Erckmann, G. H. Orians, J. O'Riordan, D. Policansky, M. H. Sadar, B. Sadler (eds) 1986. *Proceedings of the workshop on cumulative environmental effects: a binational perspective*. Ottawa, Ontario: Canadian Environmental Assessment Research Council/US National Research Council.

Bedford, B. L. & E. M. Preston (eds) 1988. Cumulative effects on wetlands. *Environmental Management* **12**(5), special edition.

Beeby, A. 1993. *Applying ecology*. London: Chapman & Hall.

Begon, M. 1979. *Investigating animal abundance: capture–recapture for biologists*. London: Edward Arnold.

Bell, S. 1993. *The elements of visual design in the landscape*. London: Spon.

Benarie, M. M. 1987. The limits of air pollution modelling. *Atmospheric Environment* **21**, 1–5.

Berkshire County Council 1985. *People in Lower Earley: results of the population survey, June*. Reading: BCC Research and Intelligence Unit.

BES (British Ecological Society) 1992. Biological flora listing. *Journal of Ecology* **80**, 879–82.

Beven, K. W. & P. Carling (eds) 1989. *Floods: hydrological, sedimentological and geomorphological implications*. Chichester: John Wiley.

BHS 1990. *Surveying for amphibians*. Conservation Committee advisory leaflet, British Herpetological Society.

Bibby, C. J. 1984. *Some guidelines for the census of birds*. Occasional Bulletin 35, Royal Society for the Protection of Birds, Sandy.

— N. D. Burgess, D. A. Hill 1992. *Bird census techniques*. London: Academic Press.

Bibby, J. S. & D. Mackney 1969. *Land use capability classification*. Technical Monograph 1 of the Soil Survey of Great Britain, Rothamsted Experimental Station, Harpenden.

Biggs, J. & P. Williams 1993. *The River Restoration Project – Phase 1 feasibility study summary*. Unpublished report, the River Restoration Project, Huntingdon.

Biggs, J., A. Corfield, D. Walker, M. Whitfield, P. Williams 1994. New approaches to the management of ponds. *British Wildlife* **5**, 273–87.

Bisset, R. 1984. Methods for assessing direct impacts. In *Perspectives on environmental impact assessment*, B. D. Clark, A. Gilad, R. Bisset, P. Tomlinson (eds), 195–212. Dordrecht: Reidel.

— & P. Tomlinson 1988. Monitoring and auditing of impacts. In *Environmental impact assessment: theory and practice,* P. Wathern (ed.), 117–28. London: Unwin Hyman.

Blower, J. G., L. M. Cook, J. A. Bishop 1981. *Estimating the size of animal populations.* London: Allen & Unwin.

Boaden, P. J. & R. Seed 1985. *An introduction to coastal ecology.* Glasgow: Blackie.

Boon, P. J. 1991. The role of Sites of Special Scientific Interest (SSSIs) in the conservation of British Rivers. *Freshwater Forum* **1**, 95–108.

— D. H. W. Morgan, M. A. Palmer 1992. Statutory protection of freshwater flora and fauna in Britain. *Freshwater Forum* **2**, 91–101.

Bornkamm, R., J. A. Lee, M. R. D Seaward 1982. *Urban ecology: 2nd European Ecological Symposium.* Oxford: Blackwell Scientific.

Bourassa, S. C. 1991. *The aesthetics of landscape.* London: Pinter (Belhaven).

Bowen, R. 1986. *Groundwater,* 2nd edn. Amsterdam: Elsevier.

Bowles, R. T. 1981. *Social impact assessment in small communities.* Toronto: Butterworth.

Box, J. D. & J. E. Forbes 1992. Ecological consideration in the environmental assessment of road proposals. *Journal of the Institution of Highway and Transportation* **39**, 16–22.

Bradshaw, A. D. & M. J. Chadwick 1980. *The restoration of land – the ecology of reclamation of derelict and degraded land.* Oxford: Blackwell Scientific.

— D. A. Goode, E. H. P. Thorpe (eds) 1986. *Ecology in design and landscape.* Oxford: Blackwell Scientific.

Brady, N. C. 1990. *The nature and properties of soils,* 10th edn. New York: Macmillan.

Branson, B., S. Foley, D. Henderson 1993. EA case study: photomontage techniques. *Landscape Design* **224**, 40.

Bras, R. L. 1990. Hydrology: an introduction to hydrologic science. Reading, Mass.: Addison–Wesley.

Brassington, R. 1988. *Field hydrology.* Milton Keynes: Open University Press.

Bratton, J. H. 1990a. Seasonal pools – an overlooked invertebrate habitat. *British Wildlife* **2**, 22–31.

— 1990b. *A review of the scarcer Ephemeroptera and Plecoptera of Great Britain.* Research and Survey in Nature Conservation 29. Peterborough: Nature Conservancy Council.

— (ed.) 1991. *British Red Data Books 3: invertebrates other than insects.* Peterborough: Joint Nature Conservation Committee.

Braun-Blanquet, J. (translated, revised and edited by C. D. Fuller & H. S. Conrad) 1965. *Plant sociology: the study of plant communities.* London: Hafner.

BRE (Building Research Establishment) 1990. *Damage to structures from ground-borne vibration.* BRE Digest 353.

Bregman, J. I. & K. M. Mackenthun 1992. *Environmental impact statements.* Chelsea, Michigan: Lewis.

Bridges, E. M. 1978. *World soils,* 2nd edn. Cambridge: Cambridge University Press.

Bronfman, L. M. 1991. Setting the social impact agenda: an organisational perspective. *Environmental Impact Assessment Review* **11**, 69–79.

Brooke, J. S. (Posford Duvivier Environment) 1992. River and coastal engineering. In *Environmental assessment: a guide to the identification, evaluation and mitigation of environmental issues in construction schemes.* Research Project 424 (ch. 4), CIRIA, Birmingham.

Brookes, A. 1988. *Channelized rivers: perspectives for environmental management.* Chichester: John Wiley.

— & K. R. Hills (in press). The impact of road developments on river corridors: lessons learnt from south-central England. In *Nature conservation and the management of drainage system habitat,* D. Harper (ed.). Chichester: John Wiley.

Brooks, S. J. 1993a. Guidelines for invertebrate site surveys. *British Wildlife* **4**, 283–6.

345

— 1993b. Review of a method to monitor adult dragonfly populations. *Journal of the British Dragonfly Society* **9**, 1-4.

Brouwer, A., A. J. Murk, J. H. Koeman 1990. Biochemical and physiological approaches in ecotoxicology. *Functional Ecology* **4**, 275-81.

Brown, A. F. & K. B. Shepherd 1993. A method for censusing upland breeding birds. *Bird Study* **40**, 189-95.

Brownrigg, M. 1971. The regional income multiplier: an attempt to complete the model. *Scottish Journal of Political Economy* **18**.

— 1974. *A study of economic impact: the University of Stirling*. Edinburgh: Scottish Academic Press.

BSI 1984. *BS5228, Part I: noise control on construction and open sites*. Milton Keynes: British Standards Institute.

— 1987. *BS8233: Sound insulation and noise reduction for buildings*. Milton Keynes: British Standards Institute.

— 1990. *BS4142: Rating industrial noise affecting mixed residential and industrial areas*. Milton Keynes: British Standards Institute.

BTO 1982. *Waterways bird survey: instructions*. Thetford: British Trust for Ornithology.

— 1983. *Common bird census: instructions*. Thetford: British Trust for Ornithology.

— 1992. *National low tide counts: detailed instructions*. Thetford: British Trust for Ornithology.

Buckland, S. T. 1992. *Review of deer count methodology*. Edinburgh: Scottish Office.

— D. R. Anderson, K. P. Burnham, J. L. Laake 1993. *Distance sampling: estimating abundance of biological populations*. London: Chapman & Hall.

Buckley, G. P. (ed.) 1989. *Biological habitat reconstruction*. London: Pinter (Belhaven).

Budd, J. T. C. 1991. Remote sensing techniques for monitoring land cover. In *Monitoring for conservation and ecology*, F. B. Goldsmith (ed.), 33-59. London: Chapman & Hall.

Burd, F. 1989. *The saltmarsh survey of Great Britain: an inventory of British saltmarshes*. Research and Nature Conservation 17. Peterborough: Nature Conservancy Council.

Burgess, J. 1990a. The influence of the media on people's relationships with, and attitudes to, nature and landscape. In *People, nature and landscape: a research review*. London: Landscape Research Group.

— 1990b. The production and consumption of environmental meanings in the mass media: a research agenda for the 1990s. *Institute of British Geographers, Transactions* **15**, 139-61.

— 1993. *Perceptions of risk in recreational woodlands in the urban fringe*. Department of Geography, University College London.

— M. Limb, C. M. Harrison 1988. Exploring environmental values through the medium of small groups, 1: theory and practice. *Environment and Planning A* **20**, 309-26.

Burgman, M. & R. Akcakaya 1993. *Risk assessment in conservation biology*. London: Chapman & Hall.

BUTT (Butterflies Under Threat Team) 1986. *The management of chalk grassland for butterflies*. Peterborough: Nature Conservancy Council.

Byfield, A. 1990. The Basingstoke Canal – Britain's richest waterway under threat. *British Wildlife* **2**, 13-21.

Calow, P. (ed.) 1993. *Handbook of ecotoxicology*, vol. 1. Oxford: Blackwell Scientific.

Canter, L. W. 1977. *Environmental impact assessment* (ch. 4). New York: McGraw-Hill.

— J. M. Robertson, R. M. Westcott 1991. Identification and evaluation of biological impact mitigation measures. *Journal of Environmental Management* **33**, 35-50.

Carley, M. J. & E. S. Bustelo 1984. *Social impact assessment and monitoring: a guide to the literature*. Boulder, Colorado: Westview.

Carter, R. W. G. 1993. *Coastal ecology: an introduction to the physical, ecological and cultural systems of coastlines*, 4th edn. London: Academic Press.

Cassettari, S. 1993. Cartography towards virtual reality. In *Survey and mapping '93: the 4th UK National Land Surveying and Mapping Conference*, 367-78. Keele: University of Keele.

CC 1987. *Landscape assessment: a Countryside Commission approach* (CCP 18). Cheltenham: Countryside Commission.

— 1988. *A review of recent practice and research in landscape assessment*. Cheltenham: Countryside Commission.

— 1990. *The Cambrian Mountain landscape*. Cheltenham: Countryside Commission.

— 1991a. *Assessment and conservation of landscape character: the Warwickshire landscapes project approach* (CCP 332). Cheltenham: Countryside Commission.

— 1991b. *Environmental assessment: the treatment of landscape and countryside issues* (CCP 326). Cheltenham: Countryside Commission and Countryside Council for Wales.

— 1992. *The Chilterns landscape*. Cheltenham: Countryside Commission.

— 1993. *Landscape assessment: a new guidance* (CCP 423). Cheltenham: Countryside Commission.

CEAA 1993. *Addressing cumulative environmental effects: a reference guide for the Canadian Environmental Assessment Act* (draft version). Hull, Quebec: Canadian Environmental Assessment Agency.

CEC (Commission of the European Communities) 1985. Council Directive on the assessment of the effects of certain private and public projects (85/337/EEC). *Official Journal of the European Communities* **L175/40**, 5 July 1985.

— 1991a. *CORINE biotopes: the design, compilation and use of an inventory of sites of major importance for nature conservation in the European Community*. Luxembourg: Office for Official Publications of the European Communities.

— 1991b. *CORINE biotopes manual: habitats of the European Community (data specifications, part 2)*. Luxembourg: Office for Official Publications of the European Communities.

— 1992. *Towards sustainability: Fifth Action Programme on the Environment*. Brussels: European Commission.

— 1993. Report from the Commission on the implementation of Directive 85/337/EEC, *COM (93) 28 final*, 2nd April 1993. Brussels: CEC.

Cernuschi, S. & M. Giugliano 1992. Air quality assessment in environmental impact studies. In *Environmental impact assessment*, A. G. Colombo (ed.), 189–209. Dordrecht: Kluwer.

Chapman, P. J. & L. L. Chapman 1982. *Otter survey of Ireland 1980–81*. London: Vincent Wildlife Trust.

Chapman, V. J. 1976. *Coastal vegetation*, 2nd edn. Oxford: Pergamon.

— 1977. *Wet coastal ecosystems*. Amsterdam: Elsevier.

CIRIA 1992. *Environmental assessment: a guide to the identification, evaluation and mitigation of environmental issues in construction schemes*. CIRIA Research Project 424, Construction Industry Research and Information Association, Birmingham.

Clapham, A. R., T. G. Tutin, D. M. Moore. 1987. *Flora of the British Isles*, 3rd edn. Cambridge: Cambridge University Press.

Clark, B. D., K. Chapman, R. Bisset, P. Wathern, M. Barrett 1981. *A manual for the assessment of major development proposals*. London: HMSO.

Clements, D. K. & R. J. Tofts 1992. *Hedgerow evaluation and grading system (HEGS) – a methodology for the ecological survey, evaluation and grading of hedgerows (test draft)*. Cirencester: Countryside Planning and Management.

Coles, S. M. 1979. Benthic microalgal populations on intertidal sediments and their role as precursors to salt marsh development. In *Ecological processes in coastal environments*, R. L. Jefferies & A. J. Davy (eds), 25–42. Oxford: Blackwell Scientific.

Collins, N. M. & S. M. Wells 1987. *Invertebrates in need of special protection in Europe*. Strasbourg: European Committee for the Conservation of Nature and Natural Resources.

Collis, I. & D. Tyldesley 1993. *Natural assets: non-statutory sites of importance for nature conservation*. Local Government Nature Conservation Initiative, Hampshire County Council.

Congdon, P. & P. Batey (eds) 1989. *Advances in regional demography: information, forecasts and models*. London: Pinter (Belhaven).

Constanza, R. (ed.) 1991. *Ecological economics*. New York: Columbia University Press.

Cooper, P. F. & B. C. Findlater (eds) 1991. *Constructed wetlands in water pollution*. Oxford: Pergamon.

Corbet, G. B. & S. Harris 1991. *The handbook of British mammals*, 3rd edn. Oxford: Blackwell Scientific.

347

Corbett, K. (ed.) 1989. *Conservation of European reptiles and amphibians*. London: Christopher Helm.

Cosgrove, D. 1984. *Social formation and symbolic landscapes*. Beckenham: Croom Helm.

— 1985. Prospect, perspective and the evolution of the landscape idea. *Institute of British Geographers, Transactions* **10**, 45–62.

— & S. Daniels (eds) 1992. *The iconography of landscape*. Cambridge: Cambridge University Press.

Cramp, S. & K. E. L. Simmons 1977 et seq. *The birds of the western Palearctic*. Oxford: Oxford University Press.

Crawford, A. K., D. Evans, A. Jones, J. McNulty 1979. *Otter survey of Wales 1977–78*. Nettleham: Royal Society for the Conservation of Nature.

Cresser, M., K. Killham, T. Edwards 1993. *Soil chemistry and its applications*. Cambridge: Cambridge University Press.

Cresswell, P., S. Harris, D. J. Jefferies 1990. *The history, distribution, status and habitat requirements of the badger in Britain*. Peterborough: Nature Conservancy Council.

Crisp, D. T. 1993. The environmental requirements of salmon and trout in fresh water. *Freshwater Forum* **3**, 176–202.

Croft, J. M., C. D. Preston, W. A. Forrest 1991. *Database of aquatic plants in the British Isles. Phase 1 (submerged and floating plants)*. NRA Project Report F028/1/N. London: National Rivers Authority.

Curtis, T. G. F. & H. N. McGough 1988. *The Irish Red Data Book 1: vascular plants*. Dublin: The Stationery Service.

Dalby, D. H. 1987. Salt marshes. In *Biological surveys of estuaries and coasts*, J. M. Baker & W. J. Wolff (eds), 38–80. Cambridge: Cambridge University Press.

Daniels, R. E. & A. Eddy 1990. *Handbook of European Sphagna,* 2nd edn. Abbots Ripton: Institute of Terrestrial Ecology.

Davidson, N. C., D. d'A. Loffoley, J. P. Doody, L. S. Way, J. Gordon, R. Key, C. M. Drake, M. W. Pienkowski, R. Mitchell, K. L. Duff 1991. *Nature conservation and estuaries in Great Britain*. Peterborough: Nature Conservancy Council.

Davies, I. M. & J. C. McKie 1987. Accumulation of total tin and tributyl tin in muscle tissue of farmed Atlantic salmon. *Marine Pollution Bulletin* **18**(7), 405–407.

Davies, K. 1992. *Cumulative environmental effects: a sourcebook*. Hull, Quebec: Federal Environmental Assessment Review Office.

De Jongh, P. 1988. Uncertainty in EIA. In *Environmental impact assessment: theory and practice*, P. Wathern (ed.), 62–84. London: Unwin Hyman.

Denzin, N. K. 1970. *Sociological methods: a source book*. Chicago: Aldine.

Digby, P. G. N. & R. A. Kempton 1987. *Multivariate analysis of ecological communities*. London: Chapman & Hall.

Disney, R. H. L., Y. Z. Erzinclioglu, D. J. de C. Henshaw, D. Howse, D. M. Unwin, P. Withers, A. Woods 1982. Collecting methods and the adequacy of attempted fauna surveys, with reference to the Diptera. *Field Studies* **5**, 607–21.

Dobson, F. S. 1992. *Lichens: an illustrated guide to the British and Irish species*. Slough: Richmond.

DOE (Department of the Environment) 1973. *Circular 10/73: Planning and noise*, HMSO.

— 1979. *Methods of biological sampling: handnet sampling of aquatic benthic macroinvertebrates, 1978*. London: HMSO.

— 1986. *Nitrate in water: a report by the Nitrate Coordination Group* (Pollution Papers 26). London: HMSO.

— 1987a *Circular 8/87: Historic buildings and conservation areas – policy and procedures*. London: HMSO.

— 1987b *Nature conservation: Circular 27/87 (Welsh Office 52/87)*. London: HMSO.

— 1988. *The Town and Country Planning (assessment of environmental effects) Regulations: Circular 15/88 (Welsh Office 23/88)*. London: HMSO.

— 1989. *Environmental assessment: a guide to the procedures* (DOE, Welsh Office). London: HMSO.
— 1990a. *Planning policy guidance note 16: development plan policies for archaeology.* London: HMSO.
— 1990b. *Integrated pollution control: a practical guide* (DOE, Welsh Office). London: HMSO.
— 1990c. *This common inheritance* (DOE, Welsh Office). London: HMSO.
— 1991a. *Policy appraisal and the environment.* London: HMSO.
— 1991b. *Environmental assessment and private bill procedures* (Consultation Paper, PDC 4). London: HMSO.
— 1991c. *This common inheritance: first year report.* London: HMSO.
— 1991d. *The potential effects of climate change in the United Kingdom, first report.* London: HMSO.
— 1992a. *Planning policy guidance note 12: Development plans and regional planning guidance.* London: HMSO.
— 1992b. *The UK environment.* London: HMSO.
— 1992c. *This common inheritance: The second year report.* London: HMSO.
— 1992d. *Planning policy guidance note 7: The countryside and the rural environment* (DOE, Welsh Office). London: HMSO.
— 1992e. *River quality: The government's proposals.* London: HMSO.
— /Welsh Office 1992f. *Planning controls over Sites of Special Scientific Interest.* Joint DOE/ Welsh Office Circular 1/92. London: HMSO.
— 1992g. *Draft planning policy guidance on nature conservation.* London: HMSO.
— 1992h. *Planning policy guidance note 20: coastal planning.* London: HMSO.
— 1993a. *Mineral planning guidance note 6.* London: HMSO.
— 1993b. *Mineral planning guidance note 11: The control of noise at surface mineral workings.* London: HMSO.
— 1993c. *Environmental appraisal of development plans.* London: HMSO.
— Annual Publication: *Digest of environmental protection and water statistics.* London: HMSO.
— MAFF (Ministry of Agriculture, Fisheries and Food) and Welsh Office Agriculture Department 1982. *Land drainage and conservation: guidance notes on procedures for water authorities, internal drainage boards, Nature Conservancy Council and Countryside Commission.* London: HMSO.
Donn, S. & M. Wade 1994. *The UK directory of ecological information.* Chichester: Packard.
Doody, P. (ed.) 1985. *Sand dunes and their management* (Focus on Nature Conservation 13). Peterborough: Nature Conservancy Council.
— & B. Barnett (eds) 1987. *The Wash and its environment* (Research and Survey in Nature Conservation 7). Peterborough: Nature Conservancy Council.
DOT (Department of Transport) 1983. *Manual of environmental appraisal.* London: HMSO.
— 1985. *Traffic flows and carriageway width assessment, TD 20/85.* London: HMSO.
— 1988. *Calculation of road traffic noise.* Department of Transport, Welsh Office. London: HMSO.
— 1989a. *Environmental assessment under EC Directive 85/337/DTp highways and traffic, departmental standard HD 18/88.* London: HMSO.
— 1989b. *Roads for prosperity.* London: HMSO.
— 1989c. *National road traffic forecasts.* London: HMSO.
— 1991. *Road accidents Great Britain 1990.* London: HMSO.
— 1992a. *Assessing the environmental impact of road schemes: response by Department of Transport to the report by the Standing Committee on Trunk Road Assessment,* London: HMSO.
— 1992b. *The good roads guide: environmental design for inter-urban roads.* London: HMSO.
— 1993. *Design manual for roads and bridges volume 11: Environmental assessment.* London: HMSO.
Dover J. 1990. Butterflies and wildlife corridors. In *The Game Conservancy review of 1989,* C. Nodder (ed.), 62–4. Fordingbridge: The Game Conservancy.
Dowdeswell, W. H. 1987. *Hedgerows and verges.* London: Allen & Unwin.
DTI (Department of Trade and Industry) 1992. *Guidelines for the environmental assessment of cross-country pipelines.* London: HMSO.
Duffey, E., M. G. Morris, J. Sheail, K. Lena, D. A. Wells, T. C. E. Wells 1974. *Grassland*

ecology and wildlife management. London: Chapman & Hall.

Duinker, P. N. 1987. Forecasting environmental impacts: better quantitative and wrong than qualitative and untestable! In *Audit and evaluation in environmental assessment and management: Canadian and international experience* (vol. 2), B. Sadler (ed.), 399–407. Canada: Beauregard Press.

Duncan, U. K. 1970. *Introduction to British lichens*. Arbroath: Buncle.

Dunne, T. & L. B. Leopold 1978. *Water in environmental planning*. San Francisco: W. H. Freeman.

Earle, R. & D. G. Erwin (eds) 1983. *Sublittoral ecology: the ecology of the shallow sublittoral benthos*. Oxford: Oxford University Press.

Earll, R. 1992. *The SEASEARCH habitat guide – an identification guide to the main habitats in the shallow seas around the British Isles*. Peterborough: Marine Conservation Society and Joint Nature Conservation Committee.

Easterbee, N., L. V. Hepburn, D. J. Jefferies 1991. *Survey of the status and distribution of the wildcat in Scotland, 1983–1987*. Edinburgh: Nature Conservation Council for Scotland.

Edington, J. M. & A. G. Hildrew 1981. *Caseless caddis larvae of the British Isles*. FBA Scientific Publication 43, Freshwater Biological Association, Ambleside.

EH 1991a. *Rescue archaeology funding: a policy statement*. London: English Heritage.

— 1991b. *Exploring our past – strategies for the archaeology of England*. London: English Heritage.

— 1992a. *Development plan policies for archaeology*. London: English Heritage.

— 1992b. *The management of archaeological projects*. London: English Heritage.

Ellenberg, H. 1988. *Vegetation ecology of central Europe*, 4th edn. Cambridge: Cambridge University Press.

Elliott, J. M. 1977. *A key to the larvae and adults of British freshwater Megaloptera and Neuroptera*. FBA Scientific Publication 35, Freshwater Biological Association, Ambleside.

— 1990. The need for long-term investigations in ecology and the contribution of the Freshwater Biological Association. *Freshwater Biology* **23**, 1–5.

— & K. H. Mann 1979. *A key to the British freshwater leeches*. FBA Scientific Publication 40, Freshwater Biological Association, Ambleside.

— & P. A. Tullett 1983. *A supplement to a bibliography of samplers for benthic invertebrates*. FBA Occasional Publication 4, Freshwater Biological Association, Ambleside.

— U. H. Humpesch, T. T. Macan 1988. *Larvae of the British Ephemeroptera*. FBA Scientific Publication 49, Freshwater Biological Association, Ambleside.

— J. P. O'Connor, M. A. O'Connor 1979. A key to the larvae of Sialidae (Insecta: Megaloptera) occurring in the British Isles. *Freshwater Biology* **9**, 511–14

Ellis A. E. 1978. *British freshwater bivalve mollusca: synopses of the British Fauna*. London: Academic Press.

Elsom, D. M. 1992. *Atmospheric pollution: a global problem*. 2nd edn. Oxford: Blackwell Scientific.

Emmet, A. M & J. Heath 1979 et seq. *The moths and butterflies of Great Britain and Ireland*, vols 1–10. Colchester: Harley.

EN 1991. *Local nature reserves in England*. Peterborough: English Nature.

— 1992. *What you should know about sites of special scientific interest*. Peterborough: English Nature.

— 1994. *Nature conservation in environmental assessment*. Peterborough: English Nature.

England, J. R., K. I. Hudson, R. J. Masters, K. S. Powell, J. D. Shortridge (eds) 1985. *Information systems for policy planning in local government*. Harlow, Essex: Longman, in association with BURISA.

Essex Development Control Forum 1992. *The Essex guide to environmental assessment*. Chelmsford: The Essex Planning Officers' Association.

Fabos, J. G. 1979. Planning and landscape evaluation. *Landscape Research* **4**(2), 4–9.

Falk, S. 1991a. *A review of the scarce and threatened flies of Great Britain, part 1*. Research and Survey in Nature Conservation 39. Peterborough: Nature Conservancy Council.

— 1991b. *A review of the scarce and threatened bees, wasps and ants of Great Britain*. Research and Survey in Nature Conservation 35. Peterborough: Nature Conservancy Council.

Feilden, B. M. 1982. *Conservation of historic buildings*. London: Butterworth.

Field, B. & B. MacGregor 1987. *Forecasting techniques for urban and regional planning*. London: Hutchinson.

Fieldhouse, K. 1993. Question time. *Landscape Design* **218**, 15–18.

Finsterbusch, K. 1980. *Understanding social impacts: assessing the effects of public projects*. Beverley Hills, California: Sage.

— 1985. State of the art in social impact assessment. *Environment and Behaviour* **17**(2), 193–221.

Fitter, R. & R. Manuel 1986. *Collins field guide to freshwater life*. London: Collins.

Forbes, J. D. & D. Heath *1990. The ecological impact of road schemes* [DOT/NCC]. London: HMSO.

Forbes, V. E. & T. L. Forbes 1993. *Ecotoxicology in theory and practice*. London: Chapman & Hall.

Forestry Commission 1992. *Community woodland design guidelines*. London: HMSO.

Foster, G. N. 1981, 1983, 1984, 1985, 1987. *Atlas of British water beetles (preliminary edition), parts 1–5. The Balfour-Browne Club Newsletter* 22, 27, 34, 35, and 40. Huntingdon: Biological Records Centre, ITE.

— & M. D. Eyre 1992. *Classification and ranking of water beetle communities* [UK Nature Conservation 1]. Peterborough: Joint Nature Conservation Committee.

— A. P. Foster, M. D. Eyre, D. T. Bilton 1989. Classification of water beetle assemblages in arable fenland and ranking of sites in relation to conservation value. *Freshwater Biology* **22**, 243–54.

Francis, C. 1992. Water supply infrastructure and wastewater treatment works. In *Environmental assessment: a guide to the identification, evaluation and mitigation of environmental issues in construction schemes* (CIRIA Research Project 424, ch. 5). Birmingham: CIRIA.

Friday, L. E. 1988. *A key to the adults of British water beetles*. Publication 189. Shrewsbury: Field Studies Council.

Frost, R. W. J. 1993. *Planning beyond environmental statements*, MSc dissertation, School of Planning, Oxford Brookes University.

Fry, R. & D. Lonsdale (eds) 1991. *Habitat conservation for insects – a neglected green issue*. Middlesex: The Amateur Entomologists' Society.

Fryer, G. 1991. *A natural history of the lakes, tarns and streams of the English Lake District*. Ambleside: Freshwater Biological Association.

Fuller, R. J. 1982. *Bird habitats in Britain*. Calton: T. & A. D. Poyser.

Garbutt, J. 1992. *Environmental law – a practical handbook*. London: Chancery.

Gauch Jr, H. G. 1982. *Multivariate analysis in community ecology*. Cambridge: Cambridge University Press.

Gibbons, B. 1993. Reserve focus: Insh Marshes, Speyside. *British Wildlife* **5**, 41–3.

Gilbertson, D. D., M. Kent, F. B. Pyatt 1985. *Practical ecology for geography and biology: survey, mapping and data analysis*. London: Unwin Hyman.

Gilpin, M. & I. Hanski 1991. *Metapopulation dynamics: empirical and theoretical investigations*. London: Academic Press.

Gimmingham, C. H. 1972. *Ecology of heathlands*. London: Chapman & Hall.

— 1975. *An introduction to heathlands*. London: Longman.

Glasson, J. 1992. *An introduction to regional planning*, 2nd edn. London: UCL Press.

— et al. 1982. *A comparison of the social and economic effects of power stations on their localities*. Power Station Impacts Team, Oxford Polytechnic.

— M. J. Elson, D. van der Wee, B. Barrett 1987. *The socio-economic impact of the proposed Hinkley Point 'C' power station*. Power Station Impacts Team, Oxford Polytechnic.

— D. van der Wee, B. Barrett 1988. A local income and employment multiplier analysis of a proposed nuclear power station development at Hinkley Point in Somerset. *Urban Studies* **25**, 248–61.

— & A. Chadwick 1988–1993. *The local socio-economic impacts of the Sizewell 'B' PWR construction project*. Impacts Assessment Unit, Oxford Brookes University.

— A. Chadwick, R. Therivel 1992. Local socio-economic impacts of the Sizewell 'B' PWR construction project: Fourth Annual Monitoring Report 1992. Impacts Assessment Unit, Oxford Brookes University.

— & D. Heaney 1993. Socio-economic impacts: the poor relations in British EISs. *Journal of Environmental Planning and Management* **36**(3), 335–43.

— R. Therivel, A. Chadwick 1994. *An introduction to environmental impact assessment*. London: UCL Press.

Gledhill, T., D. W. Sutcliffe, W. D. Williams 1993. *British freshwater Crustacea Malacostraca: a key with ecological notes*. FBA Scientific Publication 52, Freshwater Biological Association, Ambleside.

Goater, B. 1986. *British pyralid moths: a guide to their identification*. Colchester: Harley.

Godbold, D. L., E. Fritz, A. Hütterman 1988. Aluminum toxicity and forest decline. *Proceedings of the National Academy of Science, USA* **85**, 3888–92.

Goddard, J. S. & J. B. Hogger 1986. Current status and distribution of freshwater crayfish in Great Britain. *Field Studies* **6**, 383–96.

Goldsmith, F. B. 1991a. The selection of protected areas. In *The scientific management of temperate communities for conservation*, I. F. Spellerberg, F. B. Goldsmith, M. G. Morris (eds), 273–91. Oxford: Blackwell Scientific.

— 1991b. *Monitoring for conservation and ecology*. London: Chapman & Hall.

— & J. B. Wood 1983. Ecological effects of upland afforestation. In *Conservation in perspective*, A. Warren & F. B. Goldsmith (eds), 287–311. Chichester: John Wiley.

— C. M. Harrison, A. J. Morton 1986. Description and analysis of vegetation. In *Methods in plant ecology*, 2nd edn, P. D. Moore & S. B. Chapman (eds), 437–524. Oxford: Blackwell Scientific.

Golterman, H. L. 1978. *Methods for chemical analysis of fresh waters*, 2nd edn. Oxford: Blackwell Scientific.

Good, J. 1987. *Environmental aspects of plantation forestry in Wales*. London: HMSO (for ITE).

Goodchild, M. F., B. O. Parks, L. T. Steyaert (eds) 1993. *Environmental modelling with GIS*. Oxford: Oxford University Press.

Goode, D. A. 1989. Urban nature conservation in Britain. *Journal of Applied Ecology* **26**, 859–73.

Goodey, B. 1987. Spotting, squatting, sitting or setting: some public images of landscape. In *Landscape meanings and values*, E. C. Penning-Rowsell & D. Lowenthal (eds), 82–101, London: Allen & Unwin.

— 1992. Dix paysages Europeens dominants. *Paysage & Amenagement* **21**(Oct.), 8–13.

Gordon, N. D., T. A. McMahon, B. L. Finlayson 1992. *Streamflow hydrology: an introduction for ecologists*. Chichester: John Wiley.

Goudie, A. 1990. *The human impact on the natural environment*, 3rd edn. Oxford: Basil Blackwell.

Graber, M. 1992. Air pollution impact assessment of chemical plants in Israel. In *Environmental impact assessment for developing countries*, A. K. Biswas & S. B. C. Agarwal (eds), 25–35. Oxford: Butterworth–Heinemann.

Grady, S., R. Braid, J. Bradbury, C. Kerley 1987. Socio-economic assessment of plant closure: three case studies of large manufacturing facilities. *Environmental Impact Assessment and Review* **26**, 151–65.

Green, J. & R. Green 1987. *Otter survey of Scotland 1984–85*. London: Vincent Wildlife Trust.

Gregory, K. J. & D. E. Walling (eds) 1973. *Drainage basin form and process: a geomorphological approach*. London: Edward Arnold.

— 1987. *Human activity and environment processes*. Chichester: John Wiley.

Greig-Smith, P. 1983. *Quantitative plant ecology*, 3rd edn. Oxford: Blackwell Scientific.

Grime, J. P.

— & S. Lloyd 1973. *An ecological atlas of grassland plants*. Nature Conservancy Unit of Grassland Research. London: Edward Arnold.

— J. G. Hodgson, R. Hunt 1989. *Comparative plant ecology.* London: Unwin Hyman.

Gubbay, S. 1990. *A future for the coast: proposals for a UK coastal zone management plan.* Marine Conservation Society.

Gurnell, J. & J. R. Flowerdew 1990. *Live trapping small mammals. A practical guide,* 2nd. edn (Occasional Publications of the Mammal Society 3). London: The Mammal Society.

Hackett, B. 1971. *Landscape planning: an introduction to theory and practice.* Newcastle upon Tyne: Oriel.

Haigh, N. 1990. EEC *environmental policy and Britain,* 2nd edn. Harlow, Essex: Longman.

— 1993. *Manual of environmental policy: the EC and Britain.* Harlow: Longman.

Hall, M. J. 1984. *Urban hydrology.* London: Elsevier.

Halsbury's Statutory Instruments 1988. *SI 1902 – CIMAH Regulations,* 2263–81.

Hammer, D. A. 1989. *Constructed wetlands for waste water: municipal, industrial and agricultural.* London: Lewis.

Hammond, C. O. 1983. *The dragonflies of Great Britain and Ireland.* Colchester: Harley.

Hammond, P. S. 1987. Techniques for estimating the size of whale populations. *Symposium of the Zoological Society of London* **58**, 225–45.

— & P. M. Thompson 1991. Minimum estimation of the number of bottlenose dolphins (*Tursiops truncatus*) in the Moray Firth, NE Scotland. *Biological Conservation* **56**, 79–87.

Harding, P. T. 1993. *Current atlases of the flora and fauna of the British Isles.* Huntingdon: Biological Records Centre.

— (ed.) 1992. *Biological recording of changes in British wildlife.* London: HMSO (for ITE).

Hardisty, M. W. & R. J. Huggins 1975. A survey of the fish populations of the middle Severn estuary based on power station sampling. *International Journal of Environmental Studies* **7**, 227–42.

Hardman, D., S. McEldowney, S. Waite 1993. *Pollution: ecology and biotreatment.* Harlow: Longman.

Hargrave, B. T. & N. M. Burns 1979. Assessment of sediment trap collection. *Limnological Oceanography* **24**, 1124–35.

Harper, D., C. Smith, P. Barham 1992. Habitats as the building blocks for river conservation assessment. In *River conservation and management,* P. J. Boon, P. Calow, G. E. Petts (eds), 311–20. Chichester: John Wiley.

Harris, S., P. Cresswell, D. J. Jefferies 1989. *Surveying badgers.* London: The Mammal Society.

Harrop, D. O. 1986. Tackling air pollution problems with computer models. *London Environmental Bulletin* **3**(4), 11–12.

— 1994. Environmental impact assessment and incineration. In *Waste incineration and the environment: issues in environmental science and technology,* R. E. Hester & R. M. Harrison (eds), 137–53. Letchworth: Royal Society of Chemistry.

— & R. P. Carpenter 1992. Methods of assessing air quality impact. Paper presented at the National Society for Clean Air and Environmental Protection Conference on Air Quality Monitoring, Bournemouth, October 1992. Brighton: NSCA.

Haslam, S. M., C. A. Sinker, P. A. Wolseley 1975. British water plants: an illustrated key based on the vegetative features of vascular plants growing in fresh water, with notes on their ecological and geographical distribution. *Field Studies* **4**, 243–351.

Hausenbuiller, R. L. 1985. *Soil science: principles and practices,* 3rd edn. Dubuque, Iowa: Brown.

HCI 1993. *Herpetofauna workers guide.* Halesworth: Herpetofauna Conservation International.

Healey, M. J. (ed.) 1991. *Economic activity and land use: the changing information base for local and regional studies.* Harlow, Essex: Longman.

Heath, J. 1981. *Threatened Rhopalocera (butterflies) in Europe.* Nature and Environment Series 23, Council of Europe.

Hellawell, J. M. 1986. *Biological indicators of freshwater pollution and environmental management.* London: Elsevier.

Hiby, A. R., & P. S. Hammond 1989. *Survey techniques for estimating abundance of cetaceans.* Report to the International Whaling Commission, Special Issue 11, 57–80.

353

Hickman, A. J.
— & D. M. Colwill 1982. *The estimation of air pollution concentrations from road traffic.* TRRL Report LR1052. Crowthorne: Transport & Road Research Laboratory.
— & V. H. Waterfield 1984. *A user's guide to the computer programs for predicting air pollution from road traffic.* TRRL Report SR806. Crowthorne: Transport & Road Research Laboratory.
Higuchi, T. 1988. *The visual and spatial structures of landscapes*, Cambridge, Mass.: MIT Press.
Hill, M. O. 1979a. TWINSPAN – *a fortran program for arranging multivariate data in an ordered two-way table by classification of the individuals and attributes.* Section of Ecology and Systematics, Cornell University, Ithaca, New York.
— 1979b. DECORANA – *a FORTRAN program for detrended correspondence analysis and reciprocal averaging.* Section of Ecology and Systematics, Cornell University, Ithaca, New York.
— 1993. TABLEFIT *version 0.0, for identification of vegetation types.* Huntingdon: Institute of Terrestrial Ecology.
— C. D. Preston, A. J. E. Smith (eds) 1991. *Atlas of the Bryophytes of Great Britain and Ireland*, vol. 1: *liverworts (Hepaticae and Anthocerotae)*. Colchester: Harley.
— 1992. *Atlas of the Bryophytes of Great Britain and Ireland*, vol. 2: *mosses (except Diplolepideae)*. Colchester: Harley.
— 1994. *Atlas of the Bryophytes of Great Britain and Ireland*, vol. 3: *mosses (Diplolepideae)*. Colchester: Harley.
Hilton-Brown, D. & R. S. Oldam 1991. *The status of the widespread amphibians and reptiles in Britain, 1990, and changes during the 1980s.* Contract Surveys 131. Peterborough: Nature Conservancy Council.
Hiscock, K. 1990. *Marine nature conservation review: methods.* CSD Report 1072 MNCR Occasional Report MNCR/OR/05. Peterborough: Nature Conservancy Council.
Hiscock, S. 1979. *A field guide to the British brown seaweeds.* FSC Offprint 125. Shrewsbury: Field Studies Council.
— 1986. *A field guide to the British red seaweeds (Rhodophyta).* FSC Occasional Publication 13. Shrewsbury: Field Studies Council.
Hockin, D., M. Ounsted, M. Gorman, D. Hill, V. Keller, M. A. Barker 1992. Examination of the effects of disturbance on birds with reference to its importance in ecological assessments. *Journal of Environmental Management* **36**, 253–86.
Hodgetts, N. G. 1992. *Guidelines for selection of biological SSSIs: non-vascular plants.* Peterborough: Joint Nature Conservation Committee.
Holdich, D. 1991. The native crayfish and threats to its existence. *British Wildlife* **2**, 141–51.
Holling, C. S. (ed.) 1978. *Adaptive environmental assessment and management.* Chichester: John Wiley.
Holme, N. A. & A. D. McIntyre (eds) 1984. *Methods for the study of the marine benthos.* Oxford: Blackwell Scientific.
Holmes, N. T. H. 1983. *Typing rivers according to their flora.* Focus on nature conservation 4. Peterborough: Nature Conservancy Council.
— 1990. British river plants – future prospects and concerns. *British Wildlife* **1**, 130–43.
— & C. Newbold 1984. *River plant communities – reflectors of water and substrate chemistry.* Report 9. Peterborough: Nature Conservancy Council.
Hoodleff, A. & P. A. Morris 1993. An estimation of the population density of the fat doormouse (*Glis glis*). *Journal of Zoology* **230**, 337–40.
Hopkin, S. 1991. *A key to the woodlice of Britain and Ireland.* Shrewsbury: Field Studies Council.
Hopkinson, R. G. 1974. The evaluation of visual intrusion in transport situations, In *Environmental quality*, J. T. Coppock & C. B. Wilson (eds), 52–67. Edinburgh: Scottish Academic Press.
Hornung, M. & R. A. Skeffington (eds) 1993. *Critical loads: concepts and applications.* (ITE Symposium 28). London: HMSO (for ITE).
House of Commons Environment Committee 1992. *Coastal zone protection and planning.* London: HMSO.
Hubbard, C. 1984. *Grasses*, 3rd edn. London: Penguin Books.
Hudson, N. 1981. *Soil conservation*, 2nd edn. London: Batsford.
Hughes, D. 1992. *Environmental law*, 2nd edn. Edinburgh: Butterworths.

354

Hunt, D. T. E. & A. L. Wilson 1986. *The chemical analysis of water, general principles and techniques*, 2nd edn. London: Royal Society of Chemistry.

Hunt, J. C. R., R. J. Holroyd, D. J. Carruthers, A. G. Robins, D. D. Apsley, F. B. Smith, D. J. Thomson 1991. Developments in modelling air pollution for regulatory uses. In *Air pollution modelling and its application VIII*, H. van Dop & D. G. Steyn (eds), 17–59. New York: Plenum.

Hutchinson, G. E. 1975. *A treatise on limnology, vol.1, part 2: Chemistry of lakes*. Chichester: John Wiley.

Hutson, A. M. 1993. *Action plan for the conservation of bats in the United Kingdom*. London: The Bat Conservation Trust (c/o London Ecology Centre).

Hyman, P. S. & M. S. Parsons 1992. *A review of the scarce and threatened Coleoptera of Great Britain (Part 1)* (UK Nature Conservation 3). Peterborough: Joint Nature Conservation Committee.

— 1994. *A review of the scarce and threatened Coleoptera of Great Britain (Part 2)* (UK Nature Conservation 12). Peterborough: Joint Nature Conservation Committee.

Hynes, H. B. N. 1977. *A key to the adults and nymphs of British stoneflies (Plecoptera)*. FBA Scientific Publication 17, Freshwater Biological Association, Ambleside.

IEA 1993. *Guidelines for the environmental assessment of road traffic*. East Kirkby, Lincs: Institute of Environmental Assessment.

— 1994. *Guidelines for the baseline ecological input to environmental assessment in the UK: draft for consultation*. East Kirkby, Lincs: Institute of Environmental Assessment.

— & Landscape Institute 1994. *Landscape and visual impact assessment: guidelines* (draft for consultation). East Kirkby, Lincs: Institute of Environmental Assessment.

IHT 1990. *Safety auditing of highways*. London: Institution of Highways and Transportation.

— 1994. *Traffic impact assessment guidelines*. London: Institution of Highways and Transportation.

Institute of Hydrology 1979. *Design flood estimation in catchments subject to urbanisation. Flood studies supplementary report 5*. Wallingford: Institute of Hydrology.

IRPTC (International Register of Potentially Toxic Compounds) 1983. *About chemicals and chemophobia*. IRPTC Bulletin 6.

Jahns, H. M. 1983. *Collins guide to the ferns, mosses and lichens of Britain and North and Central Europe*. London: Collins.

Jefferies, R. L. & A. J. Davy (eds) 1979. *Ecological processes in coastal environments*. Oxford: Blackwell Scientific.

Jenkins, D. (ed.) 1986. *Trees and wildlife in the Scottish uplands*. London: HMSO (for ITE).

Jenkins, J. & J. R. Walker, 1985. School roll forecasting. In *Information systems for policy planning in local government*, J. R. England, K. I. Hudson, R. J. Masters, K. S. Powell, J. D. Shortridge (eds), 96–112. Harlow, Essex: Longman, in association with BURISA.

JNCC (Joint Nature Conservation Committee) 1994. *Newsletter 1: Amphibian and Reptile Recording Scheme*. Peterborough: JNCC and ITE.

Johnson, S. P. & G. Corcelli 1989. *The environmental policy of the European Communities*. London: Graham & Trotman.

Jones, C. E., N. Lee, C. Wood 1991. *UK environmental statements 1988–1990: an analysis*. Occasional Paper 29, Department of Planning and Landscape, University of Manchester.

Jones, P. 1983. *Hydrology*. Oxford: Basil Blackwell.

Jones-Walters, L. M. 1989. *Keys to the families of British spiders*. Shrewsbury: Field Studies Council.

Kapteyn, K. (ed.) 1991. *Proceedings of the First Bat Detector Workshop*. Amsterdam, Netherlands: Bat Research Foundation.

Kent, M. & P. Coker 1992. *Vegetation description and analysis: a practical approach*. London: Pinter (Belhaven).

355

Kent County Council 1991. *Kent environmental assessment handbook.* Maidstone: KCC.

Kerney, M. P (ed.) 1976. *Atlas of the non-marine mollusca of the British Isles.* Huntingdon: Institute of Terrestrial Ecology.

— & R. A. D. Cameron 1979. *A field guide to the land snails of Britain and NW Europe.* London: Collins.

Kershaw, K. A. & J. H. H. Looney 1985. *Quantitative and dynamic ecology*, 3rd edn. London: Edward Arnold.

Ketchum, B. H. (ed.) 1983. *Estuaries and enclosed seas.* Amsterdam: Elsevier.

King, D. 1987. *The Chelmer population and housing model.* Paper presented to the Regional Science Association workshop on regional demography, London School of Economics, October.

Kirby, K. J. 1988. *A woodland survey handbook.* Research and Survey in Nature Conservation 11. Peterborough: Nature Conservancy Council.

Kirby, P. 1991. *A review of the scarcer Neuroptera of Great Britain.* Research and Survey in Nature Conservation 34. Peterborough: Joint Nature Conservation Committee.

— 1992a. *Habitat management for invertebrates: a practical manual.* Sandy: Royal Society for the Protection of Birds.

— 1992b. *A review of the scarce and threatened Hemiptera of Great Britain.* UK Nature Conservation 2. Peterborough: Joint Nature Conservation Committee.

Kirpich, P. Z. 1940. Time of concentration of small agricultural watersheds. *Civil Engineering* **10**(6), 362.

Konikow, L . F. & E. P. Patten 1985. Groundwater forecasting. In *Hydrological forecasting*, M. G. Anderson & T. P. Burt (eds), 221–70. Chichester: John Wiley.

Krebs, C. J. 1989. *Ecological methodology.* New York: HarperCollins

Kroh, D. P. & R. H. Gimblett 1992. Comparing live experience with pictures in articulating landscape preference. *Landscape Research* **17**(2), 58–69.

Kruuk, H.

— & J. W. H. Conroy 1987. Surveying otter (*Lutra lutra*) populations: a discussion of problems with spraints. *Biological Conservation* **41**, 170–83.

— V. Glimmerveen, E. J. Oukerkerk 1986. The use of spraints to survey populations of otter (*Lutra lutra*). *Biological Conservation* **35**, 187–94.

Kula, E. 1992. *Economics of natural resources and the environment.* London: Chapman & Hall.

Lack, P. 1986. *The atlas of wintering birds in Britain and Ireland.* London: T. & A. Poyser.

Lambrick, G. H. 1992. The importance of the cultural heritage in a green world: towards the development of landscape integrity assessment. In *All natural things: archaeology and the green debate*, L. Macinnes & C. R. Wickham-Jones (eds), 105–126. Oxford: Oxbow.

— 1993. Environmental assessment and the cultural heritage: principles and practice. In *Environmental assessment and archaeology*, I. Ralston & R. Thomas (eds), 9–19. Birmingham: Institute of Field Archaeologists.

Lane, P. A. & R. R. Wallace 1988. *A user's guide to cumulative effects assessment in Canada.* Ottawa: Canadian Environmental Assessment Research Council.

Lang, R. & A. Armour 1981. *The assessment and review of social impacts.* Ottawa: Federal Environmental Assessment and Review Office.

Laws, E. A. 1993. *Aquatic pollution: an introductory text.* Chichester: John Wiley.

Lee N. & M. Lewis 1991. *Environmental assessment guide for passenger transport schemes. Annex 7 – air pollution.* EIA Centre, University of Manchester.

Lee, R. 1983. Ecosystem water balance. In *Disturbance and ecosystem: components of response.* H. A. Mooney & M. Gordon (eds), 99–116. Berlin: Springer.

Lenton, E. J., P. F. R. Chanin, D. J. Jefferies 1980. *Otter survey of England 1977–79.* Peterborough: Nature Conservancy Council.

Leopold, L. B. 1974. *Water: a primer.* San Francisco: W. H. Freeman.

Leventhal, B., C. Moy, J. Griffin (eds) 1993. *An introductory guide to the 1991 Census.* Henley-on-Thames, Oxfordshire: NTC Publications (in association with The Market Research Society).

356

Lewis, J. A. 1988. Economic impact analysis: a UK literature survey and bibliography. *Progress in Planning* **30**(3), 161–209.

Lewis, J. R. 1985. *Ecology of rocky shores*. London: Hodder & Stoughton.

Lipscomb, D. M. & A. C. Taylor 1978. *Noise control: handbook of principles and practices*. New York: Van Nostrand Reinhold.

Lloyd, C. S., M. L. Tasker, K. E. Partridge 1991. *The status of seabirds in Britain and Ireland*. London: T. & A. Poyser.

London Ecology Unit 1989. *Sites of metropolitan importance for nature conservation*. London: London Ecology Unit.

Long, S. P. & C. F. Mason 1983. *Saltmarsh ecology*. Glasgow: Blackie.

Lucas, G. & H. Synge 1978. *The IUCN plant red data book*. Morges, Switzerland: International Union for the Conservation of Nature.

Lucas, O. W. R. 1991. *The design of forest landscapes*. Oxford: Oxford University Press.

Luken, J. O. 1990. *Directing ecological succession*. London: Chapman & Hall.

Lyster, S. 1985. *International wildlife law*. Cambridge: Grotius Publications.

Macan, T. T. 1977. *A guide to the British fresh- and brackish water species of gastropods*, 4th edn. FBA Scientific Publication 13, Freshwater Biological Association, Ambleside.

MacIver, K. W. & A. Dickinson 1991. *Before and after studies of new superstores*. TRICS Conference, Traffic Impact Assessments, Imperial College London, September.

Mackereth, F. J. H., J. Heron, J. F. Talling 1978. *Water analysis: some revised methods for limnologists*. FBA Scientific Publication 36, Freshwater Biological Association, Ambleside.

Mader, H. J. 1984. Animal habitat isolation by roads and agricultural fields. *Biological Conservation* **29**, 81–96.

MAFF (Ministry of Agriculture, Fisheries and Food) 1988. *Agricultural land classification of England and Wales: revised guidelines and criteria for grading the quality of agricultural land*. London: HMSO.

Maguire, D. J., M. F. Goodchild, D. W. Rhind 1991. *Geographical information systems: principles and applications*. Harlow, England: Longman.

Maitland, P. S. 1972. *A key to the freshwater fishes of the British Isles with notes on their distribution and ecology*. FBA Scientific Publication 27, Freshwater Biological Association, Ambleside.

— & R. N. Campbell 1992. *Freshwater fishes of the British Isles*. London: HarperCollins.

— & A. A. Lyle 1993. Freshwater fish conservation in the British Isles. *British Wildlife* **5**, 8–15.

Malloch, A. J. C 1988 (1992). *VESPAN II: A computer package to handle and analyse multivariate species data and handle and display species distribution data*. Unit of Vegetation Science, University of Lancaster.

—1992. *MATCH: A computer program to aid the assignment of vegetation data to the communities and subcommunities of the National Vegetation Classification*. Unit of Vegetation Science, University of Lancaster.

Maltby, E. 1988. Wetland resources and future prospects – an international perspective. In *Increasing our wetland resources*, J. Zelzany & J. S. Feierabend (eds), 3–14. Washington DC: National Wildlife Federation.

Mann, K. H. 1981. *Ecology of coastal water: a systems approach*. Oxford: Blackwell Scientific.

Marmonier, P. & M. J. Dole 1986. Les amphipodes des sediments d'un bras courte-circuite du Rhône: logique de repartition et reaction aux crues. *Science del l'eau* **5**, 461–86.

Marshall, J. A. & E. C. M. Haes 1988. *Grasshoppers and allied insects of Great Britain and Ireland*. Colchester: Harley.

Martin, J. 1993. Assessing the landscape. *Landscape Design* **221**, 21–5.

Maynard-Smith, J. 1974. *Models in ecology*. Cambridge: Cambridge University Press.

McLusky, D. S. 1981. *The estuarine ecosystem*. Glasgow: Blackie.

McManus, F. 1992. In *Green's guide to environmental law in Scotland*, C. T. Reid (ed.). London: Sweet & Maxwell.

McNeely, J. A. 1988. *Economics and biological diversity: developing and using economic incentives to conserve biological resources*. Gland, Switzerland: International Union for the Conservation of Nature.

McNicholl, I. H. 1981. Estimating regional industry multipliers: alternative techniques. *Town Planning Review* **55**(1), 80–8.

Mead-Briggs, A. R. & J. A. Woods 1973. An index of activity to assess the reduction in mole numbers caused by control measures. *Journal of Applied Ecology* **10**, 837–45.

Melli, P. & P. Zannetti 1992. *Environmental modelling*. London: Chapman & Hall.

Merrett, P. 1990. *A review of the nationally notable species of spiders of Great Britain* (CSD Contract Report 127). Peterborough: Nature Conservancy Council.

Merrett, R., N. W. Moore, B. C. Eversham (in press). *Atlas of dragonflies in Britain and Ireland*. London: HMSO.

Merryweather, J. & M. Hill 1992. *The fern guide: an introductory guide to the ferns, clubmosses, quillworts and horsetails of the British Isles*. FSC Publication 211. Shrewsbury: Field Studies Council.

Meybeck, M., D. Chapman, R. Helmer (eds) 1989. *Global environment monitoring system. Global freshwater quality, a first assessment*. Oxford: Basil Blackwell.

Miller, D. H. 1977. *Water at the surface of the Earth: an introduction to ecosystem hydrodynamics*. New York: Academic Press.

Miller, R. W. & R. L. Donahue 1990. *Soils: an introduction to soils and plant growth*. Englewood Cliffs, NJ: Prentice-Hall.

Milliman, J. D. & R. H. Meade 1983. Worldwide delivery of sediment to the oceans. *Journal of Geology* **91**, 1–21.

Mills, J. 1992. *The adequacy of visual impact assessments in environmental statements*. MSc Dissertation, School of Planning, Oxford Brookes University.

Mitchell, A. 1974. *A field guide to the trees of Britain and Northern Europe*. London: Collins.

Mitchell-Jones, A. J. (ed.) 1987. *The bat worker's manual*. Peterborough: Nature Conservancy Council.

Moore, N. W. & P. S. Corbet 1990. Guidelines for monitoring dragonfly populations. *Journal of the British Dragonfly Society* **6**, 21–23.

Moore, P. D.
— & D. J. Bellamy 1973. *Peatlands*. London: Elek.
— (ed.) 1984. *European mires*. London: Academic Press.

Moore, P. G. & R. Seed (eds) 1985. *The ecology of rocky coasts*. London: Hodder & Stoughton.

Morgan Evans, D. 1985. The management of historic landscapes. In *Archaeology and nature conservation*, G. H. Lambrick (ed.), 89–94. Department of External Studies, Oxford University.

Moriarty, F. 1988. *Ecotoxicology: the study of pollutants in ecosystems*, 2nd edn. London: Academic Press.

Morris, P. A. 1993. *A red data book for British mammals*. London: The Mammal Society.

Morris, P. J. 1988. *The hydrology of Cothill Fen SSSI*. Unpublished report to the Nature Conservancy Council, School of Biological and Molecular Sciences, Oxford Polytechnic.

Moss, B. 1988. *Ecology of fresh waters: man and medium*. Oxford: Blackwell Scientific.

Moss, D., M. T. Furse, J. F. Wright, P. D. Armitage 1987. The prediction of the macro-invertebrate fauna of unpolluted running-water sites in Great Britain using environmental data. *Freshwater Biology* **17**, 41–52.

Mueller-Dombois, D. & H. Ellenberg, 1974. *Aims and methods of vegetation ecology*. Chichester: John Wiley.

Muir, R. & N. Muir 1987. *Hedgerows: their history and wildlife*. London: Michael Joseph.

Mulholland, K. A. & K. Attenborough 1981. *Noise assessment and control*. London: Construction Press.

Munday, P. K., R. J. Timmis, C. A. Walker 1989. *A dispersion modelling study of present air quality and future nitrogen oxides concentrations in Greater London* (Report LR731 (AP) M). Stevenage: Warren Spring Laboratory.

Munguira, M. L. & J. A. Thomas 1992. Use of road verges by butterfly and burnet populations, and the effect of roads on adult disperal and mortality. *Journal of Applied Ecology* **29**, 316–29.

Murdock, S. H., F. L. Leistritz & R. R. Hamm 1986. The state of socio-economic impact analysis in the USA: limitations and opportunities for alternative futures. *Journal of Environmental Management* **23**, 99–117.

Murley, L. 1991. *Clean air around the world*, 2nd edn. Brighton: IUAPPA/NSCA.

Muscutt, A. D., G. L. Harris, S. W. Bailey, D. B. Davies 1991a. *The effect of buffer strips on the run-off of pollutants from agricultural land*. *ADAS review*. (Interim Report to Ministry of Agriculture, Fisheries and Food: Initial literature survey). Cambridge: ADAS.

— 1991b. *Final review – The effect of buffer strips on the run-off of pollutants from agricultural land*. *ADAS review – final report* (Report to Ministry of Agriculture, Fisheries and Food). Cambridge: ADAS.

NCC (Nature Conservancy Council) 1983. *The ecology and conservation of amphibian and reptile species endangered in Britain*. Peterborough: Nature Conservancy Council.

— 1987. *The bat worker's manual*. Peterborough: Nature Conservancy Council.

— 1989a. *Nature conservation and estuaries in Great Britain*. Peterborough: Nature Conservancy Council.

— 1989b. *Guidelines for the selection of biological SSSIs*. Peterborough: Nature Conservancy Council.

— 1990. *Handbook for phase 1 habitat survey: a technique for environmental audit*. Peterborough: Nature Conservancy Council.

Nelson, P. 1993. Separating design from assessment. *Landscape Design* **224**, 23–24.

NERC (Natural Environment Research Council) 1975. *Flood studies report, vols 1–5*. Wallingford: Institute of Hydrology.

— 1983. *Contaminants in marine top predators* (Report to Marine Pollution Monitoring Management Group). London: DOE.

Newbold, C., J. Honnor, K. Buckley 1989. *Nature conservation and the management of drainage channels*. Peterborough: Nature Conservancy Council.

Newman, E. I. 1993. *Applied ecology*. Oxford: Blackwell Scientific.

Newman, P. J., M. A. Piavaux, R. A. Sweeting 1992. *River water quality ecological assessment and control*. Luxembourg: Commission of the European Communities.

Newson, M. D. 1975. *Flooding and flood hazard in the United Kingdom*. Oxford: Oxford University Press.

— 1979. *Hydrology: measurement and application*. London: Macmillan.

— & J. D. Hanwell 1982. *Systematic physical geography*. London: Macmillan.

Nicholls, D. C. & A. Sclater 1993. Cutting quality down to scale. *Landscape Design* **218**, 39–41.

Nordstrom, K. K. 1992. *Estuarine beaches*. London: Chapman & Hall.

North, P. M. & J. N. R. Jeffers 1991. Modelling: a basis for management or an illusion? In *The scientific management of temperate communities for conservation*, I. F. Spellerberg, F. B. Goldsmith, M. G. Morris (eds), 523–41. Oxford: Blackwell Scientific.

Novell, A. 1993. Setting the standard. *Landscape Design* **224**, 30–31.

NRA 1991a. *Water and the environment in times of drought*. Bristol: National Rivers Authority.

— 1991b. *The quality of rivers, canals and estuaries in England and Wales. Report of the 1990 survey* (Water Quality Series 4). Bristol: National Rivers Authority.

— 1992a. *Policy and practice for the protection of groundwater*. Bristol: National Rivers Authority.

— 1992b. *River corridor surveys, methods and procedures* (Conservation Technical Handbook 1). Bristol: National Rivers Authority.

— 1993a. *Otters and river habitat management* (Conservation Technical Handbook 3). Bristol: National Rivers Authority.

— 1993b. *NRA conservation strategy*. Bristol: National Rivers Authority.

NRC (National Research Council (USA)) 1992. *Restoration of aquatic ecosystems*. Washington DC: National Academy Press.

NWC 1981a. *River water quality: the 1980 survey and future outlook*. Wallingford: National Water Council.

359

— 1981b. *Design and analysis of urban storm drainage: the Wallingford procedure.* Wallingford: National Water Council.

Oke, T. R. 1987. *Boundary layer climates,* 2nd edn. London: Methuen.

OPCS 1992. *1991 Census user guide 38, local statistics: local base statistics/small area statistics explanatory notes.* Titchfield, Hampshire: Office of Population Censuses and Surveys.

— (forthcoming) *1991 Census user guide 27, guide to sources of census statistics.* Titchfield, Hampshire: Office of Population Censuses and Surveys.

Orland, B. (ed.) 1992. Data visualization techniques in environmental management. *Landscape and Urban Planning* (Special Issue) **21**.

Ormerod, S. J., S. D. Rundle, E. C. Lloyd, A. A. Douglas 1993. The influence of riparian management on the habitat structure and macroinvertebrate communities of upland streams draining plantation forests. *Journal of Applied Ecology* **30**, 13–24.

Packham, C. 1989. *Heathlands.* London: HarperCollins.

Painter, K. 1992. Different worlds: the spatial, temporal and social dimensions of female victimisation. In *Crime, policing and place: essays in environmental criminology,* D. J. Evans, N. R. Fyfe, D. Herbert (eds), 164–95. London: Routledge.

Palmer, M. A. 1989. *A botanical classification of standing waters in Great Britain* (Research and Survey in Nature Conservation 19). Peterborough: Nature Conservancy Council.

— 1992. *Trial of MATCH and TABLEFIT computer programs for placing survey data within the National Vegetation Classification* (JNCC Report 20). Peterborough: Joint Nature Conservation Committee.

– & C. Newbold 1983. *Wetland and riparian plants in Great Britain: an assessment of their status and distribution in relation to water authority, river purification board and Scottish slands areas* (Focus on Nature Conservation 1). Peterborough: Nature Conservancy Council.

— S. L. Bell, I. Butterfield 1992. A botanical classification of standing waters in Britain: applications for conservation and monitoring. *Aquatic Conservation: Marine and Freshwater Ecosystems* **2**, 125–43.

Parsons, M. S. 1993. *A review of the scarce and threatened pyralid moths of Great Britain* (UK Conservation 11). Peterborough: Joint Nature Conservation Committee.

Patmore. J. A. (ed.) 1975. Landscape evaluation. *Institute of British Geographers, Transactions* **66**, 120–61.

PCG (Pond Conservation Group) 1993. *A future for Britain's ponds.* Oxford: Pond Conservation Group.

Pearce, F. 1993. When the tide comes in. *New Scientist* (2 January), 23–27.

Pearce, D. & K. Turner 1989. *Economics of natural resources and the environment.* Brighton: Harvester Wheatsheaf.

Penning-Rowsell, E. C. 1974. Landscape evaluation for development plans. *The Planner* **60**(10), 930–4.

Perring, F. H. & L. Farrell 1983. *British Red Data Books 1: vascular plants,* 2nd edn. Lincoln: Royal Society for Nature Conservation.

— & S. M. Walters (eds) 1990. *Atlas of the British flora,* 3rd edn. London: Botanical Society of the British Isles.

Peterken, G. F. 1974. A method for assessing woodland flora for conservation using indicator species. *Biological Conservation* **6**, 239–45.

— 1993. *Woodland conservation and management,* 2nd edn. London: Chapman & Hall.

Petts, G. E. 1984. *Impounded rivers: perspectives for ecological management.* Chichester: John Wiley.

Petts, J. & G. Eduljee 1994. *Environmental impact assessment for waste treatment and disposal facilities.* Chichester: John Wiley.

Pitcher, T. J. & P. J. B. Hart 1982. *Fisheries ecology.* London: Croom Helm.

Pittman, K. 1992. A laboratory for the visualization of virtual environments. *Landscape and Urban Planning* **21**, 327–31.

Plowden, B. 1992. Sustainability criteria for minerals planning. *ECOS* **14**(4), 22–6.

Pollard, E. 1977. A method for assessing changes in the abundance of butterflies. *Biological Conservation* **12**, 115–34.

— D. O. Elias, M. J. Skelton, J. A. Thomas 1975. A method for assessing the abundance of butterflies in Monks Wood National Nature Reserve in 1973. *Entomologists' Gazette* **26**, 79–88.

— M. D. Hooper, N. W. Moore 1979. *Hedges*. London: Collins.

Polunin, O. & M. Walters 1985. *A guide to the vegetation of Britain and Europe*. Oxford: Oxford University Press.

Pond Action 1989. *Aquatic macroinvertebrates: ecology report in connection with proposed Isis Science Park, Hinksey, Oxford* (Pond Action Report 89/9). Oxford: Oxford Polytechnic.

— 1990. *The Datchet, Wraysbury, Staines and Chertsey Flood Study: aquatic biology, Part 1: a survey of the wetland plant and aquatic macroinvertebrate communities of selected gravel-pit lakes in the Datchet–Chertsey complex*. A report to the National Rivers Authority (Thames Region) [Pond Action Report 90/2], School of Biological and Molecular Sciences, Oxford Brookes University.

— 1991. *The Datchet, Wraysbury, Staines and Chertsey Flood Study: aquatic biology. A survey of the wetland plant and macroinvertebrate communities of selected gravel-pit lakes in the Datchet–Chertsey complex: Phase 2, 1991*. A report to the National Rivers Authority (Thames Region). [Pond Action Report 91/15], School of Biological and Molecular Sciences, Oxford Brookes University.

— 1994a. *National pond survey methods booklet*. Oxford: Pond Action.

— 1994b. *The Oxfordshire Pond Survey, vols 1 and 2. A report to the World Wide Fund for Nature*. Oxford: Pond Action.

Potts, G. W. & P. J. Reay 1987. Fish. In *Biological surveys of estuaries and coasts*, J. M. Baker & W. J. Wolff (eds), 342–73. Cambridge: Cambridge University Press.

Powell, M. 1989. Landscape evaluation and the quest for objectivity. *Landscape Research* **14**(2), 16–18.

Preece, R. 1991. *Designs on the landscape*. London: Pinter (Belhaven).

Priestnall, G., R. Haines-Young, N. Ward. 1993. The use of GIS in forest design. *Mapping Awareness and GIS in Europe*, **7**(2), 26–30.

Prys-Jones, O. E & S. A. Corbet 1991. *Bumblebees*. Slough: Richmond Publishing Company.

Putman, R. J. 1984. Facts from faeces. *Mammal Review* **14**, 79–97.

Purvis, O. W., B. J. Coppins, D. L. Hawksworth, P. W. James, D. M. Moore (eds) 1992. *The lichen flora of Great Britain and Ireland*. London: Natural History Museum (in association with The British Lichen Society).

QUARG (Quality of Urban Air Review Group) 1993. *Urban air quality in the United Kingdom* (First Report). London: HMSO.

Quickley, G. P. 1989. *Biological habitat reconstruction*. London: Pinter (Belhaven).

Rabinowitz, D., S. Cairns, T. Dillon 1986. Seven forms of rarity and their frequency in the flora of the British Isles. In *Conservation biology: the science of scarcity and diversity*, M. E. Soulé (ed.), 182–204. Sunderland, Massachusetts: Sinauer.

Rackham, O. 1980. *Ancient woodland: its history, vegetation and uses in England*. London: Edward Arnold.

— 1986. *The history of the countryside*. London: Dent.

— 1991. Landscape and the conservation of meaning. *Royal Society of Arts Journal* (January), 903–1013.

Ralston, I. & R. Thomas (eds) 1993. *Environmental assessment and archaeology*. Occasional Paper 5, Institute of Field Archaeologists, Birmingham.

Ranwell, D. S. 1964. *Spartina* salt marshes in Southern England, ii: rate and seasonal pattern of sediment accretion. *Journal of Ecology* **52**, 79–94.

— 1972. *Ecology of salt marshes and sand dunes*. London: Chapman & Hall.

— & R. Boar 1986. *Coast dune management guide*. London: HMSO.

Ratcliffe, D. A. (ed.) 1977. *A nature conservation review*, vols 1 & 2. Cambridge: Cambridge

University Press.

Reid, W. V. & K. R. Miller 1989. *Keeping options alive; the scientific basis for maintaining biodiversity*. Washington DC: World Resources Institute.

Relph, E. 1984. Seeing, thinking, and describing landscapes. In *Environmental perception and behavior: an inventory and prospect*, T. F. Saarinen, D. Seamon, J. L. Sell (eds), 209–23. Research Paper 209, Department of Geography, University of Chicago.

Rendel, S. 1994. Tranquility – an essential commodity? *Town and Country Planning* **63**, 31–5.

Reynoldson, T. B. 1978. *A key to the British species of freshwater triclads*, 2nd edn. FBA Scientific Publication 23, Freshwater Biological Association, Ambleside.

Rhind, D. (ed.) 1983. *A census user's handbook*. London: Methuen.

RICS 1982. *Practitioners' companion to ancient monuments and archaeological areas act*. London: Royal Institute of Chartered Surveyors.

Rieley, J. O. & S. E. Page 1990. *Ecology of plant communities: a phytosociological account of the British vegetation*. Harlow, England: Longman.

Roberts, A. & S. J. Fairman 1984. A practical method for small area population estimation. *Planning Outlook* **27**, 39–40.

Roberts, M. J. 1993. *The spiders of Great Britain and Ireland*. Colchester: Harley.

Robinson, D. G., I. C. Laurie, J. F. Wager, A. L. Traill 1976. *Landscape evaluation*. Centre for Urban & Regional Research, University of Manchester (for the Countryside Commission).

Rodwell, J. S. (ed.) 1991a. *British plant communities*, vol. 1: *woodlands and scrub*. Cambridge: Cambridge University Press.

— 1991b. *British plant communities*, vol. 2: *mires and heaths*. Cambridge: Cambridge University Press.

— 1993. *British plant communities*, vol. 3: *grasslands and montane communities*. Cambridge: Cambridge University Press.

— (forthcoming) *British plant communities*, vol. 4: *aquatic communities, swamps and tall herb fens*. Cambridge: Cambridge University Press.

— (forthcoming) *British plant communities*, vol. 5: *maritime and weed communities*. Cambridge: Cambridge University Press.

Rose, F. 1981. *The wild flower key for the British Isles and north-west Europe*. London: Warne.

— 1989. *Colour identification guide to the grasses, sedges rushes and ferns of the British Isles and north-west Europe*. London: Viking.

Rosenberg, D. M. & V. H. Resh 1993. *Freshwater biomonitoring and benthic macroinvertebrates*. New York: Chapman & Hall.

Ross, M. 1991. *Planning and the heritage: policy and procedure*. London: E. & F. Spon.

Rothwell, P. I. & S. D. Housden 1990. *Turning the tide: a future for estuaries*. Sandy, Bedfordshire: Royal Society for the Protection of Birds.

Rowell, D. L. 1994. *Soil science: methods and applications*. London: Longman.

RPS Clouston 1993. EA case study: illustrative techniques. *Landscape Design* **224**, 29.

RSPB (Royal Society for the Protection of Birds) 1991. *Coastal zone planning: evidence to the House of Commons Select Committee on the Environment*. Sandy, Bedfordshire: Royal Society for the Protection of Birds (Conservation Planning Department).

Samuelsen, G. S. 1980. Air quality impact analysis. In *Environmental impact analysis handbook*, J. G. Rau & D. C. Wooten (eds), chapter 3. New York: McGraw-Hill.

Savage, A. A. 1989. *Adults of the British aquatic Hemiptera Heteroptera*. FBA Scientific Publication 50, Freshwater Biological Association, Ambleside.

Sclater, A. & D. Nicholls 1993. Landscape beyond measure. *Landscape Design* **224**, 42–4.

Schiemer, F. & H. Waidbacher 1992. Strategies for conservation of the Danubian fish fauna. In *River conservation and management*, P. J. Boon, P. Calow, G. E. Petts (eds), 363–82. Chichester: John Wiley.

Schwab, G. O., D. D. Fangmeier, W. J. Elliot, R. K. Frevert 1993. *Soil and water engineering*, 4th edn. New York: John Wiley.

SDD 1975. *Demographic analysis for planning purposes – a manual on sources and techniques*.

Planning Advice Note 8, Scottish Development Department.

— 1986. *Scottish traffic and environmental appraisal manual*, Scottish Development Department.

Shaw, E. M. 1988. *Hydrology in practice*, 2nd edn. London: Van Nostrand Reinhold.

Shell Chemicals UK Ltd 1990. *North Western Ethylene Pipeline Environmental Statement*. Application to the Department of Energy. London: Shell Chemicals.

Shimwell, D. W. 1971. *Description and classification of vegetation*. London: Sidgewick & Jackson.

— 1983. *A conspectus of urban vegetation types*. Unpublished report to Nature Conservancy Council, Department of Geography, Manchester University.

Shirt, D. B. (ed.) 1987. *British Red Data Books 2: insects*. Peterborough: Nature Conservancy Council.

Shreeve, T. 1994. Butterfly mobility. In *Ecology and conservation of butterflies*, A. S. Pullin (ed.), 37–45. London: Chapman & Hall.

Sidaway, R. 1991. *A review of marina developments in Southern England*. RSPB/WWF Joint Report, Sandy, Bedfordshire.

Simms K. 1991. Modelling traffic pollution. *Clean Air* **21**(4), 191–6.

Simpson, B. J. 1985. *Quantitative methods for planning and urban studies*. Aldershot: Gower.

Simpson, D. & D. J. Baker 1987. New highway design flow thresholds. *Highways and Transportation* **34** (May), 5–12.

Smith, A. J. E. 1978. *The moss flora of Britain and Ireland*. Cambridge: Cambridge University Press.

— 1990. *The liverworts of Britain and Ireland*. Cambridge: Cambridge University Press.

Smith, C. J. 1980. *Ecology of the English chalk*. London: Academic Press.

Smith, I. R., D. A. Wells, P. Welsh 1985. *Botanical survey and monitoring methods for grasslands*. Peterborough: Nature Conservancy Council.

Smith M. 1951. *The British amphibians and reptiles*. London: Collins.

Sneddon, P. & R. E. Randall 1993. *The coastal vegetated shingle structures of Great Britain* (Main Report). Peterborough: Joint Nature Conservation Council.

Sorensen, J. C. 1971. *A framework for identification and control of resource degradation and conflict in the multiple use of the coastal zone*. Department of Landscape Architecture, University of California, Berkeley.

Soulé, M. E. 1987. *Viable populations for conservation*. Cambridge: Cambridge University Press.

Southwood, T. R. E. 1978. *Ecological methods*. London: Chapman & Hall.

Spellerberg, I. F. 1981. *Ecological evaluation for conservation*. London: Edward Arnold.

— 1992. *Evaluation and assessment for conservation*. London: Chapman & Hall.

— & A. Minshull 1992. An investigation into the nature and use of ecology in environmental impact assessments. *Bulletin of the British Ecological Society* **23**, 38–45.

Spencer, J. & K. J. Kirby 1992. An inventory of ancient woodland in England and Wales. *Biological Conservation* **62**, 77–93.

Spriet, J. A. & G. C. Vansteenkiste 1982. *Computer aided modelling and simulation*. London: Academic Press.

SRL (Sound Research Laboratories Ltd) 1991. *Noise control in industry*, London: Chapman & Hall.

Stace, C. A. 1991. *A new flora of the British Isles*. Cambridge: Cambridge University Press.

Staines, B. W. & P. R. Ratcliffe 1987. Estimating the abundance of red deer (*Cervus elaphus* L.) and roe deer (*Capreolus capreolus* L.) and their current status in Britain. *Symposium of the Zoological Society, London* **58**, 131–52.

Starfield, A. M. & A. L. Bleloch 1986. *Building models for conservation and wildlife management*. London: Macmillan.

Steele, D. B. 1969. *Regional multipliers in Britain* (Oxford Economic Papers 19). Oxford: Oxford University Press.

Steele, T. D. 1985. Water quality. In *Hydrological forecasting*, M. G. Anderson & T. P. Burt (eds), 271–309. Chichester: John Wiley.

Steinitz, C. 1990. Toward a sustainable landscape with high visual preference and high ecological integrity: the loop road in Acadia National Park, USA. *Landscape and Urban Planning* **19**, 213–50.

Stewart, A., D. A. Pearman, C. D. Preston (eds) (in press). *Scarce plants in Britain*. Peterborough: Joint Nature Conservation Committee.

Stewart, N. F. & J. M. Church 1992. *Red Data Book of Britain and Ireland: stoneworts*. Peterborough: Joint Nature Conservation Committee.

— (in prep.). *Red Data Book: lichens*. Peterborough: Joint Nature Conservation Committee.

— (in press). *Red Data Book: bryophytes*. Peterborough: Joint Nature Conservation Committee.

Stiles, R., C. Wood, D. Groome 1991. *Environmental assessment: the treatment of landscape and countryside recreation issues*. Cheltenham: Countryside Commission.

Strachan, R., J. D. S. Birks, P. R. S. Chanin, D. J. Jefferies 1990. *Otter survey of England 1984–1986*. Peterborough: Nature Conservancy Council.

— & D. J. Jefferies 1993. *The water vole (Arvicola terrestris) in Britain 1989–1990: its distribution and changing status*. London: The Vincent Wildlife Trust.

Stroud, D. A. 1993. Sites of special scientific interest in 1990 and 1991. In *Britain's birds in 1990–91: the conservation and monitoring review*, J. Andrews & S. Carter (eds), 7–11. Thetford: British Trust for Ornithology and the Joint Nature Conservation Committee.

Stubbs, A. & P. Chandler (eds) 1978. A Dipterist's handbook. *The Amateur Entomologist* **15**.

Suddards, R. W. 1988. *Listed buildings: the law and practice*. London: Sweet & Maxwell.

Sukopp, H. & S. Hejný (eds) 1990. *Urban ecology: plants and plant communities in urban environments*. The Hague: SPB Academic.

Swan, M. J. S. & R. S. Oldham 1989. *Amphibian communities*. Contract HF3-03-332 Year 3, Leicester Polytechnic (for the Nature Conservancy Council).

— 1993. *Herptile sites. Volume 1: National Amphibian Survey Final Report*. English Nature Research Report 38, Peterborough.

Tandy, C. 1967. The isovist method of landscape survey. In *Methods of landscape analysis, symposium report*, 9–10. London: Landscape Research Group.

Tansley, A. G. 1939. *The British Isles and their vegetation*, vols 1 & 2. Cambridge: Cambridge University Press.

ter Braak, C. J. F. 1987. The analysis of vegetation–environment relationships by canonical correspondence analysis. *Vegetatio* **64**, 69–77.

— 1988a. CANOCO – *a FORTRAN program for canonical community ordination bycorrespondence analysis (version 2.0)*. Wageningen: TNO Institute of Applied Computer Science.

— 1988b. CANOCO – an extension of DECORANA to analyse species–environment relationships. *Vegetatio* **75**, 159–60

Tett, P. B. 1987. Plankton. In *Biological surveys of estuaries and coasts*, J. M. Baker & W. J. Wolff (eds), 280–341. Cambridge: Cambridge University Press.

Therivel, R. 1994. *Environmental appraisal of development plans in practice*. Working Paper 15, School of Planning, Oxford Brookes University.

— E. Wilson, S. Thompson, D. Heaney, D. Pritchard 1992. *Strategic environmental assessment*. London: Earthscan.

Thomas, J. A. 1983. A quick method of estimating butterfly numbers during surveys. *Biological Conservation* **27**, 195–211.

Thompson, D. 1990. In defence of soil. *ECOS* **11**(1), 36–9.

Thompson, P. M. & J. Harwood 1990. Methods for estimating the population size of common seals, *Phoca vitulina*. *Journal of Applied Ecology* **27**, 924–38.

Thompson, S., J. R. Treweek, D. J. Thurling (in prep.). The potential application of strategic environmental assessment (SEA) to the farming of Atlantic Salmon (*Salmo salar* L.) in Scotland.

Thor, E. C., G. H. Elsner, M. R. Travis, K. M. Loughlin 1978. Forest environmental impact analysis: a new approach. *Journal of Forestry* **76**, 723–5.

Thornton, D. & D. J. Kite 1990. *Changes in the extent of the Thames estuary grazing marshes*. Peterborough: Nature Conservancy Council.

Tilling, S. M. 1987. *A key to the major groups of British terrestrial invertebrates*. Shrewsbury:

Field Studies Council.

Timmis R. J. & C. A. Walker 1988. *Dispersion modelling of air-quality changes due to smoke control: a case study for Leek* (Report 638 (AP)). Stevenage: Warren Spring Laboratory.

Treweek, J. R., S. Thompson, N. Veitch, C. Japp 1993. Ecological assessment of proposed road developments: A review of environmental statements. *Journal of Environmental Planning and Management* **36**, 295–307.

TRL 1990. *Traffic induced vibration in buildings* (RR246). Crowthorne: Transport Research Laboratory.

— 1991. *The appraisal of community severance.* Contractors Report 135, J. M. Clark & B. J. Hutton, Crowthorne.

Turton, P. 1993. *Who's who in the water industry.* Rickmansworth: Turret Group.

Turner, T. 1987. *Landscape planning.* London: Hutchinson.

Tutin, T. G., N. A. Burges, A. O. Chater, J. R. Edmondson, V. H. Heywood, D. M. Moore, D. H. Valentine, S. H. Walters, D. H. Webb 1964 et seq. *Flora Europaea,* vols 1–5. Cambridge: Cambridge University Press.

Twigg, G. I. 1975a. Catching mammals. *Mammal Review* **5**, 83–100.

— 1975b. Marking mammals. *Mammal Review* **5**, 101–16.

UAFHA (Urban Advisors to the Federal Highway Administrator) 1968. *The freeway in the city: principles of planning and design.* Washington DC: US Government Printing Office.

UNECE 1991. *Policies and systems of environmental impact assessment.* Geneva: United Nations Economic Commission for Europe.

Unwin, D. M. 1981. *A key to families of British Diptera.* Shrewsbury: Field Studies Council.

US Department of Commerce 1966. *A proposed program for scenic roads and parkways.* Washington DC: President's Council on Recreation and Natural Beauty.

USACE (Army Corps of Engineers) 1988. *Methodology for analysis of cumulative impacts of Corps Permit activities.* Washington DC: Institute for Water Resources Policy Study.

USDA-SCS (US Department of Agriculture, Soil Conservation Service) 1972. Hydrology. In *National engineering handbook,* section 4. Washington DC: Government Publishing Office.

— 1986. *Urban hydrology for small watersheds* (Technical Release 55). Springfield VA: National Technical Information Service.

— 1990. *Engineering field manual,* chapter 2. Washington DC: Government Publishing Office.

USDHUD (US Department of Housing and Urban Development) 1981. *Areawide environmental assessment guidebook* (Report HUD-0002384). Washington DC: Office of Policy Development and Research.

USEPA (US Environmental Protection Agency) 1987. *Industrial source complex (ISC) dispersion model user's guide,* 2nd edn (EPA Report EPA-450/4-88-002a). Research Triangle Park, North Carolina: USEPA.

— 1992. *A synoptic approach to cumulative impact assessment: a proposed methodology* (Report EPA/600/R-92/167). Washington DC: Office of Research and Development.

Usher, M. B. (ed.) 1986. *Wildlife conservation evaluation.* London: Chapman & Hall.

— 1990. Assessment of conservation values: the use of water traps to assess the arthropod communities of heather moorland. *Biological Conservation* **53**, 191–8.

van der Maarel, E. 1979. Environmental management of coastal dunes in the Netherlands. In *Ecological processes in coastal environments,* R. L. Jefferies & A. J. Davy (eds), 543–70. Oxford: Blackwell Scientific.

van der Zande, A. N.

— & P. Vos 1984. Impact of a semi-experimental increase in recreation intensity on the densities of birds in groves and hedges on a lake shore in the Netherlands. *Biological Conservation* **30**, 237–59.

— W. J. TerKeurs, W. J. van der Weijden 1980. The impact of roads on the densities of four bird species in an open field habitat – evidence of a long-distance effect. *Biological Conservation* **18**, 299–321.

365

— J. C. Berkhuizen, H. C. van Latesteijn, W. J. TerKeurs, A. J. Poppelaars 1984. Impact of outdoor recreation on the density of a number of breeding bird species in woods adjacent to urban residential areas. *Biological Conservation* **30**, 1–39.

Velander, K. A. 1983. *Pine marten survey of England and Wales 1980–1982*. London: Vincent Wildlife Trust.

Vickerman, R. 1987. Channel Tunnel: consequences for regional growth and development. *Regional Studies* **21**(3), 187–97.

Viessman, W., J. W. Knapp, G. L. Lewis, T. E. Harbraugh 1977. *Introduction to hydrology*, 2nd edn. New York: IEP.

Voorhees, W. B. 1992. Wheel-induced soil physical limitations to plant growth. *Advances in Soil Science* **19**, 73–95.

Walesh, S. G. 1989. *Urban surface water management*. New York: John Wiley.

Walker, C. H. 1990. Kinetic models to predict bioaccumulation of pollutants. *Functional Ecology* **4**, 295–301.

Wallace, I. D. 1991. *A review of the Trichoptera of Great Britain*. Research and Survey in Nature Conservation 32. Peterborough: Nature Conservancy Council.

— B. Wallace, G. N. Philipson 1990. *A key to the case-bearing caddis larvae of Britain and Ireland*. FBA Scientific Publication 51, Freshwater Biological Association, Ambleside.

Walsh, F., N. Lee, C. M. Wood. 1991. *The environmental assessment of opencast coal mines*. Occasional Paper 28, Department of Planning and Landscape, University of Manchester.

Ward, A. J., D. Thompson, A. R. Hiby 1988. Census techniques for grey seal populations. *Symposia of the Zoological Sociey of London* **58**, 181–91.

Ward, D., N. Holmes, P. José (eds) 1994. *The new rivers and wildlife handbook*. Sandy, Bedfordshire: Royal Society for the Protection of Birds.

Ward, J. V. 1992. *Aquatic insect ecology, 1: biology and habitat*. New York: John Wiley.

Ward, R. C. & M. Robinson 1990. *Principles of hydrology*, 3rd edn. London: McGraw-Hill.

Waring, P. 1994. Moth traps and their use. *British Wildlife* **5**, 137–48.

Watkins, L. H. 1991. *Air pollution from road vehicles* (TRRL State of the Art Review 1, DOT/TRRL). London: HMSO.

Waterfield, V. H. & A. J. Hickman 1982. *Estimating air pollution from road traffic: a graphical screening method* (TRRL Report SR752). Crowthorne: Transport & Road Research Laboratory.

Wathern, P. 1984. Ecological modelling in impact analysis. In *Planning and ecology*, R. D. Roberts & T. M. Roberts (eds), §3.4, 80–98. London: Chapman & Hall.

— 1988. An introductory guide to EIA. In *Environmental impact assessment: theory and practice*, P. Wathern (ed.), 3–30. London: Unwin Hyman.

Watson, A. 1993. Britain's toxic legacy. The silence over contaminated land. *The Ecologist* **23** 174–8.

Watson, A. & R. Hewson 1973. Population densities of mountain hares *Lepus timidus* in western Scottish and Irish moors and on Scottish hills. *Journal of Zoology* **170**, 151–9.

Watson, E. V. 1981. *British mosses and liverworts*, 3rd edn. Cambridge: Cambridge University Press.

WCED (World Commission on Environment and Development) 1987. *Our common future*. Oxford: Oxford University Press.

Webb, N. 1986. *Heathlands*. London: Collins.

Webster, R. 1977. *Quantitative and numerical methods in soil classification and survey*. Oxford: Clarendon Press.

Wells, S. M., R. M. Pyle, N. M. Collins 1983. *The IUCN invertebrate red data book*. Gland, Switzerland: International Union for the Conservation of Nature.

Wells, T. C. E. 1983. The creation of species – rich grasslands. In *Conservation in perspective*, A. Warren & F. B. Goldsmith (eds), 215–32. Chichester: John Wiley.

West, G. 1991. *The field description of engineering soils and rocks*. Milton Keynes: Open University Press.

Whilde, A. 1993. *Threatened mammals, birds, amphibians and fish in Ireland. Irish Red Data Book 2: vertebrates.* London: HMSO.

Whitten, A. J. 1990. *Recovery: a proposed programme for Britain's protected species* (CSD Report 1089). Peterborough: Nature Conservancy Council.

WHO (World Health Organisation) 1980. *Environmental criteria 12: Noise:.* Geneva: WHO.

— Regional Office for Europe 1987. *Air quality guidelines for Europe (*WHO Regional Publication European Series 23). Geneva: WHO.

Willing, M. 1993. Land molluscs and their conservation – an introduction. *British Wildlife* **4**, 145–53.

Wilmer, P. 1985. *Bees, ants and wasps – the British Aculeates.* Shrewsbury: Field Studies Council.

Wilson, E. M. 1990. *Engineering hydrology,* 4th edn. London: Macmillan.

Wilson, S. C. & K. C. Jones 1993. Bioremediation of soil contaminated with polynuclear aromatic hydrocarbons (PAHs): a review. *Environmental Pollution* **81**, 229–49.

Wingfield Gibbons, D., J. Reid, R. A. Chapman 1993. *The new atlas of breeding birds in Britain and Ireland: 1988–91.* London: T. & A. Poyser.

WO (Welsh Office) 1991. *Planning policy guidance note 16: archaeology and planning.* London: HMSO.

Wolf, C. P. (ed.) 1974. *Social impact assessment.* Beverley Hills, California: Sage.

Wolff, W. J. 1987. Identification. In *Biological surveys of estuaries and coasts*, J. M. Baker & W. J. Wolff (eds), 404–23. Cambridge: Cambridge University Press.

Wood, C. M. 1989a. *Planning pollution prevention: anticipating controls over air pollution sources.* Oxford: Heinemann.

— 1989b. *Environmental impact assessment: five training case studies.* Occasional Paper 19 (2nd edn), EIA Centre, University of Manchester.

— 1990. Air pollution control by land use planning techniques: a British–American review. *International Journal of Environmental Studies* **35**, 233–43.

Woodroffe, G., J. H. Lawton, W. L. Davidson 1990. Patterns in the production of latrines by water vole and their use as indices of abundance in population surveys. *Journal of Zoology* **220**, 439–45.

Woods, R. & P. Rees (eds) 1986. *Population structures and models.* London: Allen & Unwin.

World Bank 1992. *World development report 1992.* Oxford: Oxford University Press.

Wright, J. F., D. Moss, P. D. Armitage, M. T. Furse 1984. A preliminary classification of running water sites in Great Britain based on macroinvertebrate species, and the prediction of community type using environmental data. *Freshwater Biology* **14**, 221–56.

— P. D. Armitage, M. T. Furse 1989. Prediction of invertebrate communities using stream measurements. *Regulated Rivers, Research and Management* **4**, 147–55.

WWF (World Wide Fund for Nature) 1993. Endangered spaces: on the eve of destruction, *WWF News* (Autumn), 6–7.

York, D. & J. N. Speakman 1980. Water quality impact analysis. In *Environmental impact analysis handbook*, J. G. Rau & D. C. Wooton (eds), ch. 6. New York: McGraw-Hill.

Zannetti, P. 1990. *Air pollution modelling.* Ashurst, Southampton: Computational Mechanics.

Zube, E. H., R. O. Brush, J. G. Fabos 1975. *Landscape assessment: values, perceptions, and resources.* Stroudsburg, Pennsylvania: Dowden, Hutchinson & Ross.

Index

Bold text indicates EIA components and other major topics; bold page numbers refer to the glossary.

369